ENERGY

Readings From
SCIENTIFIC AMERICAN

ENERGY

With Introductions by
S. Fred Singer
University of Virginia, Charlottesville

W. H. Freeman and Company
San Francisco

Most of the SCIENTIFIC AMERICAN articles in *Energy* are available as separate Offprints. For a complete list of articles now available as Offprints, write to W. H. Freeman and Company, 660 Market Street, San Francisco, California 94104.

Library of Congress Cataloging in Publication Data

Main entry under title:

Energy: readings from Scientific American.

   Bibliography: p.
   Includes index.
   1. Power resources. 2. Power (Mechanics)
I. Singer, Siegfried Fred, 1924–   II. Scientific
American.
TJ163.2.E4842   333.7   78–31979
ISBN 0–7167–1082–X
ISBN 0–7167–1083–8 pbk.

Printed in the United States of America

9 8 7 6 5 4 3 2 1

CONTENTS

APPENDIX

Note on cross-references to Scientific American *articles:* Articles included in this book are referred to by title and page number; articles not included in this book but available as Offprints are referred to by title and offprint number; articles not included in this book and not available as Offprints are referred to by title and date of publication.

ENERGY

GENERAL INTRODUCTION

What are the most remarkable achievements of human civilization on this planet? Certainly the development of tools, the discovery of fire, the use of the wheel in transportation, the domestication of animals, and the development of agriculture are of fundamental importance, together with the invention of writing as a means of transmitting information from generation to generation. But all these developments took place more than three thousand years ago. What has happened since the flowering of Greek and Roman civilization? The answer is, relatively little until the late eighteenth century. In other words, life in the so-called civilized and advanced countries of western Europe in the eighteenth century was not so very different from life at the peak of the Roman Empire. By and large, the work was done by human labor or by draft animals, with only a small and occasional assist from such natural sources of energy as moving water and moving air—through water wheels and windmills. What was lacking was a method of concentrating large amounts of energy in one location or of transporting energy over long distances. Water energy and wind energy had to be used where they were available. Firewood, to be sure, could be carried or shipped, but it was used to produce heat, not work.

However, something happened about two hundred years ago in England to change the haphazard and spotty record of technology. Suddenly machines came to the foreground. The pump, the steam engine, and the locomotive used sources of energy that were known earlier but had not been fully utilized: firewood and coal, and later oil and gas. The relation between electricity and magnetism had been discovered by Oersted and Faraday, and crude electric motors and generators had been built, but the full impact of these developments came only with the invention of specific devices. One such important device was the electric lightbulb, perfected by Thomas Alva Edison in 1879, just one hundred years ago. This invention provided the impetus for electric generating stations, for transmission systems, and ultimately for the many industrial uses of electricity.

Clearly, technology has advanced more in the last two hundred years than it had in the previous two thousand; indeed in every small time interval since then the change has been more marked. Many of these technological developments have been related to energy sources and to electricity and electronics.

Another milestone came some forty years ago with the beginnings of nuclear energy and with the realization that even after fossil fuels were exhausted there remained a vast store of energy in uranium, thorium, and other elements. We came to realize gradually that a resource is not something that is discovered but is made by human ingenuity. Coal and petroleum were not resources to speak of until devices to use them were invented. Uranium could not be utilized until the nuclear reactor had been built.

The euphoria that accompanied the development of nuclear energy carried many into the belief that not only energy problems but many other human problems could be solved thanks to this abundant and cheap power source. Inevitable reactions set in during the last ten years—with some arguing that economic and energy growth should be limited artificially before a great ecological disaster strikes. We are probably still too close to resolve the conflict with definite, clearcut answers.

As I write these lines in 1978, the controversy centers about the National Energy Plan proposed by President Carter, with its great emphasis on energy conservation in order to limit the need for oil imports from insecure sources of supply. If history is any guide, these controversies will still be with us ten, even fifty, years from now. It may take that long before some abundant and essentially inexhaustible sources of energy to replace fossil fuels can be demonstrated, built, and integrated into the economic system of the industrialized countries.

In this collection of articles particular emphasis has been given to coal and nuclear energy, partly because these energy sources are stressed in the National Energy Plan, and partly because many suitable articles were available from recent issues of SCIENTIFIC AMERICAN.

It is my belief that solar energy, in various forms, will find increasing applications in the near future. Unfortunately, some of these developments are so new that they have not yet been described in the literature. We will be hearing about them before long.

SUGGESTED FURTHER READING

Adelman, Morris (ed.). 1975. *No Time to Confuse.* San Francisco: Institute for Contemporary Studies.

Conference on National Energy Policy. 1977. Washington, D.C.: American Association for the Advancement of Science.

Cook, Earl. 1976. *Man, Energy, Society.* San Francisco: W. H. Freeman and Company.

Energy Policy Project Staff. 1974. *A Time to Choose: America's Energy Future.* Cambridge, Mass.: Ballinger.

Energy: Use, Conservation, and Supply, vols. 1 and 2. 1974, 1978. Washington, D.C.: American Association for the Advancement of Science.

Grayson, Leslie E. (ed.). 1975. *Economics of Energy.* Princeton, N.J.: Darwin Press.

Hottel, H. C., and Howard, J. B. 1971. *New Energy Technology.* Cambridge, Mass.: MIT Press.

Lovins, Amory B. 1977. *Soft Energy Paths: Toward a Durable Peace.* Cambridge, Mass.: Ballinger.

Mancke, Richard B. 1974. *The Failure of U.S. Energy Policy.* New York: Columbia University Press.

Metz, W. D., and Hammond, A. L. 1978. *Solar Energy in America.* Washington, D.C.: American Association for the Advancement of Science.

Mitchell, Edward J. 1974. *U.S. Energy Policy: A Primer.* Washington, D.C.: American Enterprise Institute for Public Policy Research.

MIT Energy Laboratory Policy Study Group. 1974. *Energy Self-Sufficiency: An Economic Evaluation.* Washington, D.C.: American Enterprise Institute for Public Policy Research.

Nuclear Energy Policy Study Group. 1977. *Nuclear Power Issues and Choices.* Cambridge, Mass.: Ballinger.

Rowen, Henry S. (ed.). 1977. *Options for U.S. Energy Policy.* San Francisco: Institute for Contemporary Studies.

Steinhart, Carol, and Steinhart, John. 1974. *Energy: Sources, Use, and Role in Human Affairs.* North Scituate, Mass.: Duxbury Press.

ENERGY USE, CONVERSION, TRANSPORTATION, AND STORAGE

I

ENERGY USE, CONVERSION, TRANSPORTATION, AND STORAGE

INTRODUCTION

The article by Chauncey Starr of the Electric Power Research Institute sets the stage for this volume by presenting a historical overview of energy use throughout human civilization. Several phenomena are particularly striking: the switch from firewood to fossil fuels beginning about a hundred years ago, the strong correlation between the gross national product and energy consumption, and the rapid rise of the power output of machines—again, beginning about a hundred years ago, principally with the development of the water turbine, steam turbine, steam engine, and internal combustion engine. The increase in power output has been coupled with a growth in the efficiency of energy converters, but further increases will be difficult. Starr's article supports the hypothesis that the productive utilization of energy played a primary role in shaping the development of science and culture in the past century. Certainly it has led to tremendous social changes, as machines replaced farm animals and farm output per man-hour increased several times, so that more food could be produced by fewer farm workers.

Our ideas on future energy consumption are in a state of flux. Starr's article was written before the energy "crisis" of 1973, when the price of oil quadrupled. It is clear that the increased cost of all forms of energy will now lead to more emphasis on conservation and to slower growth of energy consumption. Efforts will be made to substitute for energy other production factors, including labor. Above all efforts will be made to reduce the sheer waste of energy—in industry, in the generation of electricity, and in commercial and residential applications.

What emerges clearly from the article is that the domestic fossil fuel reserves of the United States will not be sufficient to support projected energy use beyond the year 2050, even taking into account considerable energy conservation. It will be necessary to use nuclear power or some of the inexhaustible energy sources, such as solar radiation.

In this connection the reader would do well to become acquainted with the units commonly used in energy calculations. Some of these energy units and conversion factors are presented in the article by Starr. One BTU, the British thermal unit, is the energy required to raise 1 pound of water by 1 degree Fahrenheit. It is commonly used by engineers but is of course much too small to measure national requirements. On the other hand, a quadrillion BTU, or 10^{15} BTU, is a very convenient unit. It is usually called the *quad*.° Electrical

---

°A larger unit is a Q, which equals 10^{18} BTU. Hence 1 quad = 1 milli-Q (mQ).

engineers use the *watt* as a unit of power, and a *watt-second*, or *joule*, as a unit of energy. A BTU is roughly 1000 joules; and 10^{12} watt-years of thermal energy is equal to 30×10^{15} BTU or 30 quad.°

On the other hand, oil people like to deal in barrels of oil. A million barrels a day of oil has an energy equivalent of 2.12 quad per year. It is useful to keep in mind the energy content of the most common fossil fuels: 1 ton of bituminous coal represents about 25 million BTU, a barrel of oil represents approximately 6 million BTU, and 1000 cubic feet of natural gas represent just about 1 million BTU.

Extensive statistical tables are given in Robert H. Romer's book *Energy: An Introduction to Physics,* (W. H. Freeman and Company, San Francisco, 1976). Some of these tables are in the appendix to this book. It should be evident that projections of future energy consumption and of the various forms of energy to be used differ widely among authors. The reader should keep in mind also that estimates of proven reserves of fossil fuels, and especially of probable resources, are also subject to variation. The 1973–74 increase of petroleum prices has, of course, made formerly uneconomic resources much more attractive.

Starr's discussion of energy resources on the earth shows how little of the incoming solar energy is stored in plants and how tiny a fraction of the plant material is converted into fossil fuels. It is evident that the fossil fuels we are using now are the result of some 600 million years of solar radiation. When these fuels are used up they can't be replaced by natural processes in a short period of time; thus they are considered nonrenewable. Of course, the hydrocarbons in the earth's crust are more than just coal, oil, and gas. Tarsands and oil shale also exist, and these will undoubtedly be used once the more easily available and cheaper sources of fossil fuels are exhausted.

Eventually we will have to go to inexhaustible or practically inexhaustible resources, such as direct solar radiation or nuclear breeder or fusion reactors. (The present nuclear fission reactors will use up the readily available uranium unless more sophisticated devices called breeders are used.) With the breeder reactor the amount of energy obtainable from 1 gram of uranium amounts to about 80 billion joules of heat, equivalent to the energy in 2.7 metric tons of coal or 13.7 barrels of crude oil.

The important question, of course, is how soon we will need more advanced power sources such as breeders. The answer depends on the growth rate of energy use in the next fifty years and on the future price of uranium. And considerable disagreement exists in the projections of the various experts.

The conversion of energy is the subject of Claude M. Summers' article. He points out that most energy-using devices have by now reached quite respectable efficiencies and no great breakthroughs can be expected. Instead, it is necessary to concentrate on turning waste energy into useful energy. A clear candidate is the energy rejected in the generation of electricity, which amounts to nearly two-thirds of the heat generated by the burning of the boiler fuel. *Cogeneration* is a term used to denote the simultaneous generation of electricity and use of what is normally considered waste heat. But the efficient application of cogeneration may require restructuring the present utility power plants. They have increased generating efficiency (but not necessarily overall energy efficiency) by becoming larger and larger—and more distant from population centers. By properly arranging the production cycles of electricity, equalizing the load, and generally employing more sophisticated

---

° 10^{12} watt-years of electrical energy output requires an energy input of nearly 100 quad, considering an electric generating efficiency of about 30–35 percent. A large modern power plant, typically 1000 megawatts, requires a (thermal) energy input of 0.1 quad per year.

planning methods, we will achieve great energy savings—some quite soon and some in a matter of years.

In this regard, special problems are posed by potential energy sources based on renewable sources such as wind power and solar power. Because of their intermittent nature—in the case of solar power the day/night cycle—more has to be invested in energy storage systems than is the case for conventional power plants. Of course, one way to eliminate the day/night cycle is to place the solar collector in earth orbit where it will not be obscured. But transmitting the energy back to the earth in electrical form is problematical. In one particular design, microwaves are used, requiring a receiving antenna of about the same dimensions as the solar collector itself.

A very important aspect of energy is transportation from the source to the point of use, the subject of Daniel B. Luten's article. Pipelines transport oil and gas and may in the future transport coal in the form of *slurries*. Tankers transport oil and liquefied gas; coal travels in unit trains as well as ships. Electric power must be transmitted as efficiently as possible. New techniques may include even superconducting transmission lines.

Storage of energy is a vital but often overlooked aspect of the energy picture. Coal and oil are relatively easy to store, but gas is stored too, sometimes underground, sometimes in liquefied form above the ground. Even electricity can be stored, for example by converting it into the gravitational energy of water, called *pumped storage*, or by using it to electrolyze water to produce hydrogen. Batteries provide another form of storage, which may in the future become important as their efficiency rises and their cost goes down. A new mode of energy storage is compressed air in underground cavities.

Strategic storage has become of particular concern since the Arab oil embargo of 1973–74. The United States and other oil-consuming nations, as members of the International Energy Agency, are setting up strategic stockpiles of crude oil that would allow operation for at least ninety days in spite of cut-offs of oil imports. Even though stockpiles are expensive, they do furnish a considerable amount of insurance. Furthermore, their very existence can make the idea of an oil embargo quite unattractive. In addition, stockpiles furnish protection against accidental cut-offs or catastrophic interruptions of the oil supply beyond the control of the producing nation.

Energy and Power

1

by Chauncey Starr
September 1971

*Man's expanding need for energy creates difficult
economic, social and environmental problems. The
solutions call for sensible choices of technological
alternatives by the market and political process*

Between now and 2001, just 30 years away, the U.S. will consume more energy than it has in its entire history. By 2001 the annual U.S. demand for energy in all forms is expected to double, and the annual worldwide demand will probably triple. These projected increases will tax man's ability to discover, extract and refine fuels in the huge volumes necessary, to ship them safely, to find suitable locations for several hundred new electric-power stations in the U.S. (thousands worldwide) and to dispose of effluents and waste products with minimum harm to himself and his environment. When one considers how difficult it is at present to extract coal without jeopardizing lives or scarring the surface of the earth, to ship oil without spillage, to find acceptable sites for power plants and to control the effluents of our present fuel-burning machines, the energy projections for 2001 indicate the need for thorough assessment of the available options and careful planning of our future course. We shall have to examine with both objectivity and humanity the necessity for the projected increase in energy demand, its

relation to our quality of life, the practical options technology provides for meeting our needs and the environmental and social consequences of these options.

The artful manipulation of energy has been an essential component of man's ability to survive and to develop socially. Although primitive people and most animals can alter their behavior to adapt to changing environmental restrictions, the reverse ability to substantially alter the environment is uniquely man's. When primitive man learned to use fire to keep himself warm, he took the first big step in the use of an energy resource.

The use of energy has been a key to the supply of food, to physical comfort and to improving the quality of life beyond the rudimentary activities necessary for survival. The utilization of energy depends on two factors: available resources and the technological skill to convert the resources to useful heat and work. Energy resources have always been generally available, and the heating process is ancient. Power devices able to convert energy into useful work have been a recent historical develop-

ment. The prehistoric domestication of animals represented a multiplication in the power resources available to man, but not by very significant amounts. The big importance of the horse and the ox was that their fuel requirements did not deplete man's own food supply. During this period the power available limited man's ability to irrigate, cultivate and survive.

Water power for irrigation purposes, exploiting natural differences in elevation, was known in very early times. The horizontal waterwheel appeared about the first century B.C. with a power capacity of perhaps .3 kilowatt. By the fourth century the vertical waterwheel had been developed to about two kilowatts of power. These wheels were primarily used for grinding cereals and similar mechanical tasks. By the 16th century the waterwheel was by far the most important prime mover, providing the foundation for the industrialization of western Europe. By the 17th century its power output was reaching significant levels. The famous Versailles waterworks at Marly-la-Machine is said to have had a power of 56 kilowatts. The windmill first appeared in western Europe in the 12th century. It was variously used for grinding grain, for hoisting materials from mines and for pumping water. The windmill had a respectable capacity ranging from several kilowatts to as much as 12 kilowatts. The biggest disadvantage was the intermittent nature of its operation.

BAYWAY REFINERY of the Humble Oil & Refining Company in Linden, N.J., occupies most of the land area in the aerial photograph on the opposite page. Placed on-stream in 1909, when oil supplied less than 6 percent of the nation's energy (it now supplies 43 percent), the refinery has grown with the demand for petroleum products. One of five refineries operated by Humble in the U.S., the Bayway plant refines 200,000 barrels of crude oil per day. Most of it is delivered by tanker and unloaded at docks bordering Arthur Kill, a waterway that separates Staten Island, at the top of the picture, from New Jersey. The multilane highway that cuts across the photograph at an angle is the New Jersey Turnpike.

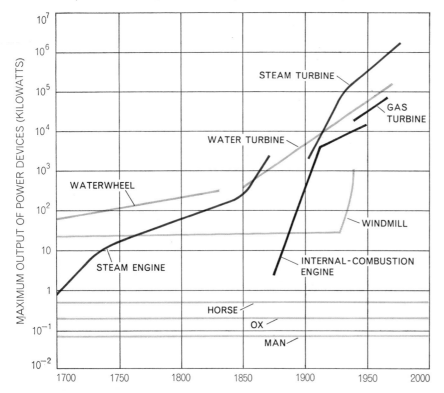

POWER OUTPUT OF BASIC MACHINES has climbed more than five orders of magnitude since the start of the Industrial Revolution (*ca.* 1750). For the steam engine and its successor, the steam turbine, the total improvement has been more than six orders, from less than a kilowatt to more than a million. All are surpassed by the largest liquid-fuel rockets (*not shown*), which for brief periods can deliver more than 16 million kilowatts.

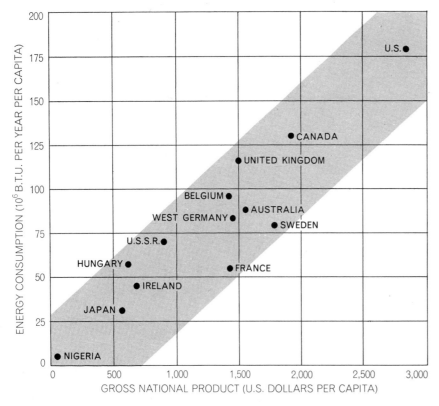

COMMERCIAL ENERGY USE AND GROSS NATIONAL PRODUCT show a reasonably close correlation.

The development of the steam prime mover is relatively modern compared with the windmill and the waterwheel. As early as the first century after Christ, Hero of Alexandria demonstrated the famous Sphere of Aeolus, a steam reaction turbine on a toy scale. Not until the 17th century was steam used effectively. The steam pump invented by Thomas Savery was a pistonless device using the vacuum of condensing steam to pump water, with a power output of about three-fourths of a kilowatt. Early in the 18th century steam engines using a moving piston were developed as power sources of several kilowatts.

The steam engine was the first mechanical prime mover to provide basic mobility. It was some time, however, before this mobility was used. The early Industrial Revolution was based on the waterwheel and the windmill as prime movers: the location of industrial centers, factories and cities was primarily determined by the availability of those power sources. It was the geographic limitation on the expansion of water power that gave the steam engine an opportunity to continue the growth of manufacturing centers. The first use of the steam engine was as an auxiliary to the waterwheel: to pump water to an elevation sufficient to increase the wheel's power. It was not until the middle of the 19th century that the steam engine became a principal prime mover for the manufacturing industry of the Western world.

The contribution of large power machines to the social development of man became important after 1700 [*see top illustration at left*]. Since 1900 a steadily growing variety of smaller power-conversion devices have been introduced whose chief virtue is mobility. From 1700 on the power output of energy-conversion devices increased by roughly 10,000 times. Most of this growth occurred in the past century, so that it has had its major impact only recently. It is this technological capability that makes our age historically one of accelerated energy utilization. The development of these prime movers required and supported the technology of iron and steel fabrication, and it involved the rise of the railroads. The consequence of these technological innovations has been an exponential increase in energy consumption.

For the millenniums preceding the 17th century the productivity of man was principally determined by his own labor and by that of domestic animals.

The growth in the world population and the manifestations of greater average affluence all appear to show significant increases in parallel with the growth in energy use. Simultaneously one witnessed rapid developments in learning, in the arts and in technologies of all kinds. Although one must be cautious when dealing with pluralistic and interacting relations, a strong case can be made for the hypothesis that the productive utilization of energy has played a primary role in shaping the science and culture of the past three and a half centuries. This hypothesis is supported by the linear relation one finds today between the per capita consumption of energy for heat, light and work and the per capita gross national product of various nations [see bottom illustration on opposite page].

As an example of the effect of power machinery on the productivity of man, the agricultural experience of the U.S. following World War I is much to the point [see illustrations on page 13]. The story is told in the following quotation from the 1960 U.S. Yearbook of Agriculture:

"Horse and mule numbers at that time [1918] were the highest in our history—more than 25 million—but the rate of technological progress had slowed down. The availability of good new land had dwindled to insignificance. One-fourth of the harvested crop acre-

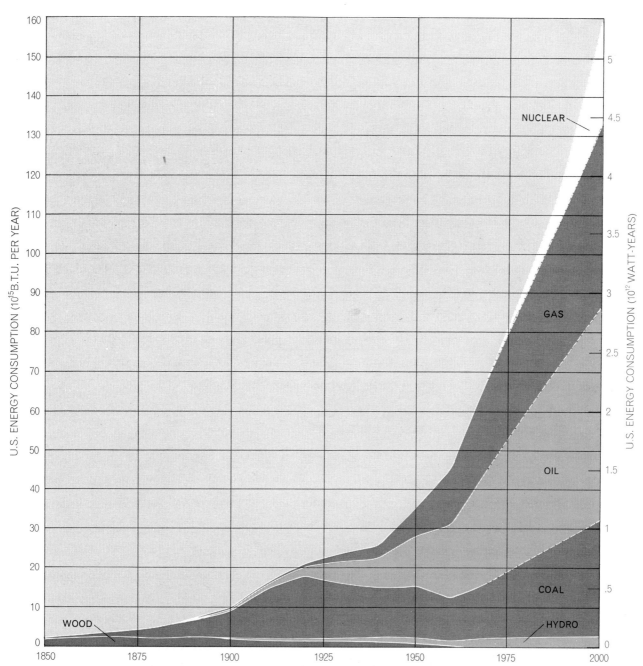

U.S. ENERGY CONSUMPTION has been multiplied some 30 times since 1850, when wood supplied more than 90 percent of all the energy units. By 1900 coal had become the dominant fuel, accounting for more than 70 percent of the total. Fifty years later coal's share had dropped to 36.5 percent and the contribution from oil and natural gas had climbed to 55.5 percent. Last year coal accounted for 20.1 percent of all energy consumed, oil and gas 75.8 percent, hydropower 3.8 percent and nuclear energy .3 percent. Energy-consumption figures are from the U.S. Bureau of Mines; projections conform to those given in the illustration on page 16.

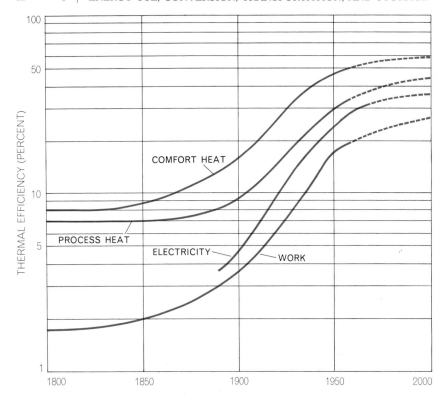

EFFICIENCY OF ENERGY CONVERTERS rose steeply from 1850 to 1950. From here on improvements will be much harder to win, partly because of thermodynamic limitations. A simple unweighted average of efficiencies in four major categories of energy use gives a value of about 8 percent in 1900, 30 percent in 1950 and a projected 45 percent in A.D. 2000.

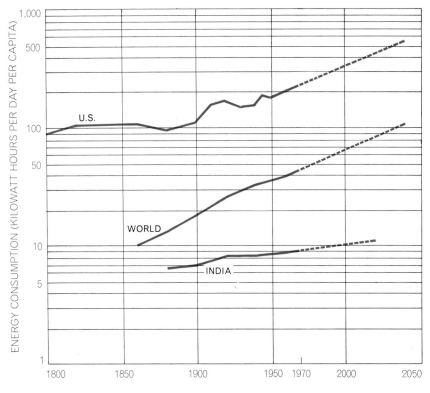

GROWTH IN ENERGY DEMAND in the U.S. is at the annual rate of about 1 percent per capita. For the world as a whole per capita consumption is growing about a third faster. Even so, the world supply of energy per capita in A.D. 2000 will be less than a fourth of the projected U.S. figure. In India the rate of increase is only about a third of the U.S. rate.

age was being used to produce feed for power animals.

"If methods had not been changed, many more horses, more men to work them, and much more land to grow feed for them would be required for today's net agricultural output. The American economy of the 1960's could not be supported by an animal-powered agriculture on our essentially fixed—in fact, slowly shrinking—land base. National progress on all fronts would have been retarded seriously had not agriculture received new forms of power and sources of energy not restricted by biological limitations.

"With the adoption of mechanical forms of power in engines, tractors, and electric motors and development of more and more types of adapted equipment to use that power, American agriculture entered a new era of sharply rising productivity."

The introduction of new hybrid grains, the use of fertilizer and pesticides, along with extensive irrigation, all contributed to the increased productivity per unit of labor. Irrigation systems and the manufacture and transportation of chemical fertilizers on a large scale all require substantial use of energy, as Earl Cook points out in Scientific American Offprint 667, "The Flow of Energy in an Industrial Society."

It is evident that the present rate of world population growth cannot be sustained indefinitely; sooner or later environmental restrictions will cause the death rate to increase substantially, and the least developed countries will be the first to suffer. The long-term alternative for the world is a controlled birthrate. Nevertheless, for some decades to come social trends will cause an inevitable increase in world population. In order to meet not only the food requirements but also a minimally reasonable quality of life, the contributions that can be made by the use of energy in various forms are essential. The issue therefore is *not* whether energy production for the world should be increased. It is rather how to increase it effectively with minimum deleterious side effects.

Because the great increase in energy consumption in the past century has taken place chiefly in the advanced countries, it is instructive to examine the trends in the U.S. The annual consumption of all forms of energy in the U.S. has increased seventeenfold in the past century, with a corresponding population increase of a little more than fivefold. During this period, in which our

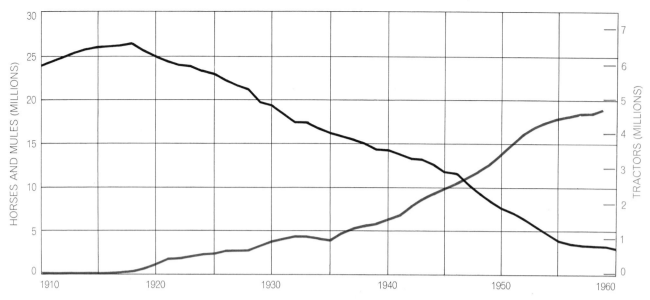

MACHINES REPLACED ANIMALS at a rapid rate on U.S. farms between 1920 and 1960. In the same period farm output more than doubled. In 1920 a fourth of U.S. farm acreage was planted in crops required to feed the nation's 25 million farm horses and mules.

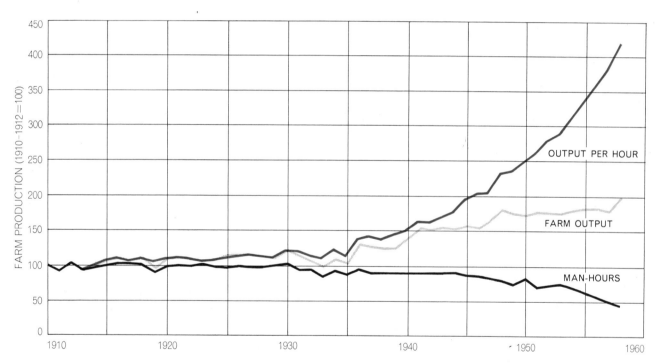

FARM OUTPUT PER MAN-HOUR approximately quadrupled between 1910 and 1958. The improvement was due not only to the internal-combustion engine but also in part to higher-yielding crops, extensive irrigation, fertilizers, herbicides and insecticides.

per capita energy use has slightly more than doubled, fuel sources have shifted steadily [see illustration on page 11]. Fuel wood was the dominant energy source in 1850; by 1910 coal accounted for about 75 percent of the total energy consumption and fuel wood had declined to some 10 percent. In the 50 years between 1910 and 1960 coal lost its leading position to natural gas and oil. Today nuclear power is emerging as a national energy source.

Thus roughly 50 years seem to be needed for the energy economy to shift substantially to a new fuel. This is determined primarily by the operating lifetime of power machinery and secondarily by the long lead time for redirecting available manufacturing and supply capability. For example, the steadily increasing demand for electric power requires construction of new power stations at a rate that exceeds the facilities of the infant nuclear-power industry; as a result fossil-fuel-burning plants must be built for many decades to come in order to meet the nation's needs. With an expected operating life of 30 years for such plants, it is evident that fossil-fuel plants will be playing a role even half a century after the change to nuclear power was initiated.

A century ago our energy resources were primarily applied to the production of heat for physical comfort. Less than a quarter of the heat was utilized for metallurgical processes and industrial activities. Today more than half of all energy consumed in the U.S. goes into useful work. Paralleling this shift in the way energy is used there has been a steady improvement in the efficiency with which energy is converted to useful forms [*see top illustration on page 12*]. There is no theoretical limit to the efficiency of energy use for heating. The theoretical limit on the conversion of heat to work is the Carnot thermodynamic efficiency [see "The Conversion of Energy," by Claude M. Summers, page 23]. The best of our power plants now operate at a thermal efficiency of 40 percent, a figure that may reach 50 percent by the year 2000. Other thermal prime movers are not as efficient. The internal-combustion engine thermal efficiency ranges from 10 to 25 percent, depending on how the engine is used. Because of the impact of efficiency on the economics of use, the motivation for improving efficiency will persist.

At present the U.S. consumes about 35 percent of the world's energy. By the year 2000 the U.S. share will probably drop to around 25 percent, due chiefly to the relative population increase of the rest of the world. The per capita increase in energy consumption in the U.S. is now about 1 percent per year [*see bottom illustration on page 12*]. Starting from a much lower base, the average per capita energy consumption throughout the world is increasing at a rate of 1.3 percent per year. It is evident that it may be another century before the world average even approaches the current U.S. level. At that time the energy gap between the U.S. and the underdeveloped world will still be large. With unaltered trends it would take 300 years to close the gap. By 2000 the world's average per capita energy consumption will have moved only from the present one-fifth of the U.S. average to about one-third of the present U.S. average. Of grave concern is the nearly static and very low per capita energy consumption of areas such as India, a country whose population growth largely negates its increased total production of energy. If the underdeveloped parts of the world were conceivably able to reach by the year 2000 the standard of living of Americans today, the worldwide level of energy consumption would be roughly 10 times the present figure. Even though this is a highly unrealistic target for 30 years hence, one must assume that world energy consumption will move in that direction as rapidly as political, economic and technical factors will allow. The problems implied by this prospect are awesome.

One can better appreciate the energy problem the world faces if one simply compares the cumulative energy demand to the year 2000—when the annual rate of energy consumption will be only three times the present rate—with estimates of the economically recoverable fossil fuels [*see illustrations on opposite page*]. The estimated fossil-fuel reserves are greater than the estimated cumulative demand by only a factor of two. If the only energy resource were fossil fuel, the prospect would be bleak indeed. The outlook is completely altered, however, if one includes the energy available from nuclear power.

There is no question that nuclear power is a saving technical development for the energy prospects for mankind. Promising but as yet technically unsolved is the development of a continuous supply of energy from solar sources. The enormous magnitude of the solar radiation that reaches the land surfaces

NUCLEAR POWER PLANT being built by the Duke Power Company near Clemson, S.C., has three 886,300-kilowatt units in various stages of completion. Unit No. 1 (*right*) is ready to be loaded with fuel; it is expected that it will be supplying power early next year. The three nuclear steam-supply systems were designed and are being manufactured by the Babcock & Wilcox Company. There are now 22 nuclear power plants with a combined capacity of 9,132 megawatts operating in the U.S. Another 99 plants with a capacity of 90,000 megawatts are under construction or on order. By A.D. 2000 nuclear fuels may be supplying half of the nation's electricity.

of the earth is so much greater than any of the foreseeable needs that it represents an inviting technical target. Unfortunately there appears to be no economically feasible concept yet available for substantially tapping that continuous supply of energy. This somewhat pessimistic estimate of today's ability to use solar radiation should not discourage a technological effort to harness it more effectively. If only a few percent of the land area of the U.S. could be used to absorb solar radiation effectively (at, say, a little better than 10 percent efficiency), we would meet most of our energy needs in the year 2000. Even a partial achievement of this goal could make a tremendous contribution. The land area required for the commercially significant collection of solar radiation is so large, however, that a high capital investment must be anticipated. This, coupled with the cost of the necessary energy-conversion systems and storage facilities, makes solar power economically uninteresting today. Nevertheless, the direct conversion of solar energy is the only significant long-range alternative to nuclear power.

The possibility of obtaining power from thermonuclear fusion has not been included in the listing of energy resources on this page because of the great uncertainty about its feasibility. The term "thermonuclear fusion," the process of the hydrogen bomb, describes the interaction of very light atomic nuclei to create highly energetic new nuclei, particles and radiation. Control of the fusion process involves many scientific phenomena that are not yet understood, and its engineering feasibility has not yet been seriously studied. Depending on the process used, controlled fusion might open up not only an important added energy resource but also a virtually unlimited one. The fusion process remains a possibility with a highly uncertain outcome.

The special environmental problems associated with generating electricity have drawn much attention, but the production of electricity is not the major environmental problem we face [see illustration on next page]. Of all the energy needs projected for the year 2000, nonelectric uses represent about two-thirds. These uses cover such major categories as transportation, space heating and industrial processes. The largest energy user at that time will be the manufacturing industry, with transportation using about half as much. These projections are based on extrapolations of present trends. One can speculate, however, on major changes in life style or

DEPLETABLE SUPPLY (10¹² WATT-YEARS)	WORLD	U.S.
COAL	670 — 1,000	160 — 230
PETROLEUM	100 — 200	20 — 35
GAS	70 — 170	20 — 35
SUBTOTAL	840 — 1,370	200 — 300
NUCLEAR (ORDINARY REACTOR)	~3,000	~300
NUCLEAR (BREEDER REACTOR)	~300,000	~30,000
CUMULATIVE DEMAND 1960 TO YEAR 2000 (10¹² WATT-YEARS)	350 — 700	100 — 140

ECONOMICALLY RECOVERABLE FUEL SUPPLY is an estimate of the quantities available at no more than twice present costs. U.S. reserves of all fossil fuels are slightly less than a fourth of the world total and its reserves of nuclear fuels are only a tenth of the world total. Fossil-fuel reserves are barely equivalent to twice the cumulative demand for energy between 1960 and 2000. Even nuclear fuel is none too plentiful if one were to use only the ordinary light-water reactors. By employing breeder reactors, however, the nuclear supply can be amplified roughly a hundredfold. (10^{12} watt-years equals 29.9×10^{15} B.t.u.)

CONTINUOUS SUPPLY (10¹² WATTS)	WORLD		U.S.	
	MAXIMUM	POSSIBLE BY 2000	MAXIMUM	POSSIBLE BY 2000
SOLAR RADIATION	28,000		1,600	
FUEL WOOD	3	1.3	.1	.05
FARM WASTE	2	.6	.2	.00
PHOTOSYNTHESIS FUEL	8	.01	.5	.001
HYDROPOWER	3	1.	.3	.1
WIND POWER	.1	.01	.01	.001
DIRECT CONVERSION	?	.01	?	.001
SPACE HEATING	.6	.006	.01	.001
NONSOLAR				
TIDAL	1.	.06	.1	.06
GEOTHERMAL	.06	.006	.01	.006
TOTAL	18+	3	1.2	.2
ANNUAL DEMAND YEAR 2000 (10¹² WATTS)	~15		~5—6	

CONTINUOUS, OR RENEWABLE, ENERGY SUPPLY can be divided into two categories: solar and nonsolar. Two sets of estimates are again presented, one for the world and one for the U.S. alone. The figure for total solar radiation includes only the fraction (about 30 percent) falling on land areas. If an efficient solar cell existed to convert sunlight directly to electric power, one could think of utilizing solar energy on a large scale. The sunlight that falls on a few percent of the land area of the U.S. would satisfy most of the energy needs of the country in the year 2000 if converted to electricity at an efficiency of 12 percent.

technology that could substantially alter these projections [see illustration on page 17]. These hypothetical shifts include all-electric homes, complete air conditioning, more use of electricity in commercial buildings, the electric automobile, the use of electricity in industrial processes, possible large-scale desalination of seawater and, finally, shifting all electricity production to nuclear plants. Such substantial changes could reduce the estimated fossil-fuel require-

ments in the year 2000 by more than 40 percent, with the greatest component being the shift from fossil to nuclear fuels in generating electricity. Even with such drastic shifts, the total fuel consumed for electricity would still represent no more than 60 percent of the national energy requirement, with the remaining 40 percent still dependent on fossil fuel.

It is clear that if in the year 2000 the U.S. were solely dependent on fossil fuels, the costs of energy would have to increase substantially [*see illustration on page 21*]. The availability of nuclear power will allow these costs to be kept reasonably low. A major reduction in cost will be achieved when the breeder reactor is successfully developed. In a breeder reactor excess neutrons from the fission of uranium 235 are used to convert nonfissionable uranium 238 and thorium 232 into the fissionable isotopes plutonium 239 and uranium 233 respectively. The breeder reactor should make it possible for nuclear fission to supply the world's energy needs for the next millennium. (If the fusion process is ever successful, its cost for electricity would be similar to that of the breeder.) The U.S. Government has recently announced that intensive development of the fast breeder reactor is now national policy. With multimegawatt fast breeders now being constructed in the U.S.S.R. and in western Europe, there appears to be little doubt about their engineering feasibility. The problems now are those of detailed engineering and performance economics.

In the past century the perceived social benefits from the uses of energy

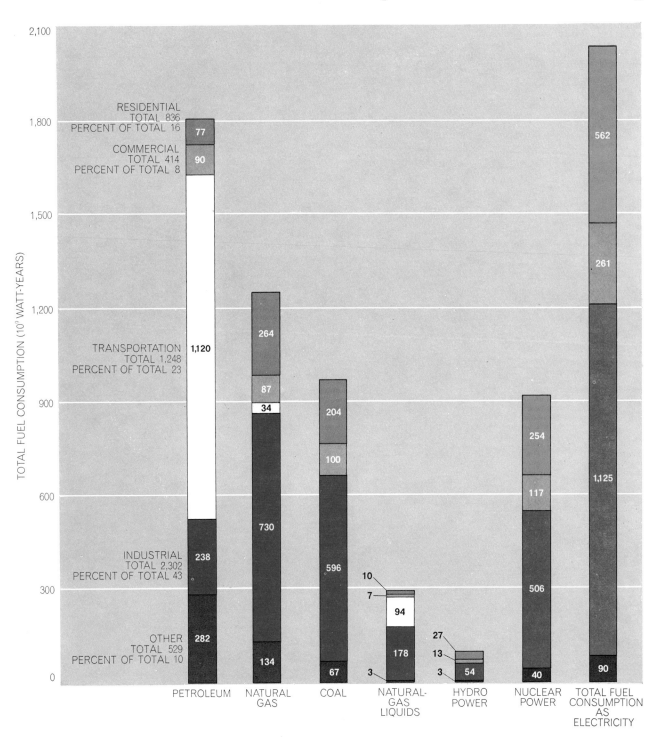

PROJECTED ENERGY SUPPLY AND USE in the U.S. in the year 2000 shows nuclear power contributing almost as much as coal to the total energy supply but both running well behind oil and natural gas. In the year 2000 the generation of electricity may consume 38 percent of the total energy input compared with 25 percent today. These estimates are a projection of present trends. One can imagine, however, a major effort to substitute nuclear for fossil fuels, with the results depicted on the opposite page.

overrode any constraint that might be set by its environmental impact. As the U.S. has grown in both population and affluence, the amount and concentration of our energy use has begun to make the deterioration of the environment serious enough to be of national concern. It is only recently that priority has been given to the technology of pollution abatement; there is little doubt that eventually control of the environmental side effects of energy utilization will be brought to socially acceptable levels. Pollution is man-made and is man-controllable. Pollution control, however, is itself a new growth industry and will create an additional energy demand. Pollution-control techniques use chemical-plant processes, all of which consume energy. For example, the proposed effluent-treatment methods for reducing pollutants from automobiles result in increased fuel consumption.

In considering the harmful effects of energy utilization it is well to distinguish between those that are short-term and geographically concentrated and those that operate over the long term, often with worldwide consequences. Of the latter there are only a few. The combustion of fossil fuels, no matter how efficiently done, must always produce carbon dioxide. Its concentration in the atmosphere has increased from some 290 parts per million to 320 within the past century and may increase to 375 or 400 parts per million by the year 2000. The mechanism for the removal of carbon dioxide is only partly understood; it is eventually absorbed by the ocean, converted into minerals or incorporated by plants in their growth. Thus the carbon dioxide ultimately but slowly returns to the biosphere in some nonpolluting form. Its effects while it resides in the atmosphere are not now predictable, although theoretically the increased carbon dioxide should cause a "greenhouse effect" by reducing the infrared heat loss from the earth and perhaps raising the mean global temperature one degree Celsius by the year 2000.

In parallel with the increase in carbon dioxide in the atmosphere there has also been a rise in suspended particulate contamination. Fine particles are released into the air not only by combustion but also by volcanic eruptions. The increased turbidity reduces solar radiation to the earth's surface. So far the observed temperature trends are not meaningful and the subject is not well understood. The meteorological data available indicate that neither the added carbon

	FOSSIL-FUEL REDUCTION (10^9 WATT-YEARS)	INCREASED ELECTRICITY PRODUCTION (10^9 WATT-YEARS)	INCREASED NUCLEAR ELECTRICITY PRODUCTION (10^9 WATT-YEARS)
ALL-ELECTRIC HOMES	230	99	99
100 PERCENT AIR CONDITIONING	NO CHANGE	14	14
INCREASED USE OF ELECTRICITY IN THE COMMERCIAL SECTOR	90	53	52
ELECTRIC AUTOMOBILES	340	92	91
REPLACE 1/3 OF INDUSTRIAL CONSUMPTION OF GAS	200	228	115
POTENTIAL NEED FOR DESALINATION IN THE WESTERN U.S.	NO CHANGE	114	115
ALL ELECTRICITY PRODUCTION SHIFTED TO NUCLEAR	1,040	NO CHANGE	515
SUBTOTALS	(−) 1,900	(+) 600	(+) 1,000
REFERENCE PROJECTION (10^9 WATT-YEARS)	4,315	1,030	515
PERCENT CHANGE	−44	+58	+194

MAJOR EFFORT TO REDUCE FOSSIL-FUEL USE by the year 2000 might conceivably eliminate $1,900 \times 10^9$ watt-years from the demand of $4,315 \times 10^9$ watt-years projected in the bar chart on the opposite page. This would amount to a reduction of 44 percent. The next column shows the amount of electrical energy needed to replace fossil fuel in each of the six categories of energy use listed at the left. The total increase in electric demand comes to 600×10^9 watt-years, or 58 percent. The reference projection of $1,030 \times 10^9$ watt-years assumes the conversion of $2,038 \times 10^9$ watt-years of fuel at a thermal efficiency of 51 percent. The reference projection also assumes that half of the electric-power production, or 515×10^9 watt-years, will be nuclear in the year 2000. The last column shows the increase in nuclear power required if all electricity were to be obtained from nuclear fuels.

dioxide nor the particulates are a serious problem yet. In any case we have at least several decades for determining the carbon dioxide pathways in our biosphere. If the carbon dioxide additions to the atmosphere were determined to be harmful, there is an ultimate but costly technological solution: we could use nuclear electric power to manufacture hydrogen by the electrolysis of water. Hydrogen would make an ideal fuel because its combustion yields water as an end product.

Other pollutants that arise from the burning of fossil fuels are in a somewhat different category. They tend to concentrate in the region where they are generated and have a relatively short life. They all eventually disappear from the atmosphere through photochemical reactions or meteorological processes such as rain. The problems they create in urban areas because of their high concen-

tration are those associated with either material damage, aesthetics, physical discomfort or public health. If one is willing to pay the cost, one can reduce the quantities of various harmful byproducts by changing combustion processes or by instituting effluent controls.

The end product of nuclear fission is an assortment of radioactive isotopes that have a wide range of lifetimes extending up to thousands of years. Although the total radioactivity decreases with time, there is no question that these radioactive substances must be carefully contained, controlled and stored. Fortunately the physical amounts involved are extremely small in bulk: about 10 cubic feet per year from a 1,000-megawatt fast-breeder power plant. The problem is one of extracting these substances during the chemical processes used for reconstituting the nuclear fuels, and then containing and storing them in a safe manner. Because of the small volume of material produced in the annual operation of a nuclear power station even elaborate handling procedures contribute only a small part to the cost of nuclear power. Although the total amount of radioactive waste today is relatively small, the amounts will be large 30 years hence. Pilot programs are needed now to develop safe handling for these future wastes.

All energy use ends up as unrecoverable waste heat. The final heat sink for the earth is radiation to space. The worldwide man-made thermal load, however, is so small compared with the solar heat load as to be insignificant on a global scale. In the year 2000 the worldwide use of energy will still be much less than a thousandth of the sun's heat input. Nevertheless, one can expect that the concentrated generation and consumption of energy in densely popu-

lated areas will be capable of affecting both the local climate and ecological systems. Since rationing of energy does not seem feasible, the only practical solution may be to limit the population density of our major cities.

While recognizing the troposphere as the ultimate heat sink, we have a number of options for influencing the flow of heat from the point where it is released to ultimate radiation into space. Of great public importance is the management of the large quantities of waste heat produced in the generation of electricity. It has been customary to locate electric-power stations on large bodies of water, rivers, lakes or oceans, for the purpose of using the available cooling water to reduce the minimum temperature of the Carnot cycle involved in the generation of power. Because of the recent growth in electric-power generation many of the inland bodies of water are approaching a natural limitation in

their ability to absorb waste heat. The most severe of such limitations is the ecological effect on marine life; the maximum temperature that can be tolerated by marine animals is not high.

One way to avoid heating inland bodies of water is to use the waste heat to evaporate a relatively small volume of water rather than to raise a large volume by only a few degrees. Evaporation is carried out by means of a "wet" cooling tower, which is now rather widely used by electric-power stations, particularly in Britain. This approach presents two problems. If the water is drawn from a small river or a small lake, the amount evaporated can reduce the amount available for other purposes. The second problem arises from the considerable amount of water vapor added to the atmosphere, which produces a sharp increase in the local humidity. In some regions of the country, valleys in particular, this would produce heavy fog

CONTROL	INDIVIDUAL SELECTION	SOCIETAL SELECTION	ECONOMIC FEASIBILITY	TECHNICAL FEASIBILITY
IMPLEMENTATION TIME (YEARS)	1	10	10 — 100	
COSTS INVOLVED (DOLLARS)	$10^2 - 10^4$	$10^6 - 10^8$	$10^9 - 10^{11}$	
	OPTIONAL USES COMFORT (HEATING, AIR CONDITIONING) ENTERTAINMENT COMMUNICATION HOME TRANSPORTATION LABOR AID CRITERIA RELATIVE COSTS PERSONAL SAFETY QUALITY OF LIFE INTANGIBLE AND SUBJECTIVE BIASES	DEVICE UTILIZATION CENTRAL STATION V. LOCAL POWER PLANT TYPE OF CONVERSION METHOD DISTRIBUTION ALTERNATIVES RESOURCE DEVELOPMENT COAL OIL AND NATURAL GAS NUCLEAR SHALE OIL COAL GASIFICATION FUSION SOLAR SITING CHOICES AT ORIGIN OF FUEL CLOSE TO USER CONSIDER AESTHETICS LAND-USE ALTERNATIVES WASTE DISPOSAL ENVIRONMENTAL DETERIORATION REGULATION AND CONTROL LEGISLATION REGULATIONS STANDARDS	SPECULATIVE RESOURCES SOLAR POWER FUSION BIOLOGICAL PHOTOSYNTHESIS FUEL CELLS, MHD, DIRECT CONVERSION ALTERNATIVE FUELS ALCOHOL LIQUID HYDROGEN AMMONIA ENVIRONMENTAL EFFECTS RECYCLE WASTES WASTE STORAGE (RADIOACTIVE) UNDERGROUND DISTRIBUTION SAFETY	

CONTROLLING FACTORS that enter into long-range energy planning are listed in this table. Some factors, such as those listed under individual and societal selection, can operate in a relatively short time. Other factors, such as those listed under economic

during much of the year, with considerable discomfort to the local population.

The next choice in heat disposal would be direct dissipation to the air from a closed-cycle heat exchanger in the form of "dry" cooling towers. This is the same technique used in an automobile radiator for cooling the engine. Although dry cooling towers obviate the need for a water supply altogether, they not only require a higher capital investment but also decrease the thermodynamic efficiency of the power station because ambient air temperatures are generally much higher than the temperatures that can be reached with a water-cooling system. Nevertheless, for inland power stations environmental considerations may force a steady increase in the use of dry cooling towers.

In many respects the most suitable location for electric-power stations is on or near the ocean. The ocean represents

NATURAL LIMITATIONS

100 — 1,000

RESOURCES
FINITE FOSSIL-FUEL RESERVES
URANIUM USAGE DEPENDS ON BREEDER

CONTINUOUS SOURCES
LIMITED EXCEPT FOR SOLAR

ENVIRONMENTAL EFFECTS
THERMODYNAMIC LIMIT ON CONVERSION EFFICIENCY

REGIONAL CLIMATIC EFFECTS

CO_2 PRODUCTION INEVITABLE FROM FOSSIL FUELS

and technical feasibility and natural limitations, involve the fate of future generations.

a heat sink of such magnitude as to be on the average unaffected by the waste heat man can introduce for the foreseeable future. There are also many areas of the ocean where local increases in temperature could even be beneficial to marine life. For this reason the location of power stations on the shores of large oceans may become increasingly popular throughout the world.

It is evident that the issues raised by the role of energy in social development fall into two broad categories: those that relate to the highly developed regions of the world and those that relate to the underdeveloped regions. Because industrialized nations now have the capability both for sustaining a modestly increasing population and for improving the average quality of life, it is likely that in the next century the per capita energy consumption in advanced countries will approach an equilibrium level.

For the underdeveloped nations, which include most of the world's population, the situation is quite different. The peoples of these nations are still primarily engaged in maintaining a minimum level of subsistence; they do not have available the power resources necessary for their transition to a literate, industrial, urban and advanced agricultural society. Such a transition will be significantly dependent on the availability of energy. It is sometimes suggested that because power production and energy consumption have harmful effects on the environment the use of energy must be arbitrarily limited. This implies the same type of social control as arbitrarily limiting the water supply, food production or population. Given the humane objective of providing the people of the world with a quality of life as high as man's ingenuity can develop, the essential role of energy must be accepted.

Within nature's limitations man has tremendous scope for planning energy utilization [see illustration on these two pages]. Some of the controlling factors that enter into energy policy depend on the voluntary decisions of the individual as well as on government actions that may restrict individual freedom. The questions of feasibility, both economic and technical, depend for their solution on the priority and magnitude of the effort applied. The time scale and costs for implementing decisions, or resolving issues, in all areas of energy management have both short-term and long-term consequences. There are so many variables that their arrangement into a

"scenario" for the future becomes a matter of individual choice and a fascinating planning game. The intellectual complexity of the possible arrangements for the future can, however, be reduced to a limited number of basic policy questions that are more sociological than technical in nature.

The first set of questions has to do with the development of energy availability. These might be succinctly stated as follows. Whose resources should be utilized? Where should power be generated? Who shall receive the polluting effluents from such activities? These questions are particularly significant because fuel resources can be shipped all over the world by inexpensive ocean transport and electric power can be transmitted as needed over grids of continental size.

Our present approach to fuel sources has resulted in an international network for the tapping of the world's oil resources. Until World War II the U.S. was a net exporter of energy supplies. Today the Middle East and Africa are the major suppliers of oil, and they possess more than half of the world's fossil-fuel reserves. Thus the underdeveloped parts of the world are exporting their natural fuel resources to the developed parts of the world. Both western Europe and Japan are unique cases of highly industrialized areas almost completely dependent on the importation of oil from underdeveloped countries. The economics of this situation has been a prime factor in discouraging the U.S. from meeting its petroleum needs with oil shale and tar sands, which are available in very large amounts. Nevertheless, both the political and the sociological consequences of depending on foreign sources of supply make it likely that the oil shale and tar sands will be tapped even at high cost. Long-range planning to prepare for this technical development obviously has to be included in any national consideration of energy policies.

A current example of the knotty issues involved in separating the source of power from the user is found in the Four Corners power-region development in the U.S. The Four Corners embrace the region where Utah, New Mexico, Arizona and Colorado meet. It is a region with abundant reserves of low-sulfur coal, plentiful cooling water from the Colorado River and a low population density. The original plan was to build six coal-fired plants to provide electric power primarily for the large cities of Los Angeles, Las Vegas, Phoenix, Al-

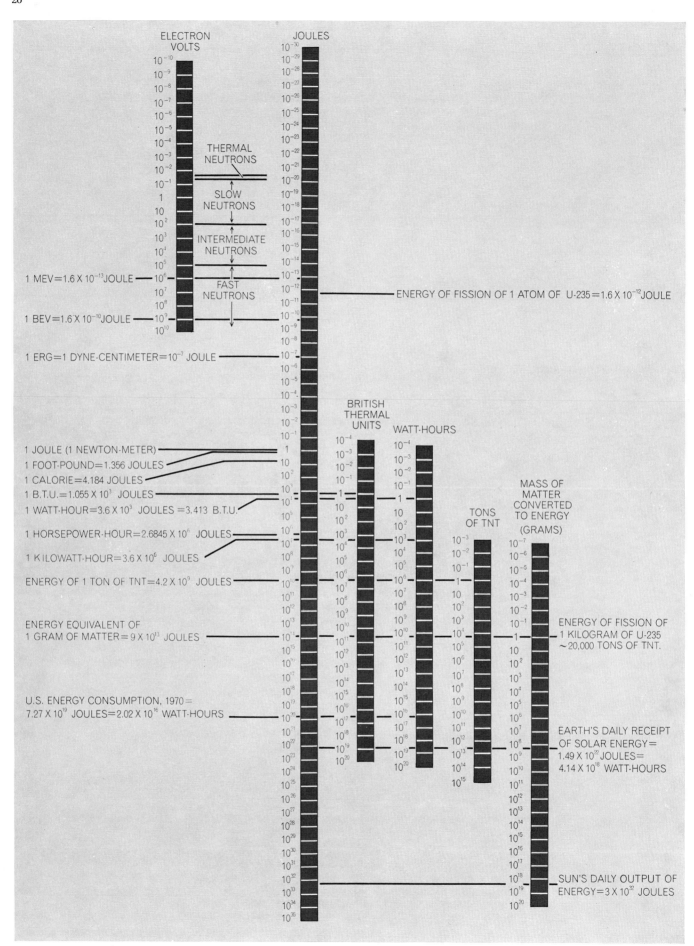

buquerque and other urban communities of the southwestern U.S.—all far distant from the Four Corners. In spite of the low population density of the Four Corners area, considerable protest has arisen over the environmental effects of intensive strip mining, the use of the Colorado for waste-heat disposal and a large-scale outpouring of stack effluents. Although some compromise of social benefits and penalties will presumably be reached to determine the acceptable levels of pollution from these plants, the Four Corners scheme epitomizes the kind of problem we can expect to encounter increasingly in planning large energy centers.

The issue of who gets the pollution, as contrasted with who gets the energy, is not only one of geographic distribution but also one of time. For example, if as a result of the rapid increase in strip-mining for coal, the acid drainage and soil erosion disrupt ecosystems over a large region, it may take decades to repair the damage in spite of the coal mining company's genuine effort to restore the local area to a semblance of its original condition. This generation of energy users will have been long gone when succeeding generations face the problem and the cost of repairing the damage of such ecological degradation. Other long-term and long-delayed problems may be associated with the effluents released by using power to do useful work. The penalties imposed on future generations are the result of social choices made today.

Another category of choices relate to the way we use energy after allocating the fraction necessary for doing useful work. Our society provides many options in which energy is used for recreation, environmental conditioning, communication and entertainment. An automobile tour of a country, a powerboat cruise, an airplane vacation trip all represent energy consumption subject to individual choice and taste. They also represent choices that produce effluents with some effect on the environment. Are we prepared to limit this freedom of choice, which implies the freedom to pollute? The answer will require a careful balancing of values.

Most significant is the allocation of

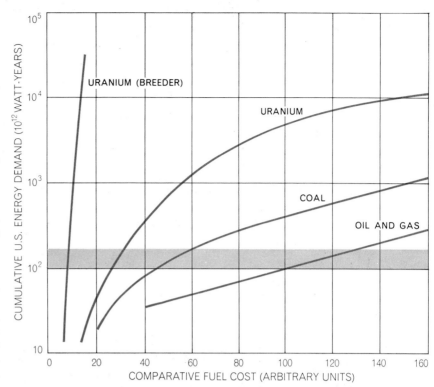

COMPARATIVE FUEL COSTS are plotted against the nation's cumulative demand for energy. The horizontal band covers the range in probable demand from 1960 to 2000. Uranium used in present-day nonbreeder reactors is already cheaper than fossil fuels. The fast-breeder reactor should hold fuel costs essentially constant far beyond the year 2000.

our national resources, manpower and technology to the improvement of the physical environment as compared with other needs. If we did not count the cost, there is little doubt we could so reduce the effluents from the utilization of energy that their effects on public health would be truly negligible. The cost, however, might be excessively large. Thus one must ask if there is an intermediate level of control that is acceptable for comfort, health and aesthetics. The continuous exposure of man to many natural pollutants (uninfluenced by man) is great enough so that there may not be much justification for reducing the pollutants of energy conversion much below the natural background levels. Since even the wealthy U.S. cannot satisfy all the demands on its resources, the level of pollution control seems bound to emerge as a major factor in the debate over national priorities. For example, it may be much more important to allocate resources to improving

public health services rather than to use that same sum to marginally reduce environmental pollution. It is unfortunately true that in a pluralistic society the value systems and priorities differ among the society's sectors. The groups seeking aesthetic and environmental improvement may be a minority compared with the much larger number seeking basic material improvements. In the energy field a decision on a national level concerning the energy system may be a determining factor in shaping the framework of our society for some generations to come.

Perhaps the most fundamental question of national policy is how we should allocate our present resources for the benefit of future generations. The development of new speculative energy resources is an investment for the future, not a means of remedying the problems of today. It is equally clear that the quality of life of the peoples of the world depends on the availability *now* of large amounts of low-cost energy in useful form. This being so, we must emphasize an orderly development of the resources available to us with present technology, and these are primarily power plants based on fossil fuels and nuclear fission.

ENERGY UNITS and conversion factors are presented on the opposite page. Physicists find it convenient to use electron volts, ergs and joules. Biologists and nutritionists think in calories. Engineers deal in British thermal units and watt-hours. Since Hiroshima energy release is commonly expressed in tons of TNT. It is less often observed that a ton of ordinary coal contains three times as much energy as a ton of TNT. The illustration is based on one that appears in *The New College Physics: A Spiral Approach,* by Albert V. Baez.

The Conversion of Energy

by Claude M. Summers
September 1971

The efficiency of home furnaces, steam turbines, automobile engines and light bulbs helps in fixing the demand for energy. A major need is a kind of energy source that does not add to the earth's heat load

A modern industrial society can be viewed as a complex machine for degrading high-quality energy into waste heat while extracting the energy needed for creating an enormous catalogue of goods and services. Last year the U.S. achieved a gross national product of just over $1,000 billion with the help of 69×10^{15} British thermal units of energy, of which 95.9 percent was provided by fossil fuels, 3.8 percent by falling water and .3 percent by the fission of uranium 235. The consumption of 340 million B.t.u. per capita was equivalent to the energy contained in about 13 tons of coal or, to use a commodity now more familiar, 2,700 gallons of gasoline. One can estimate very roughly that between 1900 and 1970 the efficiency with which fuels were consumed for all purposes increased by a factor of four. Without this increase the U.S. economy of 1971 would already be consuming energy at the rate projected for the year 2025 or thereabouts.

Because of steadily increasing efficiency in the conversion of energy to useful heat, light and work, the G.N.P. between 1890 and 1960 was enabled to grow at an average annual rate of 3.25 percent while fuel consumption increased at an annual rate of only 2.7 per-

cent. It now appears, however, that this favorable ratio no longer holds. Since 1967 annual increases in fuel consumption have risen faster than the G.N.P., indicating that gains in fuel economy are becoming hard to achieve and that new goods and services are requiring a larger energy input, dollar for dollar, than those of the past. If one considers only the predicted increase in the use of nuclear fuels for generating electricity, it is apparent that an important fraction of the fuel consumed in the 1980's and 1990's will be converted to a useful form at lower efficiency than fossil fuels are today. The reason is that present nuclear plants convert only about 30 percent of the energy in the fuel to electricity compared with about 40 percent for the best fossil-fuel plants.

It is understandable that engineers should strive to raise the efficiency with which fuel energy is converted to other and more useful forms. For industry increased efficiency means lower production costs; for the consumer it means lower prices; for everyone it means reduced pollution of air and water. Electric utilities have long known that by lowering the price of energy for bulk users they can encourage consumption. The recent campaign of the utility in-

dustry to "save a watt" marks a profound reversal in business philosophy. The difficulty of finding acceptable new sites for power plants underscores the need not only for frugality of use but also for efficiency of use. Having said this, one must emphasize that even large improvements in efficiency can have only a modest effect in extending the life of the earth's supply of fossil and nuclear fuels. I shall develop the point more fully later in this article.

The efficiency with which energy contained in any fuel is converted to useful form varies widely, depending on the method of conversion and the end use desired. When wood or coal is burned in an open fireplace, less than 20 percent of the energy is radiated into the room; the rest escapes up the chimney. A well-designed home furnace, on the other hand, can capture up to 75 percent of the energy in the fuel and make it available for space heating. The average efficiency of the conversion of fossil fuels for space heating is now probably between 50 and 55 percent, or nearly triple what it was at the turn of the century. In 1900 more than half of all the fuel consumed in the U.S. was used for space heating; today less than a third is so used.

The most dramatic increase in fuel-conversion efficiency in this century has been achieved by the electric-power industry. In 1900 less than 5 percent of the energy in the fuel was converted to electricity. Today the average efficiency is around 33 percent. The increase has been achieved largely by increasing the temperature of the steam entering the turbines that turn the electric generators and by building larger generating units [*see illustration on opposite page*]. In 1910 the typical inlet temperature was 500 degrees Fahrenheit; today the latest

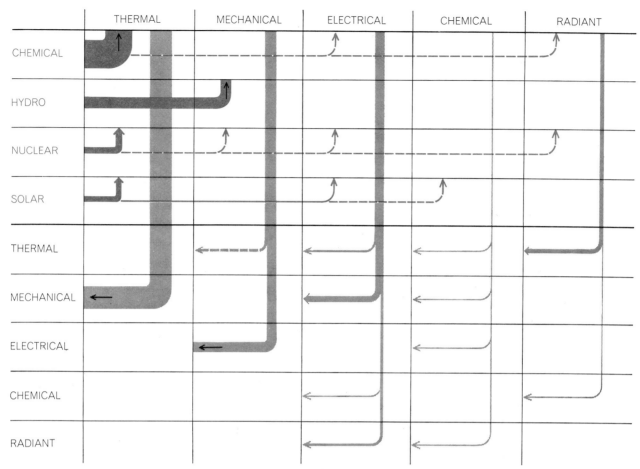

	THERMAL	MECHANICAL	ELECTRICAL	CHEMICAL	RADIANT

CONVERSION PATHWAYS link many of the familiar forms of energy. The four forms shown in color are either important sources of power today or, in the case of solar energy, potentially important. The broken lines indicate rare, incidental or theoretically useful conversions. The gray lines follow the destiny of intermediate forms of energy. Except for the thermal energy used for space heating, most is converted to mechanical energy. Mechanical energy is used directly for transportation (*see illustration below*) and for generating electricity. Electrical energy in turn is used for lighting, heating and mechanical work. As a secondary form, chemical energy is found in dry cells and storage batteries. The radiant energy produced by electric lamps ends up chiefly as heat.

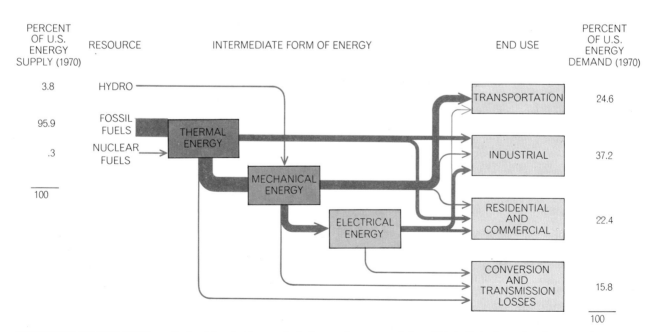

PATHWAYS TO END USES are depicted for the three principal sources of energy. The most direct and most efficient conversion is from falling water to mechanical energy to electrical energy. The energy locked in fossil and nuclear fuels must first be released in the form of thermal energy before it can be converted to mechanical energy and then, if it is desired, to electric power. Conversion and transmission losses include various nonenergy uses of fossil fuels, such as the manufacture of lubricants and the conversion of coal to coke. The biggest loss, however, arises from the generation of electric power at an average efficiency of 32.5 percent.

units take steam superheated to 1,000 degrees. The method of computing the maximum theoretical efficiency of a steam turbine or other heat engine was enunciated by Nicolas Léonard Sadi Carnot in 1824. The maximum achievable efficiency is expressed by the fraction $(T_1 - T_2)/T_1$, where T_1 is the absolute temperature of the working fluid entering the heat engine and T_2 is the temperature of the fluid leaving the engine. These temperatures are usually expressed in degrees Kelvin, equal to degrees Celsius plus 273, which is the difference between absolute zero and zero degrees C. In a modern steam turbine T_1 is typically 811 degrees K. (1,000 degrees Fahrenheit) and T_2 311 degrees K. (100 degrees F.). Therefore according to Carnot's equation the maximum theoretical efficiency is about 60 percent. Because the inherent properties of a steam cycle do not allow the heat to be introduced at a constant upper temperature, the maximum theoretical efficiency is not 60 percent but more like 53 percent. Modern steam turbines achieve about 89 percent of that value, or 47 percent net.

To obtain the overall efficiency of a steam power plant this value must be multiplied by the efficiencies of the other energy converters in the chain from fuel to electricity. Modern boilers can convert about 88 percent of the chemical energy in the fuel into heat. Generators can convert up to 99 percent of the mechanical energy produced by the steam turbine into electricity. Thus the overall efficiency is 88 × 47 (for the turbine) × 99, or about 41 percent.

Nuclear power plants operate at lower efficiency because present nuclear reactors cannot be run as hot as boilers burning fossil fuel. The temperature of the steam produced by a boiling-water reactor is around 350 degrees C., which means that the T_1 in the Carnot equation is 623 degrees K. For the complete cycle from fuel to electricity the efficiency of a nuclear power plant drops to about 30 percent. This means that some 70 percent of the energy in the fuel used by a nuclear plant appears as waste heat, which is released either into an adjacent body of water or, if cooling towers are used, into the surrounding air. For a fossil-fuel plant the heat wasted in this way amounts to about 60 percent of the energy in the fuel.

The actual heat load placed on the water or air is much greater, however, than the difference between 60 and 70 percent suggests. For plants with the same kilowatt rating, a nuclear plant produces about 50 percent more waste

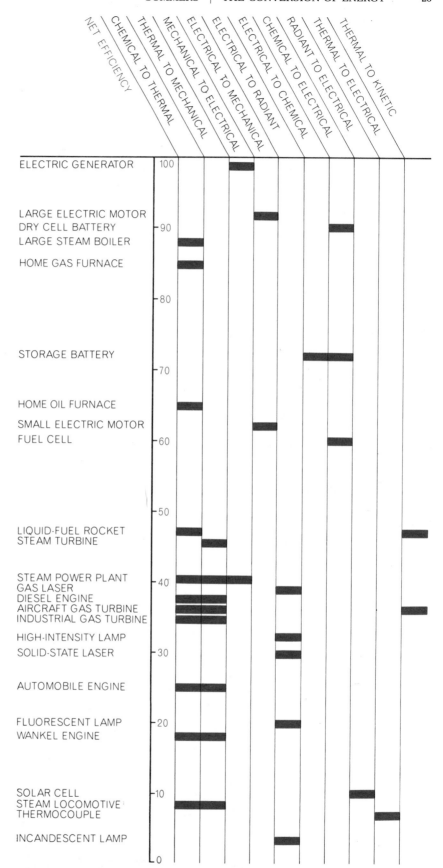

EFFICIENCY OF ENERGY CONVERTERS runs from less than 5 percent for the ordinary incandescent lamp to 99 percent for large electric generators. The efficiencies shown are approximately the best values attainable with present technology. The figure of 47 percent indicated for the liquid-fuel rocket is computed for the liquid-hydrogen engines used in the Saturn moon vehicle. The efficiencies for fluorescent and incandescent lamps assume that the maximum attainable efficiency for an acceptable white light is about 400 lumens per watt rather than the theoretical value of 220 lumens per watt for a perfectly "flat" white light.

heat than a fossil-fuel plant. The reason is that a nuclear plant must "burn" about a third more fuel than a fossil-fuel plant to produce a kilowatt-hour of electricity and then wastes 70 percent of the larger B.t.u. input.

Of course, no law of thermodynamics decrees that the heat released by either a nuclear or a fossil-fuel plant must go to waste. The problem is to find something useful to do with large volumes of low-grade energy. Many uses have been proposed, but all run up against economic limitations. For example, the low-pressure steam discharged from a steam turbine could be used for space heating. This is done in some communities, notably in New York City, where Consolidated Edison is a large steam supplier. Many chemical plants and refineries also use low-pressure steam from turbines as process steam. It has been suggested that the heated water released by power plants might be beneficial in speeding the growth of fish and shellfish in certain localities. Nationwide, however, there seems to be no attractive use for the waste heat from the present fossil-fuel plants or for the heat that will soon be pouring from dozens of new nuclear power plants. The problem will be to limit the harm the heat can do to the environment.

From the foregoing discussion one can see that the use of electricity for home heating (a use that is still vigorously promoted by some utilities) represents an inefficient use of chemical fuel. A good oil- or gas-burning home furnace is at least twice as efficient as the average electric-generating station. In some locations, however, the annual cost of electric space heating is competitive with direct heating with fossil fuels even at the lower efficiency. Several factors account for this anomaly. The electric-power rate decreases with the added load. Electric heat is usually installed in new constructions that are well insulated. The availability of gas is limited in some locations and its cost is higher. The delivery of oil is not always dependable. As fossil fuels become scarcer, their cost will increase, and the production of electrical energy with nuclear fuels will increase. Unfortunately we must expect that a greater percentage of our fuel resources (particularly nuclear fuels) will be consumed in electric space heating in spite of the less efficient use of fuel.

The most ubiquitous of all prime movers is the piston engine. There are two in many American garages, not counting the engines in the power mower, the snowblower or the chain saw. The piston engines in the nation's more than 100 million motor vehicles have a rated capacity in excess of 17 billion horsepower, or more than 95 percent of the capacity of all prime movers (defined as engines for converting fuel to mechanical energy). Although this huge capacity is unemployed most of the time, it accounts for more than 16 percent of the fossil fuel consumed by the U.S. Transportation in all forms—including the propulsion systems of ships, locomotives and aircraft—absorbs about 25 percent of the nation's energy budget.

Automotive engineers estimate that the efficiency of the average automobile engine has risen about 10 percent over the past 50 years, from roughly 22 percent to 25 percent. In terms of miles delivered per gallon of fuel, however, there has actually been a decline. From 1920 until World War II the average automobile traveled about 13.5 miles per gallon of fuel. In the past 25 years the average has fallen gradually to about 12.2 miles per gallon. This decline is due to heavier automobiles with more powerful engines that encourage greater acceleration and higher speed. It takes about eight times more energy to push a vehicle through the air at 60 miles per hour than at 30 miles per hour. The same amount of energy used in accelerating the car's mass to 60 miles per hour must be absorbed as heat, primarily in the brakes, to stop the vehicle. Therefore most of the gain in engine efficiency is lost in the way man uses his machine. Automobile air conditioning has also played a role in reducing the miles per gallon. With the shift in consumer preference to smaller cars the figure may soon begin to climb. The efficiency of the basic piston engine, however, cannot be improved much further.

If all cars in the year 2000 operated on electric batteries charged by electricity generated in central power stations, there would be little change in the nation's overall fuel requirement. Although the initial conversion efficiency in the central station might be 35 percent compared with 25 percent in the piston engine, there would be losses in distributing the electrical energy and in the conversion of electrical energy to chemical energy (in the battery) and back to electrical energy to turn the car wheels. Present storage batteries have an overall efficiency of 70 to 75 percent, so that there is not much room for improvement. Anyone who believes we

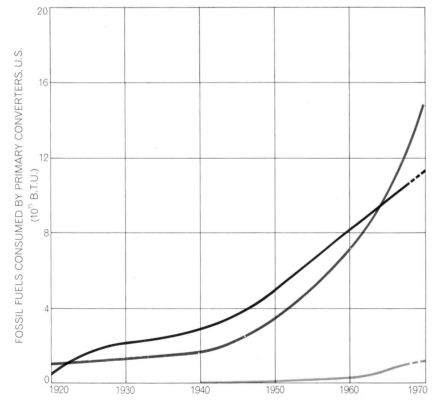

THREE OF FASTEST-GROWING ENERGY USERS are electric utilities (*color*), motor vehicles (*black*) and aircraft (*gray*). Together they now consume about 40 percent of all the energy used in the U.S. As recently as 1940 they accounted for only 18 percent of a much smaller total. The demand for aircraft fuel has more than tripled in 10 years.

would all be better off if cars were electrically powered must consider the problem of increasing the country's electric-generating capacity by about 75 percent, which is what would be required to move 100 million vehicles.

The difficulty of trying to trace savings produced by even large changes in efficiency of energy conversion is vividly demonstrated by what happened when the railroads converted from the steam locomotive (maximum thermal efficiency 10 percent) to diesel-electric locomotives (thermal efficiency about 35 percent). In 1920 the railroads used about 135 million tons of coal, which represented 16 percent of the nation's total energy demand. By 1967, according to estimates made by John Hume, an energy consultant, the railroads were providing 54 percent more transportation than in 1920 (measured by an index of "transportation output") with less than a sixth as many B.t.u. This increase in efficiency, together with the railroads' declining role in the national economy, had reduced the railroads' share of the nation's total fuel budget from 16 percent to about 1 percent. If one looks at a curve of the country's total fuel consumption from 1920 to 1967, however, the impact of this extraordinary change is scarcely visible.

Perhaps the least efficient important use for electricity is providing light. The General Electric Company estimates that lighting consumes about 24 percent of all electrical energy generated, or 6 percent of the nation's total energy budget. It is well known that the glowing filament of an ordinary 100-watt incandescent lamp produces far more heat than light. In fact, more than 95 percent of the electric input emerges as infrared radiation and less than 5 percent as visible light. Nevertheless, this is about five times more light than was provided by a 100-watt lamp in 1900. A modern fluorescent lamp converts about 20 percent of the electricity it consumes into light. These values are based on a practical upper limit of 400 lumens per watt, assuming the goal is an acceptable light of less than perfect whiteness. If white light with a totally flat spectrum is specified, the maximum theoretical output is reduced to 220 lumens per watt. If one were satisfied with light of a single wavelength at the peak sensitivity of the human eye (555 nanometers), one could theoretically get 680 lumens per watt.

General Electric estimates that fluorescent lamps now provide about 70 percent of the country's total illumination and that the balance is divided between incandescent lamps and high-

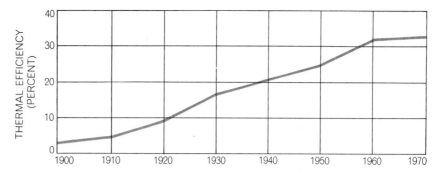

EFFICIENCY OF FUEL-BURNING POWER PLANTS in the U.S. increased nearly tenfold from 3.6 percent in 1900 to 32.5 percent last year. The increase was made possible by raising the operating temperature of steam turbines and increasing the size of generating units.

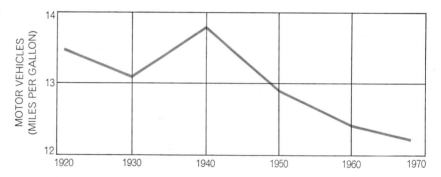

EFFICIENCY OF RAILROAD LOCOMOTIVES can be inferred from the energy needed by U.S. railroads to produce a unit of "transportation output." The big leap in the 1950's reflects the nearly complete replacement of steam locomotives by diesel-electric units.

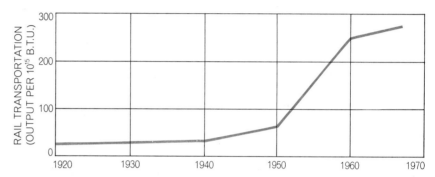

EFFICIENCY OF AUTOMOBILE ENGINES is reflected imperfectly by miles per gallon of fuel because of the increasing weight and speed of motor vehicles. Manufacturers say that the thermal efficiency of the 1920 engine was about 22 percent; today it is about 25 percent.

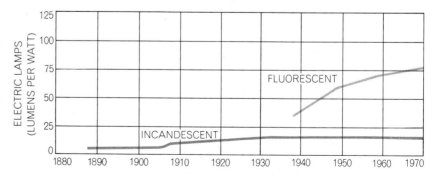

EFFICIENCY OF ELECTRIC LAMPS depends on the quality of light one regards as acceptable. Theoretical efficiency for perfectly flat white light is 220 lumens per watt. By enriching the light slightly in mid-spectrum one could obtain about 400 lumens per watt. Thus present fluorescent lamps can be said to have an efficiency of either 36 percent or 20.

intensity lamps, which have efficiencies comparable to, and in some cases higher than, fluorescent lamps. This division implies that the average efficiency of converting electricity to light is about 13 percent. To obtain an overall efficiency for converting chemical (or nuclear) energy to visible light, one must multiply this percentage times the average efficiency of generating power (33 percent), which yields a net conversion efficiency of roughly 4 percent. Nevertheless, thanks to increased use of fluorescent and high-intensity lamps, the nation was able to triple its "consumption" of lighting between 1960 and 1970 while only doubling the consumption of electricity needed to produce it.

This brief review of changing efficiencies of energy use may provide some perspective when one tries to evaluate

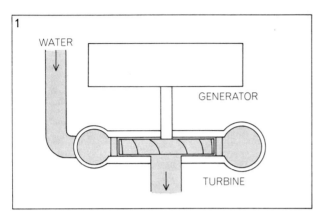

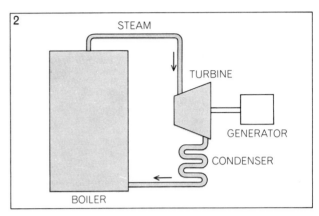

ELECTRIC-POWER GENERATING MACHINERY now in use extracts energy from falling water, fossil fuels or nuclear fuels. The hydroturbine generator (1) converts potential and kinetic energy into electric power. In a fossil-fuel steam power plant (2) a boiler produces steam; the steam turns a turbine; the turbine turns an electric generator. In a nuclear power plant (3) the fission of ura-

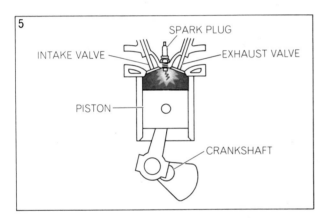

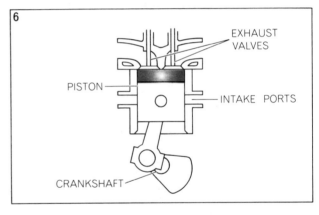

PROPULSION MACHINERY converts the energy in liquid fuels into forms of mechanical or kinetic energy useful for work and transportation. In the piston engine (5) a compressed charge of fuel and air is exploded by a spark; the expanding gases push against the piston, which is connected to a crankshaft. In a diesel engine (6) the compression alone is sufficient to ignite the charge

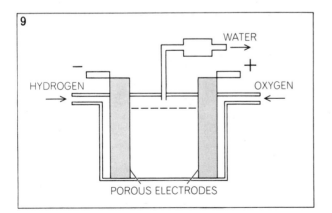

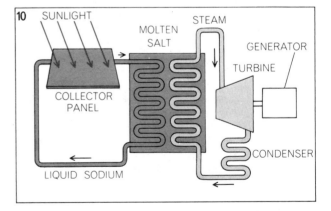

NOVEL ENERGY CONVERTERS are being designed to exploit a variety of energy sources. The fuel cell (9) converts the energy in hydrogen or liquid fuels directly into electricity. The "combustion" of the fuel takes place inside porous electrodes. In a recently proposed solar power plant (10) sunlight falls on specially coated collectors and raises the temperature of a liquid metal to 1,000 degrees F. A heat exchanger transfers the heat so collected to steam, which then turns a turbogenerator as in a conventional power plant. A salt reservoir holds enough heat to keep generating steam during the night and when the sun is hidden by clouds. In a mag-

the probable impact of novel energy-conversion systems now under development. Two devices that have received much notice are the fuel cell and the magnetohydrodynamic (MHD) generator. The former converts chemical energy directly into electricity; the latter is potentially capable of serving as a high-temperature "topping" device to be operated in series with a steam turbine and generator in producing electricity. Fuel cells have been developed that can "burn" hydrogen, hydrocarbons or alcohols with an efficiency of 50 to 60 percent. The hydrogen-oxygen fuel cells used in the Apollo space missions, built by the Pratt & Whitney division of United Aircraft, have an output of 2.3 kilowatts of direct current at 20.5 volts.

A decade ago the magnetohydrodynamic generator was being advanced as

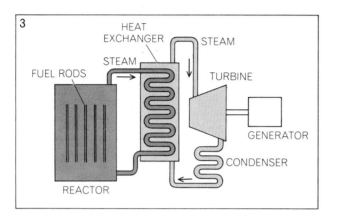

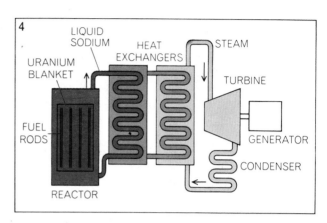

nium 235 releases the energy to make steam, which then goes through the same cycle as in a fossil-fuel power plant. Under development are nuclear breeder reactors (4) in which surplus neutrons are captured by a blanket of nonfissile atoms of uranium 238 or thorium 232, which are transformed into fissile plutonium 239 or U-233. The heat of the reactor is removed by liquid sodium.

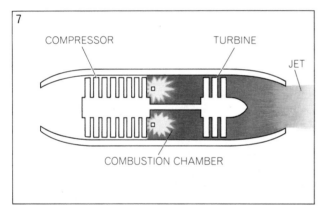

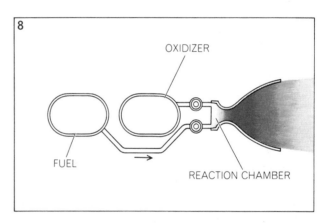

of fuel and air. In an aircraft gas turbine (7) the continuous expansion of hot gas from the combustion chamber passes through a turbine that turns a multistage air compressor. Hot gases leaving the turbine provide the kinetic energy for propulsion. A liquid-fuel rocket (8) carries an oxidizer in addition to fuel so that it is independent of an air supply. Rocket exhaust carries kinetic energy.

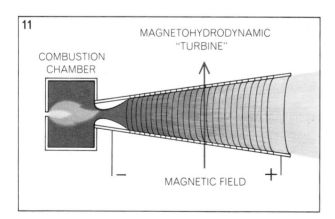

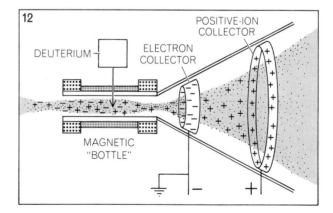

netohydrodynamic "turbine" (11) the energy contained in a hot electrically conducting gas is converted directly into electric power. A small amount of "seed" material, such as potassium carbonate, must be injected into the flame to make the hot gas a good conductor. Electricity is generated when the electrically charged particles of gas cut through the field of an external magnet. A long-range goal is a thermonuclear reactor (12) in which the nuclei of light elements fuse into heavier elements with the release of energy. High-velocity charged particles produced by a thermonuclear reaction might be trapped in such a way as to generate electricity directly.

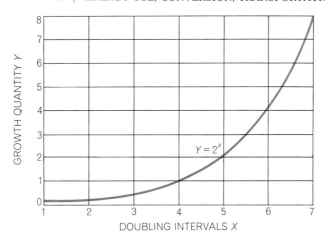

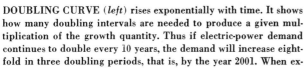

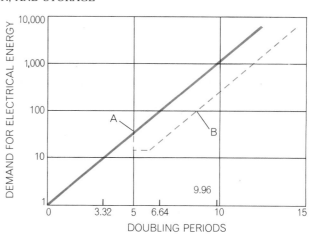

DOUBLING CURVE (*left*) rises exponentially with time. It shows how many doubling intervals are needed to produce a given multiplication of the growth quantity. Thus if electric-power demand continues to double every 10 years, the demand will increase eightfold in three doubling periods, that is, by the year 2001. When ex-

ponential growth curves are plotted on a semilogarithmic scale, the result is a straight line (*right*). If electric-power consumption were cut in half at *A*, held constant for 10 years and allowed to return to the former growth rate, time needed to reach a given demand (*B*) would be extended by only two doubling periods, or 20 years.

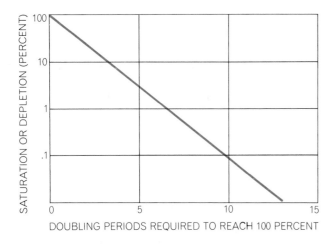

SATURATION OR DEPLETION (PERCENT)	DOUBLING PERIODS TO REACH 100 PERCENT	YEARS FROM NOW
100.0	0.0	0
10.0	3.32	33
1.0	6.64	66
.1	9.96	100
.01	13.28	133
.001	16.60	166
.0001	19.92	199
.00001	23.24	232
.000001	26.56	266
.0000001	29.88	299
.00000001	33.20	332

DEPLETION OF A RESOURCE can be read from the curve at the left. Thus if .1 percent of world's oil has now been extracted, all will be gone in just under 10 doubling periods, or 100 years if the

doubling interval is 10 years. The table at the right shows that the ultimate depletion date is changed very little by large changes in the estimate of amount of resource that has been extracted to date.

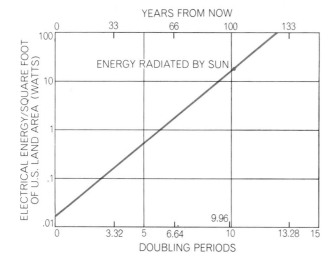

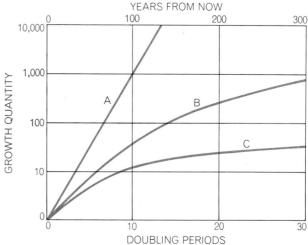

ENERGY RECEIVED FROM SUN on an average square foot of the U.S. will be equaled by production of electrical energy in roughly 100 years if the demand continues to double every 10 years.

THREE GROWTH CURVES are compared. Curve *A* is exponential. In Curve *B* each doubling period is successively increased by 20 percent. In Curve *C* the growth per doubling period is constant.

the energy converter of the future. In such a device the fuel is burned at a high temperature and the gaseous products of combustion are made electrically conducting by the injection of a "seed" material, such as potassium carbonate. The electrically conducting gas travels at high velocity through a magnetic field and in the process creates a flow of direct current [*see No. 11 in illustrations on pages 28 and 29*]. If the MHD technology can be developed, it should be possible to design fossil-fuel power plants with an efficiency of 45 to 50 percent. Since MHD requires very high temperatures it is not suitable for use with nuclear-fuel reactors, which produce a working fluid much cooler than one can obtain from a combustion chamber fired with fossil fuel.

If ever an energy source can be said to have arrived in the nick of time, it is nuclear energy. Twenty-two nuclear power plants are now operating in the U.S. Another 55 plants are under construction and more than 40 are on order. This year the U.S. will obtain 1.4 percent of its electrical energy from nuclear fission; it is expected that by 1980 the figure will reach 25 percent and that by 2000 it will be 50 percent.

Although a 1,000-megawatt nuclear power plant costs about 10 percent more than a fossil-fuel plant ($280 million as against $250 million), nuclear fuel is already cheaper than coal at the mine mouth. Some projections indicate that coal may double in price between now and 1980. One reason given is that new Federal safety regulations have already reduced the number of tons produced per man-day from the 20 achieved in 1969 to fewer than 15.

The utilities are entering a new period in which they will have to rethink the way in which they meet their base load, their intermediate load (which coincides with the load added roughly between 7:00 A.M. and midnight by the activity of people at home and at work) and peak load (the temperature-sensitive load, which accounts for only a few percent of the total demand). In the past utilities assigned their newest and most efficient units to the base load and called on their older and smaller units to meet the variable daily demand. In the future, however, when still newer capacity is added, the units now carrying the basic load cannot easily be relegated to intermittent duty because they are too large to be easily put on the line and taken off.

There is therefore a need for a new kind of flexible generating unit that may be best satisfied by coupling an industrial gas turbine to an electric generator and using the waste heat from the gas turbine to produce low-pressure steam for a steam turbine–generator set. Combination systems of this kind are now being offered by General Electric and the Westinghouse Electric Company. Although somewhat less efficient than the best large conventional units, the gas-turbine units can be brought up to full load in an hour and can be installed at lower cost per kilowatt. To meet brief peak demands utilities are turning to gas turbines (without waste-heat boilers that can be brought up to full load in minutes) and to pumped hydrostorage systems. In the latter systems off-peak capacity is used to pump water to an elevated reservoir from which it can be released to produce power as needed.

Westinghouse has recently estimated that U.S. utilities must build more than 1,000 gigawatts (GW, or 10^9 watts) of new capacity between 1970 and 1990, or more than three times the present installed capacity of roughly 300 GW. Of the new capacity 500 GW, or half, will be needed to handle the anticipated increase in base load and 75 percent of the 500 GW will be nuclear. More than 400 GW of new capacity will be needed to meet the growing intermediate load, and a sizable fraction of it will be provided by gas turbines. The new peaking capacity, amounting to some 170 GW, will be divided, Westinghouse believes, between gas turbines and pumped storage in the ratio of 10 to seven.

Such projections can be regarded as the conventional wisdom. Does unconventional wisdom have anything to offer that may influence power generation by 2000, if not by 1990? First of all, there are the optimists who believe prototype nuclear-fusion plants will be built in the 1980's and full-scale plants in the 1990's. In a sense, however, this is merely conventional wisdom on an accelerated time scale. Those with a genuinely unconventional approach are asking: Why do we not start developing the technology to harness energy from the sun or the wind or the tides?

Many people still remember the Passamaquoddy project of the 1930's, which is once more being discussed and which would provide 300 megawatts (less than a third the capacity of the turbogenerator shown on page 348) by exploiting tides with an average range of 18 feet in the Bay of Fundy, between Maine and Canada. A working tidal power plant of 240 megawatts has recently been placed in operation by the French government in the estuary of the Rance River, where the tides average 27 feet. How much tidal energy might the U.S. extract if all favorable bays and inlets were developed? All estimates are subject to heavy qualification, but a reasonable guess is something like 100 GW. We have just seen, however, that the utilities will have to add 10 times that much capacity just to meet the needs of 1990. One must conclude that tidal power does not qualify as a major unconventional resource.

What about the wind? A study we conducted at Oklahoma State University a few years ago showed that the average wind energy in the Oklahoma City area is about 18.5 watts per square foot of area perpendicular to the wind direction. This is roughly equivalent to the amount of solar energy that falls on a square foot of land in Oklahoma, averaging the sunlight for 24 hours a day in all seasons and under all weather conditions. A propeller-driven turbine could convert the wind's energy into electricity at an efficiency of somewhere between 60 and 80 percent. Like tidal energy and other forms of hydropower, wind power would have the great advantage of not introducing waste heat into the biosphere.

The difficulty of harnessing the wind's energy comes down to a problem of energy storage. Of all natural energy sources the wind is the most variable. One must extract the energy from the wind as it becomes available and store it if one is to have a power plant with a reasonably steady output. Unfortunately technology has not yet produced a practical storage medium. Electric storage batteries are out of the question.

One scheme that seems to offer promise is to use the variable power output of a wind generator to decompose water into hydrogen and oxygen. These would be stored under pressure and recombined in a fuel cell to generate electricity on a steady basis [*see illustration on next page*]. Alternatively the hydrogen could be burned in a gas turbine, which would turn a conventional generator. The Rocketdyne Division of North American Rockwell has seriously proposed that an industrial version of the hydrogen-fueled rocket engine it builds for the Saturn moon vehicle could be used to provide the blast of hot gas needed to power a gas turbine coupled to an electric generator. Rocketdyne visualizes that a water-cooled gas turbine could operate at a higher temperature than conventional fuel-burning gas turbines and achieve our overall plant efficiency of 55 percent. If the Rocketdyne concept were successful, it could use hydrogen from any source. A wind-driven hydrogen-rocket gas-turbine power plant should be unconventional enough to please the most exotic taste.

By comparison most proposals for harnessing solar energy seem tame indeed. One fairly straightforward proposal has recently been made to the Arizona Power Authority on behalf of the University of Arizona by Aden B. Meinel and Marjorie P. Meinel of the university's Optical Sciences Center. They suggest that if the sunlight falling on 14 percent of the western desert regions of the U.S. were efficiently collected, it could be converted into 1,000 GW of power, or approximately the amount of additional power needed between now and 1990. The Meinels believe it is within the reach of present technology to design collecting systems capable of storing solar energy as heat at 1,000 degrees F., which could be converted to electricity at an overall efficiency of 30 percent.

The key to the project lies in recently developed surface coatings that have high absorbance for solar radiation and low emittance in the infrared region of the spectrum. To achieve a round-the-clock power output, heat collected during daylight hours would be stored in molten salts at 1,000 degrees F. A heat exchanger would transfer the stored energy to steam at the same temperature. The thermal storage tank for a 1,000-megawatt generating plant would require a capacity of about 300,000 gallons. The Meinels propose that industry and the Government immediately undertake design and construction of a 100-megawatt demonstration plant in the vicinity of Yuma, Ariz. The collectors for such a plant would cover an area of 3.6 million square meters (slightly more than a square mile). The Meinels estimate that after the necessary development has been done a 1,000-megawatt solar power station might be built for about $1.1 billion, or about four times the present cost of a nuclear power plant. As they point out: "Solar power faces the economic problem that energy is purchased via a capital outlay rather than an operating expense." They calculate nevertheless that a plant with an operating lifetime of 40 years should produce power at an average cost of only half a cent per kilowatt hour.

A more exotic solar-power scheme has been advanced by Peter E. Glaser of Arthur D. Little, Inc. The idea is to place a lightweight panel of solar cells in a synchronous orbit 22,300 miles above the earth, where they would be exposed to sunlight 24 hours a day. Solar cells (still to be developed) would collect the radiant energy and convert it to electricity with an efficiency of 15 to 20 percent. The electricity would then be converted electronically in orbit to microwave energy with an efficiency of 85 percent, which is possible today. The microwave radiation would be at a wavelength selected to penetrate clouds with little or no loss and would be collected by a suitable antenna on the earth. Present techniques can convert microwave energy to electric power with an efficiency of about 70 percent, and 80 to 85 percent should be attainable. Glaser calculates that a 10,000-megawatt (10 GW) satellite power station, large enough to meet New York City's present power needs, would require a solar collector panel five miles square.

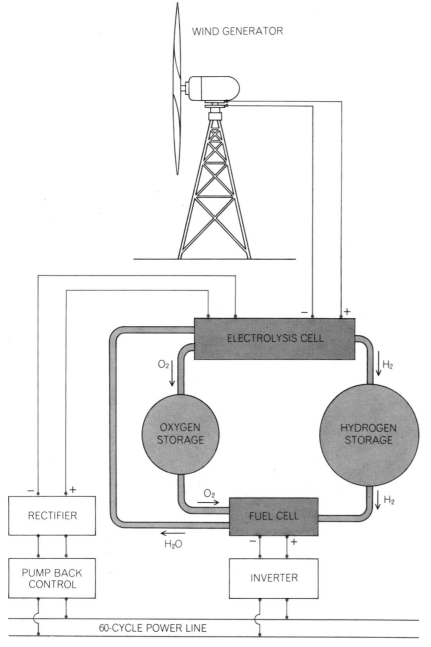

WIND AS POWER SOURCE is attractive because it does not impose an extra heat burden on the environment, as is the case with energy extracted from fossil and nuclear fuels. Unlike hydropower and tidal power, which also represent the entrapment of solar energy, the wind is available everywhere. Unfortunately it is also capricious. To harness it effectively one must be able to store the energy captured when the wind blows and release it more or less continuously. One scheme would be to use the electricity generated by the wind to decompose water electrolytically. The stored hydrogen and oxygen could then be fed at a constant rate into a fuel cell, which would produce direct current. This would be converted into alternating current and fed into a power line. Off-peak power generated elsewhere could also be used to run the electrolysis cell whenever the wind was deficient.

The receiving antenna on the earth would have to be only slightly larger: six miles square. Since the microwave energy in the beam would be comparable to the intensity of sunlight, it would present no hazard. The system, according to Glaser, would cost about $500 per kilowatt, roughly twice the cost of a nuclear power plant, assuming that space shuttles were available for the construction of the satellite. The entire space station would weigh five million pounds, or slightly less than the Saturn moon rocket at launching.

To meet the total U.S. electric-power demand of 2,500 GW projected for the year 2000 would require 250 satellite stations of this size. Since the demand to 1990 will surely be met in other ways, however, one should perhaps think only of meeting the incremental demand for the decade 1990–2000. This could be done with about 125 power stations of the Glaser type.

The great virtue in power schemes based on using the wind or solar energy collected at the earth's surface, far-fetched as they may sound today, is that they would add no heat load to the earth's biosphere; they can be called invariant energy systems. Solar energy collected in orbit would not strictly qualify as an invariant system, since much of the radiant energy intercepted at an altitude of 22,300 miles is radiation that otherwise would miss the earth. Only the fraction collected when the solar panels were in a line between the sun and the earth's disk would not add to the earth's heat load. On the other hand, solar collectors in space would put a much smaller waste-heat load on the environment than fossil-fuel or nuclear plants. Of the total energy in the microwave beam aimed at the earth all but 20 percent or less would be converted to usable electric power. When the electricity was consumed, of course, it would end up as heat.

To appreciate the long-term importance of developing invariant energy systems one must appreciate what exponential growth of any quantity implies. The doubling process is an awesome phenomenon. In any one doubling period the growth quantity—be it energy use, population or the amount of land covered by highways—increases by an amount equal to its growth during its entire past history. For example, during the next doubling period as much fossil fuel will be extracted from the earth as the total amount that has been extracted to date. During the next 10 years the U.S. will generate as much electricity as

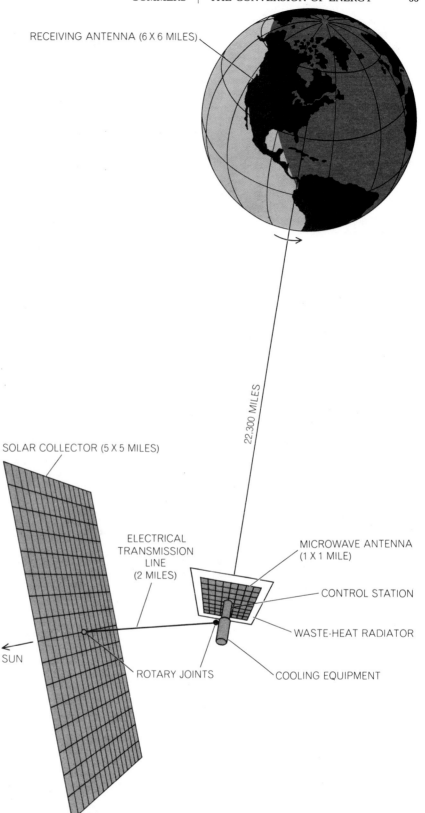

SOLAR COLLECTOR IN STATIONARY ORBIT has been proposed by Peter E. Glaser of Arthur D. Little, Inc. Located 22,300 miles above the Equator, the station would remain fixed with respect to a receiving station on the ground. A five-by-five-mile panel would intercept about 8.5×10^7 kilowatts of radiant solar power. Solar cells operating at an efficiency of about 18 percent would convert this into 1.5×10^7 kilowatts of electric power, which would be converted into microwave radiation and beamed to the earth. There it would be reconverted into 10^7 net kilowatts of electric power, or enough for New York City. The receiving antenna would cover about six times the area needed for a coal-burning power plant of the same capacity and about 20 times the area needed for a nuclear plant.

it has generated since the beginning of the electrical era.

Such exponential growth curves are usually plotted on a semilogarithmic scale in order to provide an adequate span. By selecting appropriate values for the two axes of a semilogarithmic plot one can also obtain a curve showing the number of doubling periods to reach saturation or depletion from any known or assumed percentage position [*see left half of middle illustration on page 30*]. As an example, let us assume that we have now extracted .1 percent of the earth's total reserves of fossil fuels and that the rate of extraction has been doubling every 10 years. If this rate continues, we shall have extracted all of these fuels in just under 10 doubling periods, or in 100 years. We have no certain knowledge, of course, what fraction of all fossil fuels has been extracted. To be conservative let us assume that we have extracted only .01 percent rather than .1 percent. The curve shows us that in this case we shall have extracted 100 percent in 13.3 doubling periods, or 133 years. In other words, if our estimate of the fuel extracted to this moment is in error by a factor of 10, 1,000 or even 100,000, the date of total exhaustion is not long deferred. Thus if we have now depleted the earth's total supply of fossil fuel by only a millionth of 1 percent (.000001 percent), all of it will be exhausted in only 266 years at a 10-year doubling rate [*see right half of middle illustration on page 30*]. I should point out that the actual extraction rate varies for the different fossil fuels; a 10-year doubling rate was chosen simply for the purpose of illustration.

In estimating how many doubling periods the nation can tolerate if the demand for electricity continues to double every 10 years (the actual doubling rate), the crucial factor is probably not the supply of fuels—which is essentially limitless if fusion proves practical—but the thermal impact on the environment of converting fuel to electricity and electricity ultimately to heat. For the sake of argument let us ignore the burden of waste heat produced by fossil-fuel or nuclear power plants and consider only the heat content of the electricity actually consumed. One can imagine that by the year 2000 most of the power will be generated in huge plants located several miles offshore so that waste heat can be dumped harmlessly (for a while at least) into the surrounding ocean.

In 1970 the U.S. consumed 1,550 billion kilowatt hours of electricity. If this were degraded into heat (which it was) and distributed evenly over the total land area of the U.S. (which it was not), the energy released per square foot would be .017 watt. At the present doubling rate electric-power consumption is being multiplied by a factor of 10 every 33 years. Ninety-nine years from now, after only 10 more doubling periods, the rate of heat release will be 17 watts per square foot, or only slightly less than the 18 or 19 watts per square foot that the U.S. receives from the sun, averaged around the clock. Long before that the present pattern of power consumption must change or we must develop the technology needed for invariant energy systems.

Let us examine the consequences of altering the pattern of energy growth in what may seem to be fairly drastic ways. Consider a growth curve in which each doubling period is successively lengthened by 20 percent [*see bottom illustration at right on page 30*]. On an exponential growth curve it takes 3.32 doubling periods, or 33 years, to increase energy consumption by a factor of 10. On the retarded curve it would take five doubling periods, or 50 years, to reach the same tenfold increase. In other words, the retardation amounts to only 17 years. The retardation achieved for a hundredfold increase in consumption amounts to only 79 years (that is, the difference between 145 years and 66 years).

Another approach might be to cut back sharply on present consumption, hold the lower value for some period with no growth and then let growth resume at the present rate. One can easily show that if consumption of power were immediately cut in half, held at that value for 10 years and then allowed to return to the present pattern, the time required to reach a hundredfold increase in consumption would be stretched by only 20 years: from 66 to 86 years [*see right half of top illustration on page 30*].

For long-term effectiveness something like a constant growth curve is required, that is, a curve in which the growth increases by the same amount for each of the original doubling periods. On such a curve nearly 1,000 years would be required for electric-power generation to reach the level of the radiant energy received from the sun instead of the 100 years predicted by a 10-year doubling rate. One can be reasonably confident that the present doubling rate cannot continue for another 100 years, unless invariant energy systems supply a large part of the demand, but what such systems will look like remains hidden in the future.

The Economic Geography of Energy

by Daniel B. Luten
September 1971

*The human uses of energy are reflected in patterns
on the land. The prospecting, recovery, movement and
ultimate use of energy resources are governed by
the ratio of the benefit to the cost*

All men have fire and have used it to change the green face of the earth, and those who live near fuel can have heat in abundance. Only those men who can convert heat and other forms of energy to work, and can apply that work where they will, can travel over the world and shape it to their ends. The crux of the matter is the generation of work—the conversion of energy and its delivery to the point of application. This article will explore some of the interrelations among the location of energy resources, the feasibility and cost of transporting energy commodities and the evolution of technology for converting energy.

Consider for a moment the three crucial developments of the past two centuries that have worked successive revolutions in the human utilization of energy. The first was the steam engine, invented and developed in England primarily as an answer to the flooding of deep coal mines by ground water. Removing the water was far beyond the capacity of human porters or of pumps driven by draft animals. For several centuries the task was accomplished by pumps driven by water mills. There was no realistic way, however, of conveying the action of water mills beyond the immediate site. Was coal mining to be forever confined to the streamside? The response to that challenge was the steam engine. It could operate wherever fuel could be delivered. In the 19th century its efficiency improved enough to make possible the steam locomotive, which could carry enough fuel with it to do work in transport.

The next big step came with the electric generator, the transmission of electricity and the electric motor, which freed work from its bondage to belts and shafts connected to the steam engine's flywheel; work could be provided wherever it was wanted, and in small or large amounts. The final step of this kind was the development of the automotive engine: a small power plant that was less convenient than an electric motor but was not even tied to a power line. Other fuels and conversion devices have appeared and will appear in the future, but they would seem to have less potential for working revolutions in our lives than the heat engine, small electric motors and the automobile.

Man's exploitation of an energy resource comprehends seven operations: discovery of the resource, harvest, transportation, storage, conversion, use and disposal. The discovery of the resource may be explicit and material, as in the case of a coal seam or an oil field. It may be conceptual: the idea of a reservoir or a scheme for capturing solar energy. Often it is the discovery of a conversion, as in the case of fire, the steam engine and uranium fission. And sometimes discovery comprises an entire series of technological improvements, as will be the case when shale oil is finally exploited successfully.

How much has resource discovery influenced human events? The U.S. ran on fuel wood until it had burned up the forests on croppable land as far as the prairies. England and Europe had done about the same thing, and when people ran out of wood, they turned to coal. (They complained; they preferred the old smells and smoke to the new, but they stayed warm with coal.) Whether it was the presence of coal that turned them to industry is another matter, one that is much more difficult to establish. Admittedly wood would not have sufficed, but a few lands with limited fuel have done well (notably Japan) and some with abundant fuel have not. Certainly local fuel does not seem to have been a sufficient condition, or even an entirely necessary one except for a pioneering society.

Gas and oil were adopted rather differently. It is said that as early as 1000 B.C. the Chinese drilled 3,000 feet down for natural gas, piping it in bamboo and burning it for light and heat and to evaporate brine for salt. Elsewhere candles persisted for millenniums and were only slowly succeeded by fatty oils in lamps. Coal had little to offer as an illuminant, but the coking of coal provided gas as well as coke. Handling gas required innovation, which came through the chemical studies of the late 18th century. In England "town gas" soon undercut the price of fatty oil for lamps in the new factories; in the less urban U.S. oil lamps persisted until kerosene appeared in the mid-19th century.

Discovery comes first in the exploitation of a resource; use and then disposal are the next to last and last steps. The sequence of the intervening steps can vary depending on the resource and on the economics and geography and the specific set of operations they dictate. In some cases a preliminary conversion step is introduced: wood may be made into charcoal or coal into coke and petroleum must be refined.

A commodity can move by land either in a continuous process in a conduit or as a batch in a vehicle; shipping by sea must be by batches in vessels. The batch shipper has freedom of destination; a conduit constrains shipment to the chosen destination. The batch shipper, however, needs terminal storage facilities at both ends of every trip so that he can pick up and deliver his cargo with minimum lost time. For some commodities there are many possibilities; for others there are few choices or none.

The constraints on transport have had a significant effect on the adoption of new energy technologies. Primitive people can carry wood easily, coal less handily. The handling of liquids calls for pots and baskets; gases are uncooperative and elusive. Before the advent of simple and efficient equipment for containing and pumping fluids at high pressures, a development largely of recent decades, petroleum moved in barrels or in wood vats on flatcars, and long-distance transmission of gas was impractical. At sea, however, there were tankers, which began carrying oil from the Caucasus almost a century ago.

The combination of tankers and pipelines brought the fossil-fuel industry to a momentarily stable condition in the

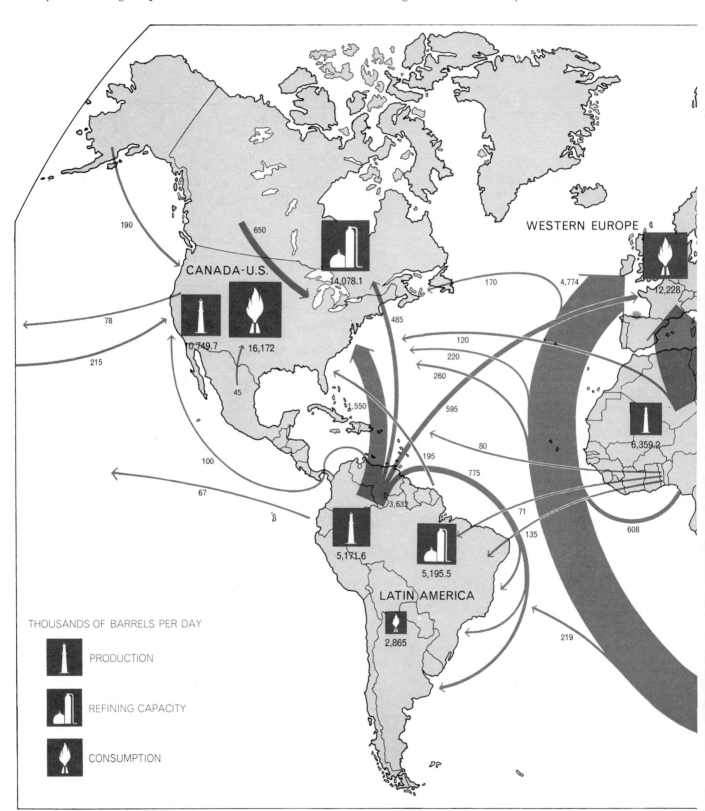

WORLDWIDE PATTERNS of oil production, refining, shipping and consumption are summarized by this map based on maps from the *International Petroleum Encyclopedia*. The data are for 1970. All quantities are in thousands of barrels per day. Export figures

years after World War II. All the possibilities seemed to have been exploited. Now, with competition tightening, innovations are again being pressed hard and marginal improvements are being squeezed for any advantage. Oil brought great distances is made competitive by

increasingly large tankers, but the million-ton supertankers now being proposed must be near the limit. Larger pipelines also shave costs, but most of the pipeline routes that have enough potential also raise international political issues; the proposed trans-Alaska pipe-

line has become a domestic political issue.

As technology advances, the feasibility of transporting some commodities improves. The fact remains that most commodities that can be transported at an acceptable cost today could also be

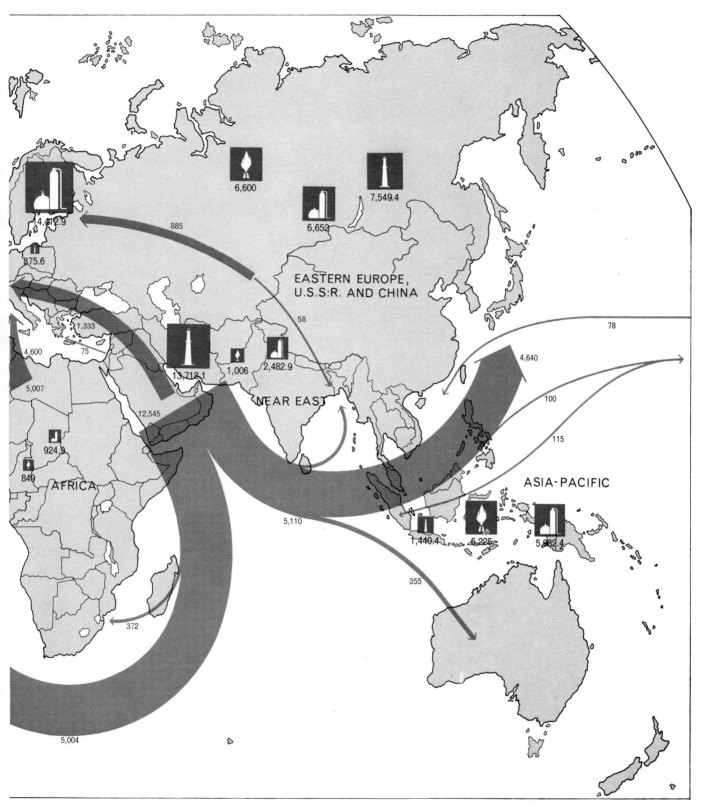

for eastern Europe, the U.S.S.R. and China refer only to exports from those areas to other parts of the world. The arrows indicate the origins and destinations of the principal international oil movements, not the specific routes. The U.S. is a heavy net importer.

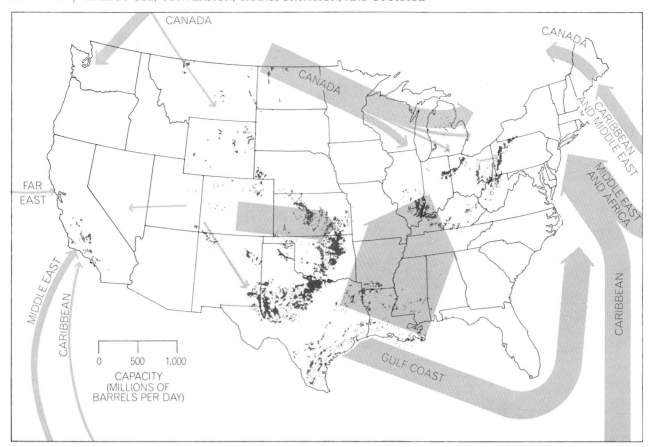

TRANSPORTATION OF CRUDE OIL to the U.S. and within the country is shown by a map adapted from the *National Atlas.* Data are for 1966. Arrow widths are proportional to movements by pipeline (*land*) and tanker (*sea*). Areas in solid color are oil fields.

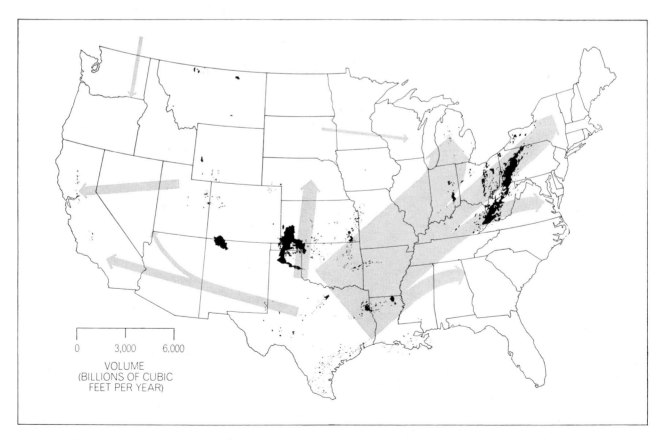

NATURAL-GAS MOVEMENTS are charted, based on figures for 1965. The development of techniques for transporting gas at high pressures in pipes has led to the sharp increase in the use of natural gas since World War II. Areas that are shown in black are gas fields.

transported economically long ago, although admittedly the distances have grown a great deal in this century. Some movements are still impossible; we do not know how to move electricity by sea, for example [*see bottom illustration on page 44*]. The only recent real innovations in transport (except for the appearance of nuclear sources, with their trivial costs of transportation) are the movement of natural gas by sea as a refrigerated liquid and the development of new technologies for electrical transmission.

The power provided to any electrical-conversion unit is the product of the drop in voltage within the unit and the flow of current; the loss of energy as heat in a transmission line is the product of the square of the current and the resistance of the wire. Lower currents, higher voltages and larger wires (less resistance) therefore reduce waste. There are limits to the size of a wire, and so improvements in transmission were achieved primarily by utilizing alternating current (which could be transformed easily) and stepping up the voltage. Transmission voltages have increased as demands have grown and as transmission technology (insulation, for example) has improved, but the gains have required successive doublings of voltage rather than incremental increases, and the end of the road seems to have been reached for alternating current at less than 500,000 volts.

These gains, combined with the high growth rate of the electric-power industry in the U.S. and with the large economies of scale in the construction of power plants, have changed the look of the land. The oldest power plants were small and widely scattered about the cities; the countryside had no electricity and no prospect of having it. Today power plants have become even larger; they are moving out of the cities, and high-voltage lines dominate miles of the countryside. Electricity is provided where it is wanted; transmission is not as cheap as moving fuel, and yet it is attractive to build big power plants and move electric power more than 100 miles to consumers. Still, the pressure for innovation continues. The privately owned public utilities that provide most of our electric power, even though they are entitled to prices that guarantee them a "fair profit" and are therefore in a sense free to rest on their laurels, are driven by their own imperatives to seek every possible increase in operating efficiency. (For one thing, as utilities lower their costs the public-utility commissions that

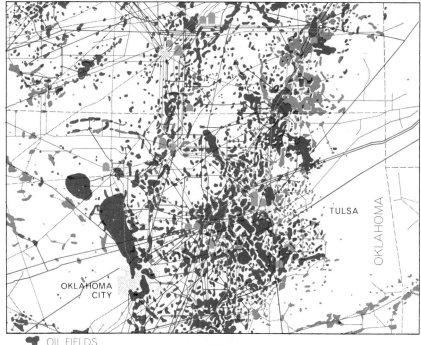

OIL FIELDS
CRUDE-OIL PIPELINES
REFINERY
PRODUCT PIPELINE
GAS FIELDS
GAS PIPELINES

PIPELINES radiate from the rich oil fields and natural-gas fields of Oklahoma. Crude oil is piped from wells to refineries in the region or farther away; petroleum products from the refineries are piped to industrial and commercial centers, primarily in the Middle West.

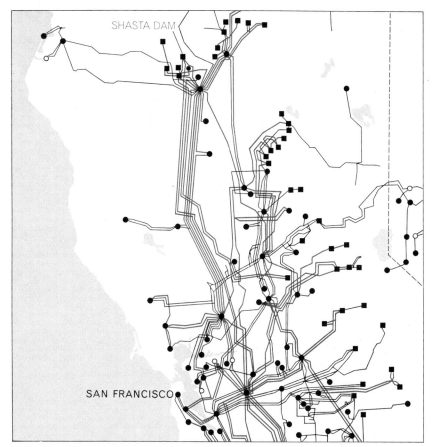

MORE THAN 189,000 VOLTS
LESS THAN 189,000 VOLTS
HYDRO
FUEL
SUBSTATION

TRANSMISSION LINES radiate from power plants in northern California, the largest of which is Shasta Dam plant. Most of the electric power goes to the San Francisco area.

set utility rates seem to lag in lowering the prices of electricity.)

The result is that even a trivial innovation may earn thousands of dollars a day, and the tendency is to judge its value by that potential rather than by its capacity to initiate a substantial revolution. Thus power companies adopt small improvements to alleviate some of the following inherent problems: (1) The demands of customers vary systematically by the time of day and the season, but unpredictable demands also arise and emergency shutdowns do occur. An isolated system must have enough spare equipment to handle such contingencies, but linking systems together with lines of high capacity makes some of the spare equipment unnecessary. (2) Peak demands are closely related to urban time schedules as well as to the sun. When a time-zone boundary is crossed, the period of peak demand shifts by an hour. Bringing in electricity from a neighboring time zone broadens the peak, reduces its magnitude and thereby again reduces the amount of generating equipment needed. (3) Some of the great hydropower facilities—Grand Coulee is the best example—can sell power very cheaply; others were built with the intention of selling power for premium prices, mostly at the hours of peak daily demand. Outlets for such peak-hour power may be many hundreds of miles away.

For all these reasons the power grids of the 48 states are now fairly well interlinked. It must be doubted that the resulting savings come to as much as 10 percent. Still, the interest in ever cheaper transport persists. Recently the devices of solid-state physics have provided means for transforming voltage (and current) with direct current. Because direct current is more tractable than alternating current at high voltages, utilities are now turning back from alternating to direct current and are beginning long-distance power transmission at extrahigh voltages (EHV) of 750,000. The next step may be the use of superconductors. All metals, when cooled to near the boiling point of helium, become supercon-

ductive, or quite without resistance to the flow of current. The use of superconductors could change the technical task involved in transmission dramatically, from the reduction of energy loss as heat to the operation of an elongated ultralow-temperature refrigerator. The first commercial application of superconductor transmission may be to bring power into urban areas too crowded for the wide corridors required for conventional high-voltage lines.

Storage presents its own set of constraints. Electricity is hardly storable as such in commercial quantities. Instead we resort to a subterfuge: building artificial reservoirs into which water can be pumped electrically and from which electricity can be retrieved by reversing the flow of water and letting the motors and pumps act as generators and turbines. Although this is quite efficient, it is a clumsy sort of thing; still, it is the best we can do. Storage batteries are not a substitute because they are expensive and have little capacity. One would think that about as much electricity could be stored in a battery as oil can be stored in a tank, because the same kinds of forces are being manipulated. Unfortunately reliable storage batteries are very heavy because they use chemical elements at the heavy end of the periodic table, notably lead, and provide only about as much energy as would result from an equal number of atoms at the light end of the series. (Clearly what is needed is a good storage battery in which lithium is oxidized and reduced instead of lead!) The same phenomenon gives electric automobiles an unsatisfactory performance and cruising range compared with automobiles that depend on hydrocarbon fuels.

The difficulties of storing electricity impose exacting constraints on the operation of electric-utility systems, as residents of many U.S. cities have learned in recent years. When customers demand more electricity by switching on lights or air conditioners or other machinery, the production of power must be increased to meet the demand. Little flexibility

exists; electricity does not stretch or squeeze easily. To keep a system in balance requires minute-by-minute attention; at least, since electricity moves at a notably high velocity, increased production does reach the customer without delay.

If gas customers ask for a greater flow, on the other hand, gas will simply expand to a considerable degree within the pipeline and so meet the increased demand. Minute-by-minute flow is therefore no problem. The other side of the coin is that gas comes down the pipeline rather slowly, and so if the neighborhood supply runs short, it may take a day or two to make up the deficiency. Accordingly the marketers of gas usually have to arrange some kind of local storage. The large gasholders one sees on the outskirts of cities do not hold enough. To meet the possible peak demand for a day in the San Francisco Bay area, for example, would take a gasholder equivalent in volume to a cube 1,000 feet on a side; existing ones have perhaps 1 percent of that capacity. A common provision is therefore storage in depleted gas fields or in the transmitting pipeline itself. Gas is compressible, and if the upstream pressure is increased, not only can the gas be sent along faster but also larger amounts can be stored in the pipeline near the demand. A pipeline three feet in diameter running at 400 pounds per square inch contains about a million cubic feet of gas per mile, or a billion cubic feet per 1,000 miles—equal to the capacity of the 1,000-foot cube.

The third case is that of the supplier of liquid fuels. Here storage is so easy and so much of it is provided all along the distribution chain that no real technological problem remains. It is easier than keeping grocery shelves stocked.

In the synthesis of these unit operations both technological advances and economies of large-scale operation have contributed to lowering the cost of the alternatives for meeting demands. In general great economies of scale result only from the phenomena of liquid flow, which cause the capacity of a pipeline to increase as a high power of its diameter. One very different example of such economy is seen in strip-mining: the stripping away and movement of overburden is now being handled by outsized equipment, making operations economically attractive that would have been unacceptable a generation ago. Yet one suspects that here too, as in the case of supertankers, the end of the road of increasing scale is close at hand.

Energy is almost never harvested in

COAL FOR EXPORT passes through the yards of the Norfolk and Western Railway at Norfolk, Va. It is primarily high-grade metallurgical coal from fields in Virginia, West Virginia and Kentucky; Japan is the largest single customer. The yard can accommodate 11,520 coal hopper cars; a nearby yard handles another 9,880 cars. The two adjacent piers at lower left handle about 1,000 vessels a year. The pier at left, extending 1,870 feet into the Elizabeth River, is said to be the largest and fastest coal-loading facility in the world. It has two traveling loaders, each as high as a 17-story building, that can handle up to 8,000 tons of coal an hour. The system, consisting of car dumpers, conveyor belts and the loaders, combines coal of different kinds and grades, blending the shipments to order for the individual customers.

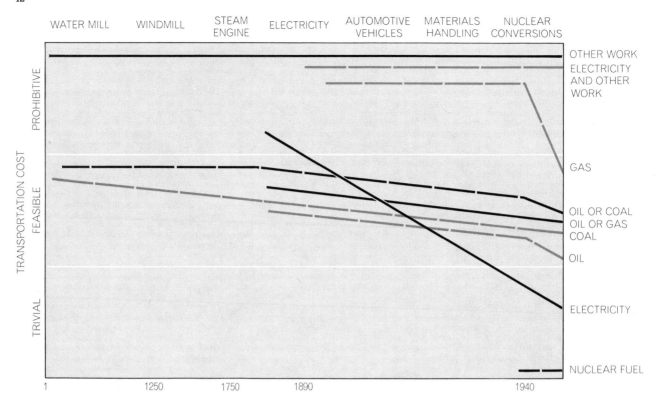

WATER MILL   WINDMILL   STEAM ENGINE   ELECTRICITY   AUTOMOTIVE VEHICLES   MATERIALS HANDLING   NUCLEAR CONVERSIONS

TRANSPORTATION COST

PROHIBITIVE

FEASIBLE

TRIVIAL

OTHER WORK
ELECTRICITY AND OTHER WORK

GAS

OIL OR COAL
OIL OR GAS
COAL

OIL

ELECTRICITY

NUCLEAR FUEL

1   1250   1750   1890   1940

TRANSPORTATION COSTS may make it impossible to move some forms of work, such as wind or water power, from the site where they are developed. The costs of other commodities have varied through history; in many cases technological changes make a cost feasible that was once prohibitive. The curves relate the general level of costs for transportation by sea in batches (*broken colored lines*), by land in batches (*broken black lines*) and by continuous methods such as power lines or pipelines (*solid black lines*).

the form in which it is to be used, and therefore it must ordinarily go through a conversion step [see "The Conversion of Energy," by Claude M. Summers, page 23]. The most significant conversions are those of latent energy to heat through combustion, and of heat to electricity. Once energy is in the form of electricity all the gates are open, even though the toll through some gates is excessive.

Centuries of development, innovation and growth have built up an intricate pattern of physical facilities and economic relations that connect discovered and harvested resources with sites of conversion, use and disposal [*see illustrations on pages 36 through 37*]. For the most part the patterns reflect the movement of energy from resource sites to the homes and places of work of growing populations at the various times and in the various amounts and forms that are needed.

In a sense the customer has been king; he has received what he wanted when and where he wanted it. It might be argued, as a matter of fact, that societies in which energy costs have been excessive have simply not prospered. In the U.S. the consumer has usually paid the

price and paid little attention to paying; in return the energy industry has ordinarily met his demands while asking for a very minor share of his income. To estimate that share is difficult because so much of it is paid indirectly and because one scarcely knows whether to apply retail or wholesale prices, what to do about gasoline taxes and so on. Very roughly, every American consumes each day about 15 pounds of coal (200,000 B.t.u.) for 10 cents; two and half gallons of petroleum, half of it as gasoline (350,000 B.t.u.), for 50 cents; 300 cubic feet of

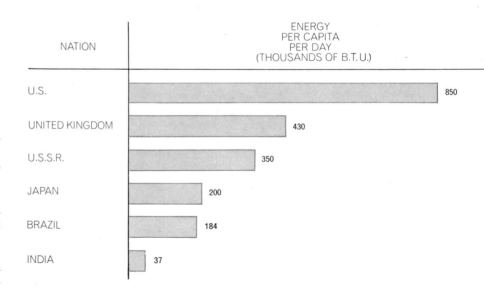

NATION	ENERGY PER CAPITA PER DAY (THOUSANDS OF B.T.U.)
U.S.	850
UNITED KINGDOM	430
U.S.S.R.	350
JAPAN	200
BRAZIL	184
INDIA	37

ENERGY PATTERNS are revealed by some international statistics. Energy per capita is about as expected, with a large advantage in the developed nations. (B.t.u. figures for Brazil and India would be about 150,000 and 22,000 respectively if "primitive" fuels were in-

natural gas (300,000 B.t.u.) for 20 cents, and 24 kilowatt-hours of electricity for 45 cents. If one subtracts the 250,000 B.t.u. of the fuels that are used in generating the electricity, or 15 cents, the total comes to $1.10. Marked up to retail level, that would be about a tenth of the U.S. per capita personal income. Other inquiries have arrived at lower estimates, such as 4 percent or 7 percent, for the share of personal income spent on energy, but my own feeling is that a figure of 10 percent more nearly represents the situation from the consumer's point of view.

The resistance of the consumer varies. Two-thirds of his consuming is done for him in industry, commerce and transportation other than his own, and is beyond his direct control. It is hard to tell how much he resists buying industrial products, but his interest has been turning toward spending for services and it does seem that in some vague way his resistance to the purchase of industrial energy is increasing. In his home he behaves differently. No one can measure the extent to which he turns down the heat, turns off electric lights or skimps on gasoline, but the general impression is: not much. (Has anyone under the age of 30 ever turned off an electric light?) He pays a good deal more for electricity than he does for fuels but is easily persuaded to use electricity as a fuel, even though it costs him many times as much per unit of heat.

How about the rest of the world? First, it seems plausible, since fuels have long been available to men, that a highly technological society should show a high ratio of work to total energy, as expressed perhaps by kilowatt-hours per million B.t.u. Second, it can be argued that the construction of a thermal power system requires an intricate structure extending from mining through diverse forms of consumption, whereas the construction of a hydropower system (perhaps with assistance from a more technological society) can precede and is often intended to initiate development. Accordingly a high fraction of hydropower should be common in developing societies. Certainly the general experience is that the fraction of hydropower diminishes in the highly technological societies. Third, growth rates of electricity, for example, should be higher in the developing societies.

Such patterns do appear in the statistics but are far from infallible [see illustration below]. The U.S. is by no means the highest in kilowatt-hours per million B.t.u. In fact, it uses 35 percent of the world's electricity, just as it does with total energy. The reason is at least partly obvious: it is our excessively high consumption of gasoline for private automobiles. Brazil and India come in too high on the kilowatts-to-B.t.u. ratio, but the formalized statistics on which these numbers are based take no account of contributions from "primitive" sources: notably fuel wood in Brazil and cow-dung fuel in India. If these sources are counted in, the ratio drops from 32 for Brazil to six; for India it falls from 15 to six. (Perhaps as energy economies evolve the ratio should pass through a maxi-mum and then decline.) The electrical growth rates are much what one would expect, except that Brazil's seem low. The hydropower percentages are generally in line, but they remind one not to forget climate and topography: Japan and the United Kingdom are both insular, mid-latitude and humid, but the former is mountainous, with a great many hydropower sites, and the latter is flat.

These, to be sure, are only the most superficial of the patterns associated with energy. Close examination of any society will reveal the influences on it of its particular experience with energy resources and energy conversion. The patterns one finds depend not only on such physical factors as the waxing and waning of resources but also on cultural variables: the development of technologies, changes in social patterns and the constraints of tradition, governmental policies and local fads and preferences. The resulting patterns are seldom simple, and it is particularly difficult to foresee the future. I should like merely to raise a few questions: Is the correlation between increasing use of energy and human welfare good enough, and is the hypothesis that more energy means a better life plausible enough, to warrant any hopeful extrapolation? Where on the rising consumption curve is the breaking point between gains and losses? Are we likely to find that point by encouraging growth until the customer—no longer interested in more energy or unable to afford it—finally offers resistance, and growth ends?

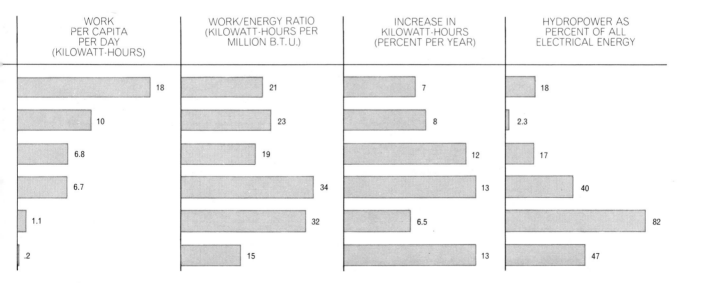

WORK PER CAPITA PER DAY (KILOWATT-HOURS)	WORK/ENERGY RATIO (KILOWATT-HOURS PER MILLION B.T.U.)	INCREASE IN KILOWATT-HOURS (PERCENT PER YEAR)	HYDROPOWER AS PERCENT OF ALL ELECTRICAL ENERGY
18	21	7	18
10	23	8	2.3
6.8	19	12	17
6.7	34	13	40
1.1	32	6.5	82
.2	15	13	47

cluded.) One would expect kilowatt-hours per million B.t.u., a measure of the ratio of work to energy, to reflect technical expertise in the advanced countries, but the inefficiency of gasoline engines reduces the ratio there instead. The figures for hydropower's share of total electrical energy reflect not only the state of development (hydropower comes early) but also the geography of the countries.

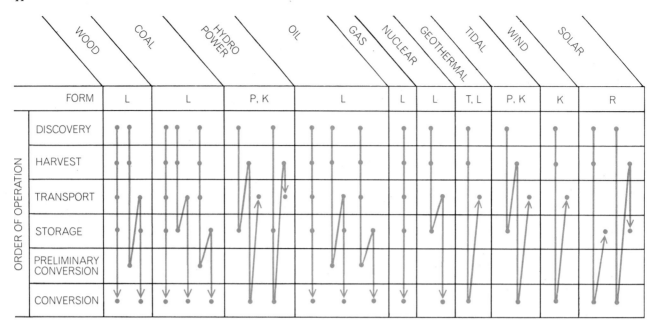

SEQUENCE OF OPERATIONS between the discovery and the use of an energy commodity is diagrammed. Energy is discovered in various forms: latent (*L*), potential (*P*), kinetic (*K*), thermal (*T*) or radiant (*R*). The resource is harvested, transported, stored and converted; sometimes there is a preliminary conversion step. The sequence of these steps varies for different commodities; in some cases there are alternate sequences. Arrows indicate the order in which the operations can be accomplished for 10 kinds of energy.

		WOOD	COAL	PETROLEUM	GAS	HEAT (STEAM)	ELECTRICITY	HYDRO POWER
LAND	BATCH	ARMLOAD	▨						
		PACK	▨						
		BASKET		▨					
		POT			▨				
		WAGON	▨	▨					
		TRUCK	▨	▨					
		RAIL	▨	30	>15				
		VEHICLE FUEL TANK			▨				
	CONTINUOUS	AQUEDUCT							▨
		PIPELINE			10	20	▨		
		TRANSMISSION LINE						50	
		SLURRY PIPELINE		30					
SEA	BATCH	CARGO SHIP	▨	<30					
		COLLIER		<30					
		BARGE		<30					
		TANKER			5				
		LNG TANKER				>20			
		SUPERTANKER			<5				

TRANSPORTATION of energy commodities can be by land or sea; on land it can be in batches or continuous, by sea only in batches. The colored boxes in the matrix indicate the feasible means of transport for each commodity. The numbers in some of the boxes give the approximate lowest cost for the major means of transport in cents per British thermal unit for a 1,000-mile haul.

II

COAL

II COAL

INTRODUCTION

The dependence of the United States economy on imported oil is a worrisome matter. Even with the existence of a strategic stockpile, a long-term interruption could cause serious damage to our industrial output and employment. One can easily conceive of situations in which the major oil fields of the Arabian peninsula are put out of production for many months. For this reason, industrialized countries have been searching for alternatives to oil in order to make themselves as energy-independent as possible. For most countries, nuclear energy appears to be the only answer, and even the dependence on imported uranium can be reduced by the use of nuclear breeder reactors. The United States probably has more options available than any other industrialized country, except perhaps the Soviet Union. The United States not only has large native uranium resources but also one of the largest reserves of coal.

Coal, of course, was the fuel in common use until oil and gas, because of their low cost and convenience, displaced it in its many applications, particularly for home and commercial heating, fueling locomotives, and even generating electricity. Still, coal has managed to hold a certain percentage. With the current significantly increased oil and gas prices, coal has again become a desirable fuel, but only in applications where its handling cost, and particularly its pollution-control cost, are not prohibitive. For all practical purposes, therefore, coal in its original form can be used only as a boiler fuel for the production of electricity, and in industry for the production of steam.

But coal is also an important raw material that can furnish the feedstock for synthetic oil and gas. This conversion can be accomplished by a number of well-developed commercial processes and by improved processes now under development. The article by Neal P. Cochran describes several of these processes, including the most important methods for turning coal into synthetic oil and into both low-BTU and high-BTU gas. Cochran even describes the concept of a "coal refinery," which furnishes not only a variety of fuels but also important industrial raw materials, such as benzene, xylene, and toluene.

Should the conversion of coal become economic, the use of coal in the United States is likely to increase by a very large factor, from its present 600 million tons per year to double or triple this amount. For example, 20 typical conversion plants producing a total of 1 million barrels a day of liquid products and 4 billion cubic feet of gas would require 165 million tons of coal per year, about a quarter of the present production. Even so, the United States has nearly 2 trillion tons of recoverable coal, leading to the possibility that some day these plants may become commonplace. This large reserve of coal assures an essentially constant price, assuming that the productivity of mining can keep pace with the rise in the cost of labor and machinery. But since the price of oil must rise as this resource is gradually depleted, a point will be reached at which synthetic fuels will become competitive. In this respect the role of the federal government is of crucial importance. By guaranteeing the investments, by agreeing to purchase synthetic fuels at a certain price, or by underwriting the necessary research, it can bring closer the time when synthetic fuels will be generally available.

If the large-scale use of coal should come to pass, particularly for the production of synthetic fuels, then the cheapest available source will be strip-mined western coal. Incredibly rich coal seams exist in Wyoming, Montana, and North Dakota, as well as in Utah, Arizona, and New Mexico.

It has been estimated that if all of the U.S. coal were to be mined in Wyoming's Powder River Basin, the amount of land disturbed per year would only be 10 to 20 square miles. Unfortunately, all coal needed cannot always be mined at such productive sites. The transportation cost of coal makes eastern coal more desirable, even though the seams in the east are much thinner and therefore require more disturbance of land. In addition, the mining of steeper slopes, particularly in the Appalachian region, can lead to a great deal of land damage from strip-mining. Another problem associated with coal mining is water pollution from underground mining.

One of the crucial problems the nation faces is how to manage its coal resources without the spoiling of land. The issue has been highly emotional. Extreme views have been prevalent, ranging from the belief in the need for a complete moratorium on all strip-mining, particularly on agricultural land, to the claim that land restoration would improve upon the original quality of the land. The true situation is somewhere in between, as pointed out in the article by Genevieve Atwood. Some special areas should probably not be strip-mined under any circumstances.

How much money should be devoted to land reclamation? Money is simply a means of buying resources, and the nation's resources are limited and should be devoted to the most important social goals. The complete restoration of land may be an unrealistic goal, particularly if the restoration costs exceed by a large factor the actual value of the land, at least as expressed in its price. These issues have not yet been faced, but they will undoubtedly come to the foreground as strip-mining becomes more prevalent because of its lower production costs.

4

Oil and Gas from Coal

by Neal P. Cochran
May 1976

The conversion can be accomplished by several tested processes. The present effort is to combine such processes in a large-scale system that will manufacture the oil and gas at reasonable cost

For both political and economic reasons it seems clear that sooner or later the U.S. will come to rely much more on coal as a source of energy than it has over the past few decades. Except in large installations such as power stations, however, coal is not the ideal fuel both because it is not fluid and because it usually burns less cleanly than either oil or gas. Moreover, since coal-driven locomotives and ships have almost disappeared from the scene, coal can scarcely serve at all as a fuel for vehicles. Success in exploiting the nation's huge reserves of coal therefore depends on the development of a technology that will convert coal into oil and gas on a large scale. The principles of the technology already exist, as do a number of pilot and demonstration plants where the conversion is being accomplished on a small scale. The problem, then, is to mobilize the financial and industrial resources that are needed to put the technology on a commercial basis.

The fuller exploitation of coal will entail a substantial shift in the economics of energy. For the past decade or so the sources of energy in the U.S. have been predominantly oil and gas (44 and 31 percent respectively), with coal accounting for 21 percent and all other sources, including hydroelectric and nuclear plants, accounting for 4 percent. In contrast, coal accounts for 75 percent of the nation's fossil-fuel resources.

It is also instructive to look at the markets for energy. Industrial and commercial activities take about 40 percent of the energy produced in the country each year, transportation 25 percent, electric utilities 20 percent and homes 15 percent. The direct consumption of coal is virtually ruled out in transportation and seems unlikely to expand much in industrial, commercial and residential markets because of restrictions against air pollution. To reach those markets coal must be converted into oil and gas.

The conversion of coal into gas was an established commercial technology in the U.S. as early as the 1820's. The "gas works" became a familiar feature of cities of the Northeast and the Middle West, manufacturing gas from coal for illumination and cooking. The gashouses disappeared after World War II as natural gas came to be widely distributed by pipeline, and the technology that sufficed then would be inadequate now because the gas it made from coal could not match the heating value of natural gas.

Although the conversion of coal into oil has not been accomplished commercially in the U.S., it was a large-scale operation in Germany during World War II and is being pursued on a substantial scale in South Africa today. The German production of synthetic gasoline from coal reached the level of 12,000 barrels per day, with the largest plant turning out nearly 4,000 barrels per day. The Sasol synthetic-fuel plant in South Africa, which has operated for 20 years, converts coal into more than 300 million cubic feet of gas per day and then converts the gas into liquids that are similar to petroleum. Neither the German undertaking nor the South African one would be considered large by U.S. standards; the German plants at the time of peak production were processing about 600 tons of coal per day, and the input to the Sasol plant is about 3,500 tons per day. The plants envisioned for the U.S. would process upward of 25,000 tons of coal per day, a level of operation to which the German or South African processes cannot be economically adapted.

The amount of work done to study and develop processes for converting coal into oil has tended to rise and fall with estimates of how adequately the known and projected reserves of petroleum would supply the projected markets for gasoline and other petroleum products. Until recently most of the research in the U.S. was done by two agencies of the Department of the Interior: the

Bureau of Mines and the Office of Coal Research. Those programs are being continued by the Energy Research and Development Administration (ERDA), which was established in 1975 to coordinate the Government's research and development efforts to make the U.S. independent of foreign energy resources.

Enough work has been done to make it possible to predict the development of a scaled-up technology that can convert coal into oil and gas at prices that are competitive with the current prices of such products in the U.S. In terms of actual cost, of course, oil and gas from coal cannot now compete with crude oil from the Middle East. One should remember, however, that the cost and the price of Middle East crude are not related, as was demonstrated by the embargo and price increases the exporting countries of the Middle East imposed starting in 1973. In many ways those actions may prove to have benefited the U.S., since they could encourage the development of domestic sources of oil and gas and hasten the production of synthetic fuels from coal. The price of coal can be expected to rise with the price of oil, but the price of synthetic oil and gas will decline with respect to the price of the natural product because the capital charges for plants making the synthetic fuels will stabilize.

In the simplest terms the conversion of coal into oil or gas calls for adding hydrogen to the coal. (The ratio of hydrogen atoms to carbon atoms in coal is .8 to 1; in oil it is 1.75 to 1.) The source of the hydrogen is water (as steam). The energy for the process by which hydrogen is separated from water must be obtained from the coal itself if the economics of the conversion technology are to be favorable. The production of hydrogen is a major cost in the conversion of coal into oil.

It follows that every conversion process

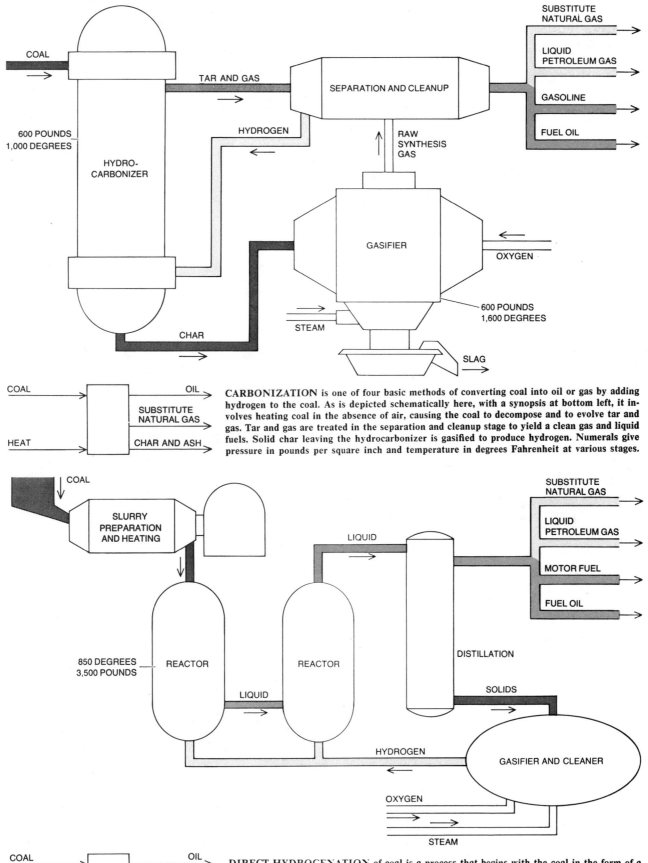

COAL

600 POUNDS
1,000 DEGREES

HYDRO-
CARBONIZER

TAR AND GAS

SEPARATION AND CLEANUP

SUBSTITUTE
NATURAL GAS

LIQUID
PETROLEUM GAS

GASOLINE

FUEL OIL

HYDROGEN

RAW
SYNTHESIS
GAS

GASIFIER

OXYGEN

600 POUNDS
1,600 DEGREES

STEAM

CHAR

SLAG

COAL — OIL

SUBSTITUTE
NATURAL GAS

HEAT — CHAR AND ASH

CARBONIZATION is one of four basic methods of converting coal into oil or gas by adding hydrogen to the coal. As is depicted schematically here, with a synopsis at bottom left, it involves heating coal in the absence of air, causing the coal to decompose and to evolve tar and gas. Tar and gas are treated in the separation and cleanup stage to yield a clean gas and liquid fuels. Solid char leaving the hydrocarbonizer is gasified to produce hydrogen. Numerals give pressure in pounds per square inch and temperature in degrees Fahrenheit at various stages.

COAL

SLURRY
PREPARATION
AND HEATING

LIQUID

DISTILLATION

SUBSTITUTE
NATURAL GAS

LIQUID
PETROLEUM GAS

MOTOR FUEL

FUEL OIL

850 DEGREES
3,500 POUNDS

REACTOR

REACTOR

LIQUID

SOLIDS

HYDROGEN

GASIFIER AND CLEANER

OXYGEN

STEAM

COAL — OIL

SUBSTITUTE
NATURAL GAS

HYDROGEN — CHAR

DIRECT HYDROGENATION of coal is a process that begins with the coal in the form of a slurry. The slurry is fed into a reactor, where it is reacted with hydrogen at high pressure. The reactor usually contains a catalyst, such as cobalt molybdenum, that facilitates the process. After hydrogenation the liquid yielded by the reaction is distilled to remove solids, which are gasified to provide hydrogen for the operation. In this process, as in others, water in the form of steam is the source of the hydrogen that must be added to the coal to convert it into oil and gas.

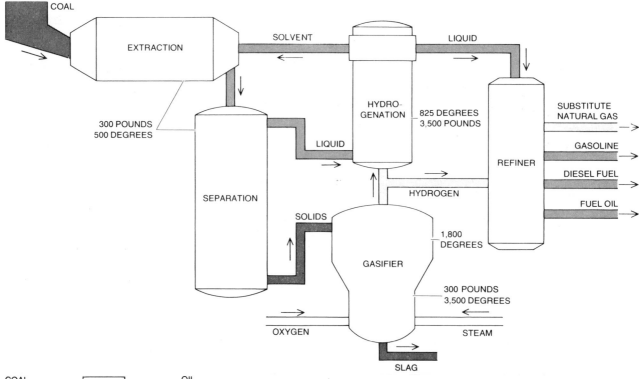

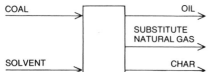

EXTRACTION PROCESS for converting coal into oil and gas involves dissolving the coal in a solvent. The hydrogen-donor process, which is one of two methods of extraction, is depicted here. The coal is dissolved by being mixed with the solvent at low pressure. The liquid that results is hydrogenated, yielding both a synthetic crude oil and the solvent for extraction. Since the solvent is rich in hydrogen, it transfers hydrogen to the coal during the extraction process. Part of the coal remains undissolved; it is gasified, as in other processes, to yield hydrogen.

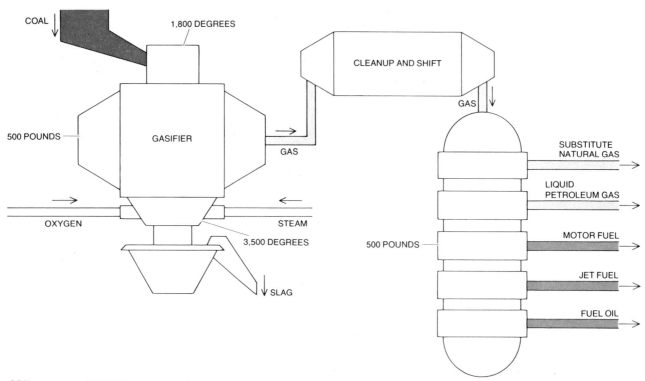

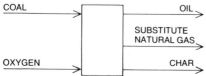

FISCHER-TROPSCH SYNTHESIS is the fourth basic method of converting coal to oil and gas. Coal goes into a gasifier, where it is burned in the presence of oxygen and steam. The combustion generates a gas consisting mainly of carbon monoxide and hydrogen. In the cleanup-and-shift stage the gas is purified. Then it is passed over a catalyst, producing not only substitute natural gas of pipeline quality but also a variety of liquid products. Fischer-Tropsch synthesis is employed in a South African plant that processes some 3,500 tons of coal per day.

must involve a gasification step, in which the coal is reacted with steam. The aim is not primarily to get pipeline gas but to make a synthesis gas ($CO + H_2$) that can be modified with more steam ($CO + H_2O \rightarrow CO_2 + H_2$) to obtain more of the hydrogen needed to convert coal into oil and hydrocarbon gas. Moreover, all conversion processes yield a mixture of hydrocarbon gases, including methane (CH_4), which is a substitute for natural gas. Since the national economy requires both gas and oil, the aim of a conversion technology should be to conserve the gas generated in the liquefaction of coal and to regard it as a coproduct with the oil.

The simplest of the four basic processes for producing oil from coal is carbonization, which consists in heating coal in the absence of air. The heat causes the coal to decompose, evolving tar and gas and leaving a solid residue (coke). Carbonization has long been employed in making coke. Even though it also yields a liquid and a gas, it is not economically practical for the production of oil because the capital charge for coke ovens is high in relation to the unit cost of the oil produced.

Research is therefore focused on improved carbonization processes. A consortium of companies is developing the COGAS (coal-oil-gas) process, in which carbonization is accomplished in a fluidized bed of coal at comparatively low temperature. The fluidized bed improves the efficiency of decomposition of the coal. The next steps are the combustion of the solid char with air and the reaction of steam with the hot char. Typical products are tar and pipeline gas.

A hydrocarbonization process is the basis of a demonstration plant that will be built in Illinois under a $237-million contract awarded to the Coalcon Company by ERDA. The process yields products ranging from substitute natural gas to a high-quality (low-sulfur) fuel oil by heating coal to about 1,000 degrees Fahrenheit in the presence of hydrogen at a pressure of about 600 pounds per square inch. In the basic carbonization step a moderate hydrogen pressure is established to improve the yield and quality of liquid products (by facilitating the reaction of hydrogen with the products of pyrolysis) and to reduce costs. Material leaving the top of the reactor as a gas is cooled to condense a heavy oil and is then treated to remove sulfur compounds, so that the gas coproduct is clean. The liquid product is distilled to yield a light motor-fuel fraction and a heavy fuel-oil fraction. The gas is distilled at cryogenic (very low) temperature to separate hydrogen, which is recycled to the reactor. The char leaving the reactor is gasified to produce the hydrogen required by the process, and energy for the entire operation is obtained by burning part of the char or the excess gas. The process can be adapted to the recovery of chemicals, but that is not the objective of the present work.

A recent report issued by Coalcon indicates that a full-scale plant would produce the substitute natural gas to sell at a price of $2.40 per million cubic feet. Fuel oil would be priced at $15 per barrel and gasoline at 33 cents per gallon (not including tax). The prices are comparable to current prices charged by producers for natural products and are typical of the prices of synthetic oil and gas.

The second basic conversion process is hydrogenation. It calls for reacting coal with hydrogen at high pressure, usually in the presence of a catalyst. (This was one method employed to convert coal into oil in Germany during World War II.) The coal can be hydrogenated directly by feeding it into a reactor in the form of a slurry. In a system that is being developed for ERDA by Hydrocarbon Research, Inc., the reactor contains a catalyst of cobalt molybdenum, which is kept in motion by the recycling of liquid in the reactor. Intimate contact is achieved between the solid (the catalyst), the liquid (oil) and the gas (hydrogen).

Another system of direct hydrogenation is being investigated by the Pittsburgh Energy Research Center of ERDA. In this system the catalyst bed is fixed, which simplifies the construction of the reactor but complicates the removal of heat and thus the control of the temperature in the reactor. (Recent work seems to indicate that the catalyst is not required.) Control of the temperature in the reactor is crucial in both systems if the formation of carbon and the ultimate plugging of the reactor is to be avoided. The temperature of the reactor also affects the nature and quality of the products.

These systems and other methods of hydrogenation call for a temperature of about 850 degrees F. and pressures ranging from 2,000 to 4,000 pounds per square inch. A lower pressure and a shorter time for reaction limit the reaction between the coal and the hydrogen and favor the production of heavy fuel oil; a higher pressure and a longer time for reaction favor the production of lighter fractions. Following the hydrogenation step the liquid fraction of the product is distilled and decompressed so that solids can be removed. The heaviest oil fraction contains the unreacted solid coal, which must be separated from the liquid. The solid fraction is gasified with additional coal to provide hydrogen for the operation. Direct hydrogenation provides more liquid than any of the other processes discussed here. The selling price of the product will range from $14 to $18 per barrel.

The third basic conversion process is extraction, in which coal is partly or completely dissolved. Two systems are being investigated commercially and in the program of ERDA. They differ in the method of bringing hydrogen to the coal. In the solvent-refined-coal system coal is dissolved in an organic liquid in the presence of hydrogen gas that is under high pressure (about 2,500 pounds per square inch). The process dissolves nearly all the coal and in subsequent steps filters the resulting slurry, distills the liquid fraction to recover solvent and re-forms the coal by cooling. If the starting coal has a high sulfur content (about 3 percent), the remaining solid fuel will contain from .5 to .8 percent sulfur. The products of the system can be improved by further hydrogenation in the presence of a catalyst.

In the hydrogen-donor process coal is dissolved by being mixed with the solvent at a low pressure (about 300 pounds per square inch). The resulting liquid is separated from the undissolved coal and then hydrogenated to obtain a synthetic crude oil and the solvent required for extraction. In other words, the solvent is derived from the process; because it is rich in hydrogen, it transfers hydrogen to the coal during extraction. The hydrogen-donor system has two key features: the extraction process is carried out at low pressure and the hydrogenation step employs a clean feed. In terms of 1975 dollars the selling price of the products of extraction systems is about $2.30 per million cubic feet for the substitute natural gas and $15 per barrel for the oil.

The fourth basic method of liquefying coal is usually called Fischer-Tropsch synthesis, after the German chemists Franz Fischer and Hans Tropsch, who originally developed it. Coal is burned in the presence of oxygen and steam, generating a gas composed mostly of carbon monoxide and hydrogen. The gas is purified and then passed over a catalyst, yielding liquid products ranging from methanol ("wood alcohol") to hydrocarbons of high molecular weight, including waxes and oils. The process can be directed primarily toward the production of motor fuel and substitute natural gas. This process is employed in the Sasol plant to produce waxes, oils, motor fuel and chemicals. It has been estimated that a Fischer-Tropsch plant in the U.S. with the capacity to produce 50,000 barrels per day would be able to sell its output at about $2.25 per million cubic feet of substitute natural gas and $13.05 per barrel (31 cents per gallon) for gasoline.

It is useful to compare the four basic processes in various ways. One is by yield per ton of coal. For carbonization the yield is from one barrel to 1.5 barrels of liquid products and from 4,000 to 5,000 cubic feet of gas products; for direct hydrogenation it is 2.5 to 3.5 barrels and 2,000 to 3,000 cubic feet; for extraction-hydrogenation it is two to three barrels and 3,500 to 4,500 cubic feet, and for Fischer-Tropsch synthesis it is 1.5 to two barrels and 8,000 to 10,000 cubic feet. Another measure is the pressure at which the process operates, since the cost of the equipment rises with the pressure. Carbonization is done at from atmospheric pressure (14.7 pounds per square inch at sea level) to 70 atmospheres, direct hydrogenation at 200 atmospheres, hydrogen-donor extraction at 20 atmospheres followed by hydrogenation at 200 atmospheres and Fischer-Tropsch synthesis at 30 atmospheres. In thermal efficiency the range for carbonization is from 55 to 65 percent, for direct hydrogenation 60 to 65 percent, for the extraction processes 60 to 70 percent and for Fischer-Tropsch synthesis 55 to 70 percent.

Although each of the four basic conversion processes will work by itself, it may be that the most successful commercial operation will be one that involves a combination of processes. Such a plant could be described as a coal refinery. One attractive possibility is a combination of hydrogen-donor extraction and Fischer-Tropsch synthesis. This possibility arises because the synthesis step in the Fischer-Tropsch process proceeds at a pressure of from 350 to 500 pounds per square inch, which is also the range of pressure in the extraction phase of the hydrogen-donor process.

Coupling the extraction and synthesis steps in the low-pressure zone also couples them in the sections of the plant handling coal solids. It is always difficult to handle such solids, and the low-pressure feature therefore simplifies the design of such components as feeders, reactors, piping, pumps, valves and instruments. Low pressure also reduces the cost of the equipment for handling gas.

The steps in a combined process could go as follows [see illustration below]. Coal from the mine, which is treated as an integral part of the plant, is handled directly, with little or no storage. (If oil and gas made from coal are to compete with natural oil and gas supplies, a conversion plant must have a high capacity. Coal-handling is expensive in any case, and the cost rises if the coal has to be put into and withdrawn from a storage pile. Hence it is economical to gear the plant to a large and close supply of coal and to put the coal through the plant at a high rate with minimal storage.)

Leaving the mine, the coal is ground fine and made into a slurry with a solvent that has been produced in the plant and recy-cled. The solvent, a mixture of compounds that are saturated with hydrogen, serves as a source of hydrogen to aid the dissolving of the coal. The slurry of coal and solvent is heated to about 500 degrees F. in a fired heater and then is pumped to the extractor, which can be visualized as two simple agitated tanks. The time the mixture spends in the extractor depends on the feed rate of the plant and the level of slurry maintained in the vessels. With a long residence time the extraction can dissolve more than 90 percent of the coal, but the economics of the process is most favorable when the rate of extraction is between 60 and 75 percent.

The ratio of solvent to coal in this step is about 2 : 1 by weight. Hydrogen can be added to the reactor to enhance the transfer of hydrogen from the solvent to the coal. After extraction the mixture of solvent, liquid extract and undissolved coal is pumped to a filter for separation. The filter, with a diatomaceous earth as the medium, works well

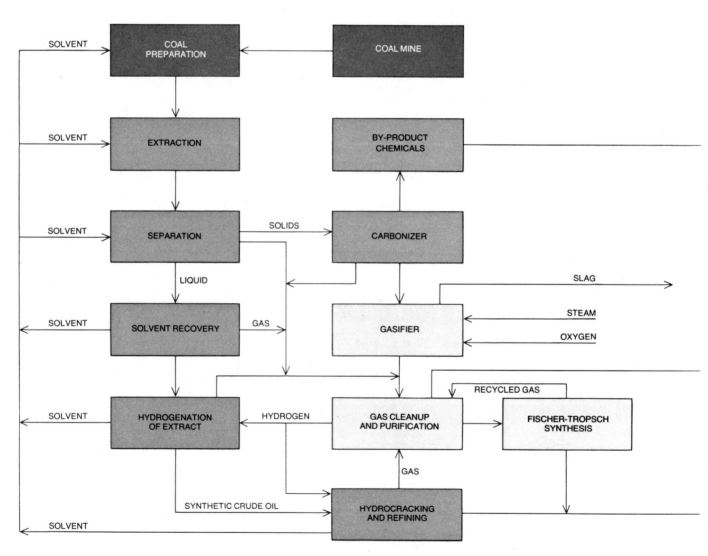

COAL REFINERY as envisioned by the author combines extraction, hydrogenation and Fischer-Tropsch synthesis to yield a large variety of products. Coal moving directly from a nearby mine is ground fine and made into a slurry with a solvent. That is an extraction process.

In the next step the liquids and undissolved coal are separated. Solvent is recovered and the remaining liquid is hydrogenated. Solids remaining after extraction go as a sludge to a carbonizer, where the sludge is heated with recycled gas to generate gas and a char. The

for the separation of solids from a liquid that would clog conventional filter cloth.

The filter cake, consisting of undissolved coal and ash-producing minerals, leaves the filter as a wet sludge. The sludge is moved to a fluid-bed carbonizer, where it is heated with recycled gas to recover tar acids and also gases of low molecular weight. Another product of the carbonizer is a solid char similar to the char produced in conventional carbonization. This char serves as a source of energy for the entire process and of hydrogen for the hydrogenation step.

The char is gasified (with oxygen) in a high-temperature, entrained-bed gasifier operating under slag-forming conditions. The gasifier is designed to conserve much of its own waste heat, which is recovered as high-pressure steam. The gasifier also achieves a total utilization of carbon by producing a slag that contains no carbon. Without carbon the slag is inert, which simplifies its disposal as waste.

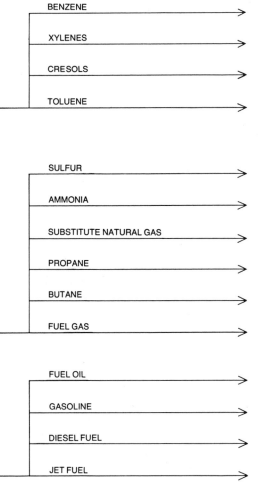

char is gasified and the gas is purified. The resulting clean gas proceeds to a Fischer-Tropsch step. The liquid extract (*bottom left*) yields a synthetic crude oil that can be refined.

After the gas from the gasifier has passed through the waste-heat recovery system it goes to a purification section of the plant, where acid gases, hydrogen sulfide and carbon dioxide are removed. The ratio of carbon monoxide to hydrogen is adjusted by the addition of hydrogen recovered from gases generated in other processes of the plant. Hydrogen is separated from those gases in a cryogenic separation step. The adjusted char gas then goes to the Fischer-Tropsch synthesis, where it passes over a catalyst. The primary products of this step are naphtha, diesel fuel and premium fuel oil. In addition the synthesis manufactures large quantities of gases of low molecular weight, which become the primary source of substitute natural gas and liquid petroleum gas from the plant.

The liquid extract from the filtration step, representing about 60 percent of the coal brought in from the mine, is distilled to recover solvent, which is employed in extraction and as the wash in the filtration step. Further distillation yields a primary material, the coal extract, which becomes the feed material for the hydrogenation step. Hydrogenation is conducted at a temperature of from 750 to 850 degrees F. and a pressure of about 3,000 pounds per square inch. Both the temperature and the pressure can be varied to alter the proportions of gas and liquid in the product.

The hydrogenated liquid is distilled to make both a solvent fraction, which is recycled, and a synthetic crude oil, which serves as the primary feed for a "cracking" step that yields gasoline. In locations that already have petroleum refineries the synthetic crude oil could be diverted to them. The gasoline produced by cracking would be combined with motor-fuel fractions produced in the Fischer-Tropsch section of the plant.

As I have indicated, a plant making oil and gas from coal must be located at the mine; otherwise the cost of moving from 25,000 to 100,000 tons of coal per day would make the plant uneconomical. Coal varies considerably from one area to another. For example, in the Appalachian region the coal is bituminous and high in volatiles; on the northern Great Plains it is lignite. Energy markets also vary from region to region. The coal refinery I have described, combining extraction, hydrogenation and Fischer-Tropsch synthesis, can be adapted to fit the characteristics of any coal, and the mixture of products can be adjusted to any market. In the Appalachian region the main products might be fuel oil for utility plants and substitute natural gas for residential and commercial consumption. A plant on the northern Great Plains would probably produce a higher proportion of liquids, since they could be transported economically by pipeline.

Suppose such a plant was designed to produce 50,000 barrels per day of liquid products. The plant would require about 25,000 tons of coal per day. The total capi-

tal investment would be about $1.5 billion: $200 million for equipment to mine and prepare coal, $200 million for extraction and separation, $300 million for gasification and synthesis, $100 million for equipment to handle by-products, $200 million for a utilities and services unit, $100 million for the air-separation plant, $250 million for contingencies and $150 million for engineering. Once the plant was operating it would manufacture daily 205 million standard cubic feet of gas, 5,000 barrels of liquid petroleum gas, 40,000 barrels of motor fuels, 5,000 barrels of fuel oil and (as by-products) 500 tons of sulfur and 3,000 tons of slag. The sulfur could be sold or stockpiled; some of the slag could be sold for making concrete blocks and other construction materials, but much of it would have to be returned to the mine as waste.

Twenty plants would produce a million barrels per day of liquid products and 4.1 billion cubic feet of gas. In a year, operating at 90 percent of capacity, the plants would make some 330 million barrels of nonpolluting liquid fuel and 1.353 trillion cubic feet of substitute natural gas. The output would meet about 6 percent of the nation's present demand for energy.

The coal requirement for 20 plants would be 165 million tons per year, which is about 25 percent of the national production of coal in 1975. The U.S. has nearly two trillion tons of recoverable coal. Therefore 100 plants, producing about 30 percent of the nation's energy requirement (stated in terms of consumption in 1975), could be operated for 30 years on approximately 1 percent of the coal resources. In Eastern locations the system would entail a land use of about 10 acres of coal per day.

The total capital investment for this array of plants would be about $30 billion, which is a large figure but a modest one on a national scale. If construction were spread over 10 years, expenditure would be at a maximum of $4 billion per year from the third year through the sixth. Plants could be added at the rate of four per year by increasing the expenditure to $6 billion per year. With this investment and production costs of from $160 to $180 million per year, the output of the plant would sell for $2.10 per million British thermal units to yield a rate of return of 12 percent on a discounted-cash-flow basis. The availability of nonpolluting fuels of high quality at $12.20 per barrel is an attractive alternative to imported oil at steadily rising prices.

To achieve a capacity of a million barrels per day of synthetic fuel calls for a vigorous Government program. Plants would be funded by the Government and operated by industry, which ultimately would buy the successful plants. The creation of a synthetic-oil industry would assure the nation of ample supplies of oil and gas. Moreover, the synthetic fuels would act as a ceiling on the price of crude oil from abroad. At the same time the country would have the assurance that it was no longer dependent on foreign supplies of oil.

The Strip-Mining of Western Coal

5

December 1975

*If the U.S. is to become self-sufficient in energy terms,
it will have to take huge amounts of coal from the
thick shallow deposits of the Western states. Can it be
done without despoiling the land?*

The availability of low-cost natural resources within the U.S. has been a major contributing factor to the country's economic and social growth. With oil and gas no longer plentiful, coal has become the mainstay of Project Independence: the effort to make the U.S. self-sufficient in energy terms. The coal industry is hoping to increase production from the 1974 level of 603 million tons to 1.3 billion tons by 1980 and to 2.1 billion tons by 1985. Since a ton of coal has the approximate heating value of four and a half barrels of oil, such an achievement would substitute coal for billions of barrels of oil.

Although people in the coal industry are delighted by the prospect, they are concerned that it may prove to be just another "boom and bust" cycle of the kind that has characterized the industry's history. Prosperity in the coalfields has never been predictable, since it fluctuates according to the availability of other fuels and the development of new energy technologies. There are currently no real alternatives to oil, gas and coal as energy sources; a large-scale role for such sources as oil shale, nuclear energy, geothermal and solar energy is a long way off, and many such sources are subject to attack by environmentalists. Just as steel has had its ups and downs in the market but has never been replaced, so coal may become essential to the U.S. economy.

Thus it seems almost certain that the country will turn increasingly to the vast coal deposits of the Western states, where 150,000 square miles of land are underlain by an estimated 2.9 trillion tons of coal. These deposits constitute 72 percent of the country's identified and hypothetical coal resources. Two distinctive things about minable Western coal reserves (198 billion tons) are that most of the coal has a low sulfur content and that a good deal of it (43 percent, or 86 billion tons, underlying 24,000 square miles) is close enough to the surface for strip-mining, in which the overburden is removed and coal is excavated from an open trench. Strip-mining in the West and in other parts of the country, however, has proved to have catastrophic effects on the land. The questions that arise are: Can the land be restored to its original capacity? If it can be, at what cost? Should the cost be borne by the public that needs the fuel or be covered by the mine operator and incorporated in the cost of the coal? Interested citizens and mine operators in the Western states with large deposits of coal accessible to strip-mining are actively debating these questions.

The amount of coal mined in the Western states (Montana, North Dakota, Wyoming, Utah, Colorado, Arizona and New Mexico) has increased substantially in recent years. The increase, from 75,000 tons per mine in 1961 to an average of a million tons in 1972, was brought about by the escalating market for low-sulfur coal and made possible by advances in mining technology.

Although the production of Western coal has risen sharply, the number of mines operating at any one time has remained fairly constant. Whereas in the past 40 years the number of operating strip mines in the Appalachian region of the East increased from 29 to 2,089, the number of strip mines in the West has remained at about 50 at any given time. Fewer than 35 of those mines account for more than 90 percent of Western coal production. Most Western strip mines are new: three-fourths of the area's tonnage comes from mines that are less than 10 years old.

Each mine has its special characteristics, but most of the mines follow the same six-step mining process. First, scrapers remove the topsoil and other unconsolidated material and put it on spoil piles. Second, the consolidated strata overlying the coal are drilled and blasted. Third, this overburden is removed by a huge dragline or power shovel. Fourth, the coal is drilled and blasted. Fifth, it is loaded into trucks. In the final step the spoil piles are graded by bulldozers.

In the East and the Middle West, where strip mines occupy areas smaller than those of the Western mines, the mining machinery tends to be smaller and more versatile. In the large Western mines the machines are of heroic size. The major workhorse is the dragline, which is employed to remove the overburden and put it in the pit opened by previous mining. The largest dragline now in operation stands more than 300 feet high and weighs 13,000 tons.

Nonetheless, the dragline is the limiting factor in coal production. In most Western mines these machines work full time, whereas the machines used for loading, hauling and reclamation work part time. Production can be increased only by increasing the rate of removing overburden. Bigger machines are not the answer; some draglines are already so large and heavy that the ground can barely support them as they move on to the next cut.

The trend from underground mining to surface mining has resulted in an enormous increase in productivity. The output per man-hour at a strip mine is approximately eight times the output at a fairly modern underground mine. The same trend has affected the geographic distribution of coal production in the Western states. From 1950 to 1967 Utah was the leading producer of coal west of the Mississippi, with an output of from four to five million tons per year from underground mines. In 1961 Utah International, Inc., opened the first big strip mine, the Navajo Mine in New Mexico. The production capacity of that mine alone is more than seven million tons per year. As a result of the trend toward strip-mining, Utah, where production is from underground mines only, has slipped almost to the bottom of the list of Western coal-producing states. First place has been taken over by Wyoming, where several huge strip mines are in operation.

In Wyoming, Montana and much of the northern Great Plains the conditions for surface mining approach the ideal. Coal seams from 50 to 75 feet thick, with an overburden of a mere 30 to 40 feet, are common. In Eastern and Middle Western mines a 10-foot seam is regarded as being exceptionally thick. There is also more overburden at those mines; at one mine in Oklahoma 95 feet of material must be removed to reach an 18-inch seam of coal.

The level of coal production from Western mines is expected to rise steeply in the near future. In Montana alone it is foreseen that 50 million tons of coal per year will be taken from five large mines by the early 1980's. Burlington Northern, Inc., expects to be hauling 90 million tons of coal per year out of the area by the same time. Plans are under consideration for thousands of miles of slurry pipelines to move millions of tons of coal per year across the country: from Wyoming to Arkansas, Montana to Washington, Wyoming to the Great Lakes and Utah to Nevada.

Although the projections for coal production in the West have virtually no ceiling, the current fraction of the nation's energy supplied by coal has risen only slightly (from 17.6 percent in 1973 to 17.8 percent in 1974). Total coal production rose only 1.4 percent (to about 603 million tons) from 1973 to 1974. Even the present low rate of coal production has subjected Western strip-mining to environmental constraints; sharp increases in production can only result in the multiplication of such constraints. Among the principal environmental concerns, of course, is whether or not Western land disturbed by strip-mining can be restored to something approximating its original condition.

Even though each mine site is different from all the others, a few generalizations can be made about the physical characteristics of Western mines and their potential for reclamation. Part of what can be said about reclamation is based on academic research and part on what the surface-mining industry has done (intermittently for 30 years and quite vigorously for the past five years) toward reclaiming mined land.

Geology and climate have the most significant bearing on a site's potential for reclamation. On the basis of their complex geological history the Western coal lands can be divided into two provinces: the Rocky Mountain province and the northern Great Plains province. The Rocky Mountain province includes the coal lands of the cordillera from Idaho to New Mexico, which lie between forested, granite-cored mountain ranges in broad basins that are comparatively flat but have been dissected by streambeds. The basins are covered by grasses and other dry-environment plants, shrubs and trees. The southwestern section is generally drier than the rest of the province, and the slopes of its mountains are less rugged and less forested than those in Utah, Idaho and Colorado.

The coal laid down in the Rocky Mountain province originated some 100 million years ago in the Cretaceous period. During later epochs of mountain building, forces that compressed the deposits and drove out volatile components upgraded some of the deposits from lignite to bituminous coal. After several million years coal swamps once again formed in the area on the flatlands between emerging highlands. Eventually material eroded from the highlands covered those coal deposits with thick layers of sediment.

The northern Great Plains province adjoins the Rocky Mountain province. During Paleocene time, some 65 million years ago, extensive beds of lignite were laid down over vast swampy areas. The mountains were low, and moisture-bearing winds from the west reached far into the interior of the continent. As the Rockies continued to be uplifted the winds were gradually cut off and the

STRIP COAL MINE is operated by the Western Energy Co. at Colstrip in southeastern Montana. The prominent ribbed ridge at lower right is a spoil pile made up of ma-

swamps gave way to grasslands. Today coal underlies thousands of square miles of basins covered at lower elevations by mixed-grass prairies and at higher ones by ponderosa pine.

Since Western coals formed under such a variety of geological conditions, they include all grades of coal (the main grades are lignite, subbituminous coal,

bituminous coal and anthracite). They are buried under a variety of overburden materials and call for different techniques of both mining and reclamation. Much of the coal underlying several large structural basins is accessible only to underground mining and will not be discussed here. In the current state of the technology the coal has to be within

about 225 feet of the surface to be accessible to strip-mining methods.

The present climate of the Western coal lands is arid or semiarid. The annual mean precipitation is low, ranging from four inches or less in the Four Corners area (where the boundaries of Utah, Colorado, Arizona and New Mexico meet) to 20 inches or more in some of

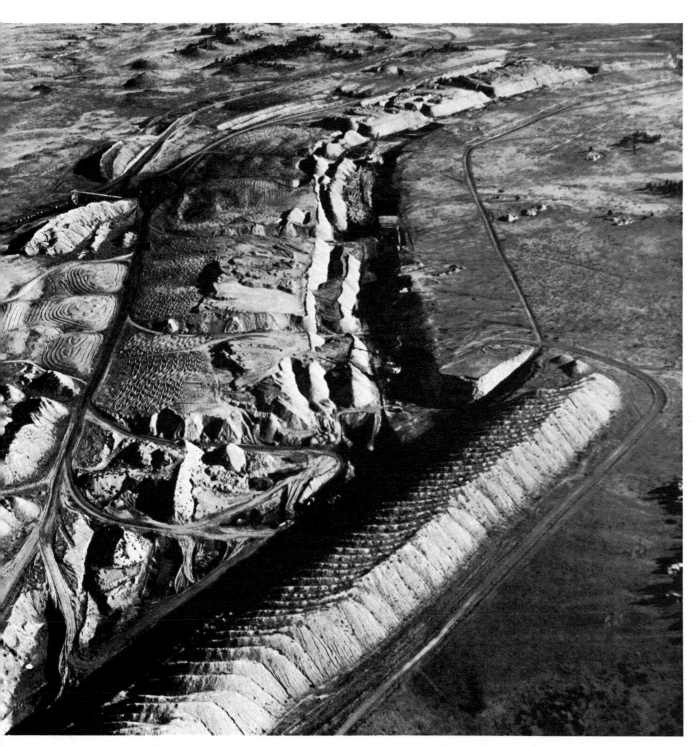

terial that lay over the seam of coal. The conspicuous rills are caused by erosion. Where the trench running toward the top of the photograph bends to the right, a dragline is removing overburden from the coal seam. In the trench a large shovel loads coal into hauling trucks. Running diagonally at left center is a railroad where coal cars are being loaded through a conveyor and a hopper. In the area below the hopper are several old piles of spoil that have been graded as a preliminary step in the reclamation process.

the Colorado coalfields. Droughts are common, and when precipitation does occur, it may come as a cloudburst. The temperature in summer reaches levels that desiccate seedling plants, so that only the hardiest organisms, tolerant to both summer heat and winter cold, survive.

The soils at arid and semiarid Western sites are poorly developed. Rocks weather slowly, and what sparse vegetation

there is adds little organic matter to the soil. Only terrain where glaciers have deposited material has a supply of soil that is more than barely adequate for reclamation. Wind, unimpeded by vegetation, dries the soil and drives sand and soil before it, killing tender plants. The rates of erosion are among the highest in North America.

Under such conditions natural processes are slow. It might take decades or

centuries for a disturbed site in a desert to be restored without assistance. Indeed, no successful reclamation after stripping has yet been achieved at any of the more arid sites, although considerable success has been achieved in parts of Colorado and in the more favorable conditions of the semiarid northern Great Plains. Fortunately glaciated and semiarid conditions are more typical of areas where the largest increases in the sur-

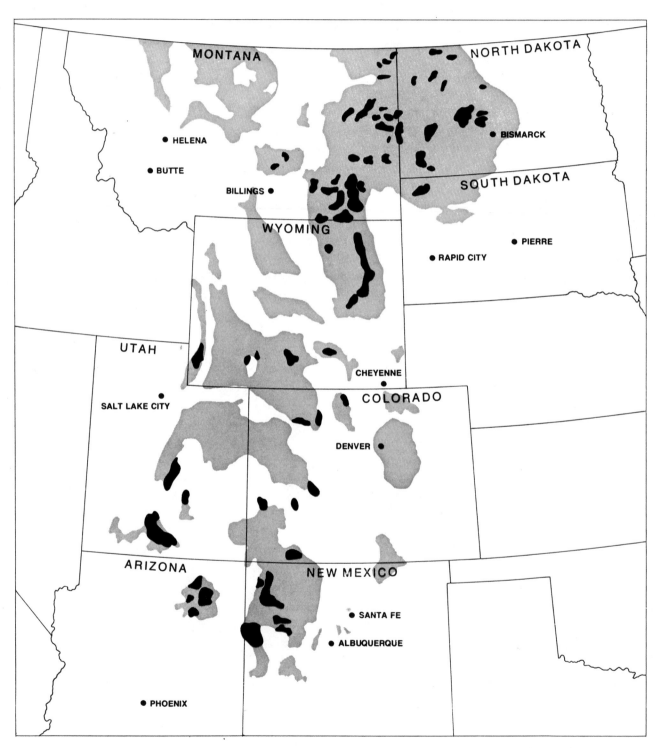

LOCATION OF RESERVES of Western coal is indicated. The parts of the map in color indicate the areas in the region that have major reserves of coal. The areas where the coal is close enough to the surface to be accessible by strip-mining are indicated in black.

face production of coal can be expected.

On the northern Great Plains rainfall ranges between 12 and 16 inches per year; the terrain is gently rolling; the overburden consists of alkaline shales and sandstones, and most of the land is grazed. Reclamation can be successful at most of these sites if enough planning, management, money and time are put into it. The prerequisites to successful reclamation include stability of the site, a nontoxic soil medium capable of holding moisture, proper plant-seeding techniques (which generally require slopes that can be traversed by farm machinery), occasional supplementary water and adequate management of grazing animals.

In the Appalachian region much of the damage has been done by water; the land has been scarred by acid drainage, instability of slopes, erosion and sedimentation. Water is scarce in most of the West, where the most difficult reclamation problems are revegetation and the maintenance of the hydrologic conditions in and around the mine site. Let us examine more closely how strip-mining can give rise to such problems.

Nearly all the Western lands underlain by coal can be classified according to some productive purpose and are capable of serving a range of uses: grazing, farming, recreation, watershed management or wildlife management. Surface mining without reclamation removes the land forever from productive use; such land can best be classified as a national sacrifice area. With successful reclamation, however, surface mining can become just one of a series of land uses that merely interrupt a current use and then return the land to an equivalent potential productivity or an even higher one.

Before 1950 most surface mines in the West (for all minerals, not just coal) operated according to the unwritten principle that the mined land would be treated as a sacrifice area. Today some states have strict regulations requiring reclamation. The regulations are not uniform, and they are not uniformly enforced. Moreover, none of them has been in effect for more than five years. Accordingly it cannot yet be determined whether the reclamation efforts made so far have either succeeded or failed.

Mining an area for coal must always disturb the land to some extent, but detrimental changes in topography brought about by mining can be remedied in most areas by employing earth-moving machines to grade and reshape the spoil. The Four Corners region may prove to

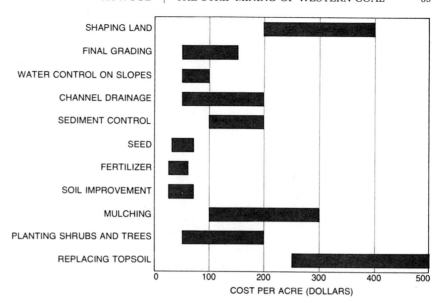

COST OF RECLAMATION of Western land where coal has been removed by strip-mining is given as a range of estimates. The figures cover the direct cost of reclamation and are based on data in a report prepared by the National Academy of Sciences for the Ford Foundation. Cost estimates made by mining companies range from $500 to $5,000 per acre.

be an unfortunate exception. The mesa topography of the area has largely been created by flash floods that have cut gullies through layers that are otherwise resistant to erosion. Mining operations involving blasting and the overturning of strata leave a fairly homogeneous nonresistant material. This material is subject to rapid erosion, and if the erosion is widespread, badlands will be formed. In any event the material cannot give rise to the mesa topography that was displaced.

Strip-mining usually buries topsoil under subsurface unconsolidated material. Natural processes will not soon regenerate such soils. Although certain soil characteristics can be artificially created, some mine operators have found it cheaper simply to segregate the topsoil during strip-mining operations. Then when the coal has been removed and the other overburden has been put back, the original topsoil can be spread over it. In a semiarid area reclamation is virtually impossible unless the ground is covered with a nontoxic material capable of holding water.

The changes in surface material brought about by strip-mining may result in changes in vegetation that are in themselves detrimental to the productive capacity of the land. Even in relatively nonproductive regions the impact can be heavy. For example, halogeton, a toxic weed, is among the first and most tenacious plants to revegetate disturbed desert land, to the dismay of ranchers. In more productive areas much can be done to restore or even to enhance the

productivity of mined areas by irrigating them. That, however, calls for a long-term commitment to land management that many mining companies are unwilling or unable to make.

Water, in fact, is the key to reclamation in the West. Mined areas, unlike those in the Appalachians, cannot be regarded as having been reclaimed when they have simply been reshaped and revegetated. Their hydrologic function must also be restored. In areas where productive activities, such as ranching, depend on a barely adequate supply of water, any disturbance of that supply can be critical.

The hydrologic measures normally taken after mining can usually reduce such surface-water problems as erosion, sedimentation, silting, ponding and changes in the quality of the water. A variety of techniques for handling material, impounding water and treating water (developed at Appalachian and Middle Western mines and at a strip mine in Washington) can be applied in Western mines. After mining, surface water usually infiltrates the spoil material rapidly, which tends to promote revegetation and the recharge of aquifers. If a soil material has not been successfully established over the spoil, however, a layer of hardpan can form. Then the surface runoff increases, quite likely giving rise to undercutting and the erosion of gullies downstream. Such effects are almost always irreversible. They are similar to the widespread damage to streambeds caused at the turn

STRIP-MINING OPERATION is shown at the Peabody Coal Company's Black Mesa Mine in Arizona. At top a dragline removes overburden from the coal seam. The seam is visible in the trench below the dragline, as are two large trucks that haul the coal out of the trench. At bottom center is a drilling machine employed to prepare the overburden for blasting.

REMOVAL OF COAL proceeds at the Black Mesa Mine. At left a large power shovel digs coal from the seam and stands by to load it into a haulage truck. The rock wall with the drilling machine at its lip (*far right*) is the overburden that must be removed from coal.

of the century by the excessive runoff brought on by overgrazing.

Changes in the quality of the surface water caused by the mining operation can be controlled by treatment at the site. Subtler and more serious are the consequences of leaching and the mechanical erosion of the highly alkaline overburden. An excess of dissolved salts, a high content of trace metals and an increased load of sediment are the commonest symptoms. Although such effects may not be detected at first, they can eventually cause the groundwater to become contaminated.

Groundwater can also be contaminated by the percolation of water through the spoil, the residue of chemicals used in the mining process, the mixing of groundwater layers during the mining cycle and the seepage of low-quality groundwater from one stratum to another. Effects of this kind, which are difficult to prevent and almost impossible to reverse, may not be detected for decades. The groundwater at Colstrip, Mont., is only now being found to have been contaminated by mining operations of 50 years ago.

Pumping water out of a mine so that the mining itself can proceed often gives rise to another set of problems. The water table in the vicinity of the mine is drawn down and thereby lowers the water level in nearby wells. In Gilette, Wyo., for example, it is estimated that intensive mining could result in the lowering or dewatering of some 200 wells used for livestock.

Mining may also change the characteristics of aquifers. It is almost axiomatic that mining alters the porosity and permeability of the overburden. On the northern Great Plains the most valuable aquifer may be the coal seam itself, and the overburden may also be an aquifer. In addition water of poor quality may be contained in pockets in the overburden and in the strata just below the coal. By disturbing these components of the aquifer and by introducing their contents into the rest of the aquifer, mining can lower the quality of the water in it.

The problem of maintaining the aquifer is even more difficult when mining is done on an alluvial valley floor, which is by definition an area of unconsolidated overburden and a high water table. In the West such areas act as buffers to the seasonal fluctuations of surface water and provide the naturally irrigated land where winter hay is grown for cattle. Where surface mining interrupts the alluvial aquifer and reduces the amount of water in it, the surface-water table downstream is lowered and vegetation with short roots is desiccated. Without its protective cover of vegetation the unconsolidated material of the valley rapidly erodes, downstream areas are undercut and gouged and the productivity of the area can be lost for decades.

If the essential hydrologic functions of alluvial areas could be maintained during mining and restored after mining, there would be no reason not to mine them, particularly since the handling of overburden tends to be cheaper in such areas than it is in the hillier country around them. It is by no means certain, however, that the hydrologic functions can be maintained during mining. Until reliable methods of doing so are devised, surface mining will inevitably endanger the long-term productivity of an area for a cash crop of coal that can be harvested only once.

The environmental damage caused by strip-mining can be viewed in at least four ways. One is to accept the damage, on the basis that the cost to society of controlling it is excessive. The second is to insist that enough remedial work be done to reclaim the area after mining. The third is to forestall the problems by requiring preventive measures during the mining cycle. The last is to avoid the problems by not mining the area at all.

Some threats to the environment, such as the slowing of photosynthesis in leaves coated by the dust stirred up at the mine site, seem hardly worth the trouble involved in eliminating the cause, but others are clearly quite serious and must be balanced against the importance of coal to the national economy. Society can, either consciously or by default, treat strip mines as national sacrifice areas.

Most citizens and most coal-mining companies find the concept of sacrifice areas unacceptable, particularly in view of the fact that most mined areas can be reclaimed. For the past 20 years or so the general rule has been remedial action; when an environmental problem arose after the mining cycle had been completed, it was patched up. Usually this meant covering the mined area with a suitable soil material.

This "add on" method of reclamation proved to be inefficient, expensive and frequently unsuccessful. Accordingly the trend has been toward preventive measures such as segregating spoil, burying toxic material and incorporating grading operations into the mining cycle. As a result revegetation has been more successful and certain of the hydrologic problems have been avoided.

The policy of not mining areas that cannot be reclaimed is regarded as a

RECLAMATION OF STRIP MINE is in progress at the Big Sky Mine of the Peabody Coal Company in Montana. Here the spoil banks have been shaped and partly seeded with oats and alfalfa, which serve as cover crops for a mixture of native grasses. The view is southeast toward hills and coniferous trees that were not affected by the strip-mining operation.

harsh one by coal companies. It would be devastating to a couple of existing mines and would create hardships for companies that have invested heavily in such operations. Yet so much Western coal is easily available for mining that the nation can afford in the future to avoid mining areas that are in one way or another irreplaceable.

Based on estimates made by the Western mining companies themselves it can be concluded that the cost of reclamation (from $500 to $5,000 per acre) has not jeopardized the competitive position of coal in the market. The higher estimate would add, to strike a rough average, less than 10 cents per ton to the price of Western coal. Moreover, most operators of Western coal mines hold the view that reclamation does not interfere with production, although they do believe that it reduces productivity. (Production, which is output in tons per day, should not be confused with productivity, which is output in tons per man-day. Production will not necessarily decrease when productivity decreases.)

Certain issues other than reclamation must be considered in any discussion of Western coal mining. One is the concern being voiced in the mining states over the fact that much of the money for and the direction of the mining activity will come from outside the region. Another has to do with the shortage of water in the West. Numerous objections to what may be in effect an allocation of scarce water resources for the production of energy for other states have been made, and more can be expected as coal production increases. Although there appears to be ample water for reclamation, the conversion of coal into electricity requires large quantities of water and will force the reallocation to energy production of water now dedicated to agriculture and recreation. At the same time voices are heard in the East warning that the shift to Western mines will be a severe blow to Eastern mines.

Such issues are more political than economic. So too is the question of how much reclamation should be undertaken and what timetable should be followed in the work. The choices that will shape the development of coal mining in the West and the reclamation of the mined lands will be made less in the marketplace than in the political arena. It seems clear, however, that large amounts of coal can be removed from a relatively few large mines in the Western states, that such operations can be limited to sites that can be reclaimed within an acceptable period of time after mining and that the cost of reclamation can be absorbed in the price of the coal.

III

NUCLEAR ENERGY

NUCLEAR ENERGY III

INTRODUCTION

Nuclear energy may have arrived just in the nick of time. Our supplies of low-cost fossil fuels clearly will be depleted within a few decades. For general central-power generation solar and geothermal energy will not be economically feasible for some time. Nuclear power is beginning to fill the gap in a very significant way, especially in resource-poor countries such as Korea and Taiwan. Currently, nuclear energy generates heat from the fission of the uranium nucleus; soon the fission of thorium or plutonium may become commercial. Breeder reactors may "breed" fissile material. Fusion reactors may be developed that create essentially an inexhaustible energy supply. For the time being, nuclear energy is confined to the indirect generation of electricity through steam, simply replacing another boiler fuel, such as oil. But such applications as high-temperature chemical processing or even direct generation of electricity may become possible.

The problems and prospects of nuclear power are discussed in the article by H. A. Bethe, who makes a strong case for fission reactors. He argues that the release of radioactivity from reactors is negligible, and that the possibility of a reactor accident is so small as to be negligible as well, in relation to other types of risks that face us in modern society. It is surprising, for example, that the health hazards posed by a nuclear plant, even when the possibility of accidents is accounted for, are considerably fewer than those posed by a coal-burning power plant of equal electric output. Additionally, it now appears that nuclear energy is less expensive than electricity from coal plants when one includes pollution-control equipment.

Aside from the fear of accidents, the principal concerns in people's minds seem to be the disposal of radioactive wastes from the reactors and the reprocessing of nuclear fuels generally. Both of these problems turn out to be not particularly difficult. We have adequate time to look for the cheapest and safest solutions. These conclusions emerge from the articles by Bernard L. Cohen and William P. Bebbington.

Clearly, civilian nuclear energy is still a very controversial topic in the United States and in many other democratic countries. A flavor of this controversy is conveyed by the letters and replies to the articles by Bethe and Cohen.

More to the point is the possibility of nuclear theft, nuclear terrorism, and nuclear proliferation. These topics are somewhat related, and the article by David J. Rose and Richard K. Lester deals with this complex of issues and relates it to United States energy policy.

The key to nonproliferation as well as to the theft of nuclear material by terrorist groups may lie in the reprocessing of spent nuclear fuels. The "Purex process" in current use (see Bebbington) was specifically developed during

World War II to extract pure plutonium for weapons manufacture. (Plutonium-239 is not especially radioactive and can therefore be hand-assembled into a weapon.) But reprocessing techniques exist for producing theftproof material—by leaving the unfissioned uranium and plutonium together and keeping in a proportion of the radioactive wastes. The result is fuel material too "hot" to handle except by remote processing behind heavy concrete shields. Such fuel, however, would be quite suitable for use in nuclear breeder reactors.

Because the development and commercialization of breeders is now proceeding in many countries, it becomes urgent to institute agreements on reprocessing and develop the necessary technology. The article by Georges A. Vendryes describes the construction of a full-scale breeder reactor in France, part of a joint European project.

Breeder reactors would, of course, greatly extend the energy potential of uranium and thorium. Current fission reactors use only between 1 and 2 percent of the potential energy content of uranium. But an alternative concept, fusion reactors, would produce an essentially inexhaustible energy source—although not necessarily an inexpensive one. The principles of nuclear fusion, as well as the bewildering number of devices currently being tested, are described in the article by William Gough and Bernard Eastlund. Although basically optimistic, the authors cannot set a time-table because a laboratory demonstration of a self-sustained fusion reactor has not yet been achieved. Clearly, nuclear fusion is proving to be a challenging scientific problem, as well as an engineering one.

6

The Necessity of Fission Power

by H. A. Bethe
January 1976

*If the U.S. must have sources of energy other than fossil fuels,
the only source that can make a major contribution between
now and the end of the century is nuclear fission*

The quadrupling of the price of oil in the fall of 1973 came as a rude but perhaps salutary shock to the Western world. It drew attention to the fact that oil is running out, and that mankind must turn to other fuels, to strict energy conservation or to both.

The price increase was not entirely unjustified. From 1950 to 1973 the price of oil, measured in constant dollars, had declined steadily. Moreover, it has been estimated that if world oil production were to continue to increase at the same rate that it has in the past two decades, the upward trend could persist only until about 1995; then the supply of oil would have to drop sharply [*see **illustration on next page**]. Accordingly the oil-producing countries must see to their own economic development while their oil lasts so that they can rely on other sources of revenue thereafter. At the same time the rest of the world must take measures to become less dependent on oil—particularly imported oil—while there is still time.

What would it take for the U.S., which currently gets more than 15 percent of its energy in the form of imported oil, to become "energy independent?" In a report issued last June the Energy Research and Development Agency (ERDA) outlined its plans for the U.S. to achieve this goal. The ERDA projections are expressed in terms of quads, or quadrillions (10^{15}) of British thermal units (B.t.u.). According to ERDA, the drive to achieve energy independence calls for a two-pronged approach. First, the U.S. must be technologically geared not only to expand the production of its existing principal energy resources (oil, gas, coal and uranium) but also to develop several new energy sources. Second, a major energy-conservation effort must be initiated both to reduce total energy consumption and to shift consumption to sources other than oil. Only if both remedies are successfully applied can energy independence be achieved—and then it can be achieved only by 1995 [*see illustration on page 69*]. Without any new initiatives the need for imported oil will rise steadily from about 12 quads at present to more than 60 in the year 2000. At current prices the importation of that much oil would cost about $120 billion, compared with $25 billion in 1974, an increase of $95 billion.

Now, $95 billion may not sound like a gigantic sum when this fiscal year's Federal budget deficit is projected to be about $70 billion. The economics of international trade, however, is a different matter. Even a $10 billion trade deficit has a major effect on the stability of the currency. It is almost impossible to think of exports that could bring in an additional $95 billion. Besides, if current trends are allowed to continue, the U.S. would take about 30 percent of the world's oil production when that production is at its maximum. Clearly it is critical that the U.S. not follow this course.

What is critical for the U.S. is a matter of survival for Japan and the countries of western Europe. After all, the U.S. does have substantial amounts of oil and gas and plenty of coal. Japan and Italy have none of those fuels. England and Norway will have a limited domestic supply of oil in a few years, but other countries of western Europe have no natural oil resources of their own and have limited amounts of coal. If the U.S. competes for scarce oil in the world market, it can only drive the price still higher and starve the economies of western Europe and Japan. The bankruptcy of those countries in turn would make it impossible for the U.S. to export to them and thus to pay for its own imports.

For the next five years or so there is only one way for the U.S. to make measurable progress toward the goal of energy independence, and that is by conserving energy. There are two kinds of energy conservation. One approach is to have the country lower its standard of living in some respects, for example by exchanging larger cars for smaller ones. This measure has been widely accepted, probably at some cost in safety. To most Americans, however, it appears undesirable to continue very far in this direction.

The other approach to conservation is to improve the efficiency with which energy is consumed. A number of useful suggestions have been made, such as insulating houses better, increasing the efficiency of space-heating and water-heating systems, improving the way steam is generated for industry and upgrading other industrial processes. Conversions of this type require substantial investment, and their cost-effectiveness on a normal accounting scheme is not clear. Much leadership, public education and tax or other incentives will be needed to realize the potential for increased efficiency. If all these things are provided, the total energy consumption of the U.S. in the year 2000 could be reduced from 166 quads to 120.

ERDA predicts that if at the same time the generation of electricity from coal and nuclear fuel is allowed to expand as it is needed, the U.S. can achieve an intermediate trend in oil imports: a satisfactory decline in the first 10 years, followed by a rise until oil imports are higher in 2000 than they are now. Energy independence will not have been achieved by that course either.

In all three ERDA projections it is assumed that the U.S. will move gradually from liquid fuels (oil and gas) to solid fuels (coal and uranium). For example, in President Ford's State of the Union Message in January, 1975, the actual contribution of various fuels to our energy budget in 1973 was presented along with the President's aims for 1985 and the expected situation in 1985 if no action is taken [see illustration on page 70]. The latter situation would require the importation of 36 quads of oil, in fair agreement with ERDA's prediction of 28 quads for 1985.

The Ford projection envisions a total U.S. consumption of 103 quads in 1985, 28 quads more than in 1973. Since much of the added energy would go into the generation of electricity, with a thermal efficiency of 33 to 40 percent, however, consumable energy would increase by only 17 quads, or 26 percent. Taking into account an expected 22 percent increase in the working population during that period, the consumable energy per worker would stay roughly constant.

The Ford message projects that domestic oil production will increase by seven quads by 1985 and that natural-gas production will decrease by only two quads, in spite of the fact that in the U.S. oil production has declined in the past two years and natural-gas discoveries have run at less than half of consumption for the past eight years. The ERDA report agrees that by stimulating the domestic production of oil and gas the U.S. could attain just about the total production figure used by the President, 53 quads, with gas somewhat higher than his estimate and oil lower.

Of course, the country would be depleting its resources more rapidly and would have to pay for it by having less domestic oil and gas in the years after 1985. The proposed stimulation of domestic oil and gas production, however, would provide the breathing space needed to bring other forms of energy into play. The only energy resources the U.S. has in abundance are coal and uranium. Accordingly President Ford calls for a massive increase in coal production, from 600 million tons in 1973 to 1,000 million tons in 1985. Meanwhile the Administration's energy program calls for the building of 200 nuclear-fission reactors with an energy output equivalent to about 10 quads.

Coal should certainly be substituted for oil and gas in utilities and in other industrial uses wherever possible. The conversion of coal into synthetic gas or oil is essential; demonstration plants for these processes and price guarantees should be given the highest priority. The same applies to oil from shale.

Coal cannot do everything, however, particularly if it is used intensively for making synthetic fuel. The U.S. needs another, preferably nonfossil, energy source. The only source that is now sufficiently developed to play any major role is nuclear fission. Thoughtful people have raised a number of objections to nuclear-fission reactors, which I shall discuss below, but first let me review some of the alternative energy sources that have been suggested.

Nuclear fusion is the energy source that has most strongly captured the imagination of scientists. It is still completely unknown, however, whether useful energy can ever be obtained from the fusion process. It is true that both stars and hydrogen bombs derive their energy from the fusion of light atomic nuclei, but can such energy be released in a controlled manner on the earth? The requirements for accomplishing the task are tremendous: a mixture of heavy-hydrogen gases must be brought to a temperature of about 100 million degrees Celsius and kept there long enough for energy-releasing reactions between the hydrogen nuclei to take place at a rate sufficient to yield a net output of energy.

The most obvious way to try to satisfy this condition is by magnetic confinement. At 100 million degrees hydrogen is completely ionized, and the positively charged nuclei and negatively charged electrons can be guided by magnetic

FINITE SUPPLY OF OIL is responsible for the shape of this curve representing world oil production over a two-century span. The projection is based on the work of M. King Hubbert of the U.S. Geological Survey, who estimates that if world oil production were to continue to increase at the same rate that it has in the past two decades, output would peak in about 1995 and then drop sharply. Energy content of various fuels discussed in this article is expressed in quads, short for quadrillions (10^{15}) of British thermal units (B.t.u.).

fields. Since the early 1950's physicists in many countries have designed many intricate magnetic-field configurations, but they have not succeeded in attaining the break-even condition. Great hopes have alternated with complete frustration. At present the prospects seem better than ever before; a few years ago Russian experimenters developed the device named Tokamak, which has worked at least roughly according to theoretical expectations. This device has been reproduced in the U.S. with comparable success. More than $200 million has now been committed by ERDA for a much larger device of the Tokamak type, to be built at Princeton University; if that machine also fulfills theoretical expectations, we may know by the early 1980's whether or not power from fusion is feasible by the Tokamak approach.

There have been too many disappointments, however, to allow any firm predictions. Work on machines of the Tokamak type is also going forward in many other laboratories in the U.S., in the U.S.S.R. and in several countries of western Europe. If the problem can be solved, it probably will be. Money is not the limiting factor: the annual support in the U.S. is well over $100 million, and it is increasing steadily. Progress is limited rather by the availability of highly trained workers, by the time required to build large machines and then by the time required to do significant experiments. Meanwhile several alternative schemes for magnetic confinement are being pursued. In addition there are the completely different approaches of laser fusion and electron-beam fusion. In my own opinion the latter schemes are even further in the future than Tokamak.

Assume now that one of these schemes succeeds in the early 1980's. Where are we then? The problem is that the engineering of any large, complex industrial plant takes a long time, even after the principle of design is well known. Since preliminary fusion-power engineering is already under way, however, it is a reasonable hope that a prototype of a commercial fusion reactor could operate in about 2000, and that fusion might contribute a few percent of the country's power supply by 2020.

Solar power is very different. There is no doubt about its technical feasibility, but its economic feasibility is another matter. One should distinguish clearly between two uses of solar power: the heating of houses and the production of all-purpose power on a large scale.

Partial solar heating of houses may become widespread, and solar air-conditioning is also possible. ERDA is spon-

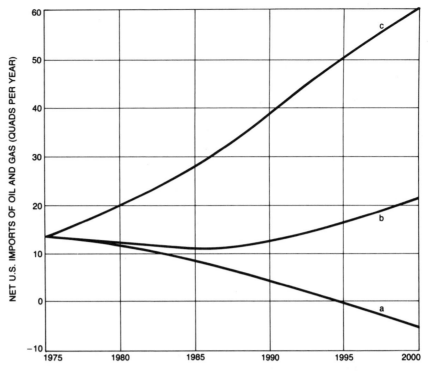

THREE PROJECTIONS of the extent to which the U.S. will continue to be dependent on imported oil and gas for the rest of this century were made in a 1975 report by the Energy Research and Development Administration (ERDA). Curve *a*, which shows the U.S. ending its dependence on imported fuel in about 1995, assumes that all the recommendations of ERDA's National Plan for achieving "energy independence" are put into effect. Curve *b* shows the effect of an intermediate approach emphasizing energy conservation. Curve *c* shows the expected trend in oil imports in the absence of any major new initiatives. All three curves assume that the energy efficiency of automobiles will improve by about 40 percent over the next five years or so as a result of the trend toward smaller cars and that the number of coal-burning and nuclear power plants will expand to meet rising demand for electricity. Any slowdown on nuclear plants would shift all three curves sharply upward.

soring the development of model solar-heated houses. Private estimates for solar-heating systems, for a "standard" house of 1,200 square feet, run between $5,000 and $10,000 in mass production, compared with about $1,000 for conventional heating systems. With such an installation one might expect to supply about 50 percent of the house's heating requirements (more in the South, less in the North, particularly in regions of frequent cloud cover). In any case an auxiliary heating system supplied with gas, oil or electricity must be provided; otherwise the cost of the solar-heating system becomes exorbitant.

ERDA estimates that 32 million new households will be established between 1972 and 2000, and that they will then comprise about a third of all dwelling units. If all the new units are equipped with solar heating, it would require a private investment of $150 to $300 billion. The heating requirement for all residential units in 1973 was close to 10 percent of the country's total energy consumption, and that fraction is likely to remain about the same. Some of the new

dwelling units will not use solar energy, but let us assume (optimistically) that an equal number of older houses will be converted to solar heat. In that case a third of all houses would derive on the average about half of their heat from the sun, which would then supply somewhat less than 2 percent of the country's total energy needs. This contribution would be helpful but clearly would not be decisive.

The use of solar heat on a large scale for power generation is something else again. (Here I shall assume electric power, but the situation would not be essentially different if the energy were to be stored in fuels such as hydrogen.) Of the many proposals that have been made, the most practical in my opinion is to have a large field (perhaps a mile on a side) covered by mirrors, all of which reflect sunlight to a central boiler. The mirrors would be driven by a computer-controlled mechanism; the boiler would generate electricity in the conventional manner. At least three separate groups, supported by ERDA, are working on this kind of project. The best estimates I have heard give about $2,500 per installed kilowatt (power averaged over the 24-hour day) exclusive of interest and the effects of inflation during construction. On the same basis nuclear-fission reactors cost about $500 per kilowatt, so that solar power is roughly five times as expensive as nuclear power.

That cost estimate may sound high, but a little thought will show that it is not. First of all, the sun shines for only part of the day. On a sunny winter day in the southern U.S. one square mile of focused mirrors is just about enough to generate an average of 100 megawatts of electric power at a cost of about $250 million. To achieve that output the full heat of the sun must be utilized whenever it shines. At noon such a system would generate about 400 megawatts; near sunrise and sunset it would generate correspondingly less; at night it would generate none. To get an average of 100 megawatts one must have equipment to generate 400 megawatts, so that the generating equipment (boilers, turbines and so on) would cost roughly four times as much as they would in a comparable nuclear or fossil-fuel power plant. To this total cost must be added the cost of storing the energy that will be needed at night and on cloudy days. (The means of storage is so far a largely unsolved problem.)

Assume now that half of the cost is allotted to the mirrors and their electronic drive mechanisms; that would amount to $125 million for a plant of one square mile, or less than $5 per square foot. It is hardly conceivable that the mirrors and their drives could be built that cheaply, even in mass production, when a modest house costs $30 a square foot. I conclude therefore that all-purpose solar power is likely to remain extremely expensive.

Although it seems clear that solar power can never be practical for western Europe and Japan, the countries that need power most urgently, it might be just the right thing for certain developing countries, provided that the capital-cost problem can be solved. Many of those countries have large desert areas, rather modest total energy needs and abundant cheap manpower, which is probably required for the maintenance of any solar-power installation.

In addition to the alternative energy sources discussed above, a variety of other schemes have been suggested, such as harnessing the wind or the tides, burning garbage or agricultural wastes, converting fast-growing plants into fuels such as methane or tapping the earth's internal heat. Each of these approaches presents its own special difficulties, and at best each can make only a minor contribution toward the solution of the energy problem.

I do not mean to imply that work on alternative-energy projects is worthless. On the contrary, I believe that research and development on many of them should be pursued, and in fact ERDA is stepping up this type of work. I want to emphasize, however, that it takes a very long time from having an idea to proving its value in the laboratory, a much longer time for engineering development so that the process can be used in a large industrial plant and a still longer time before a major industry can be established. Certainly for the next 10 years and probably for the next 25 years the U.S. cannot expect any of the proposed alternative energy schemes to have much impact.

For all these reasons I believe that nuclear fission is the only major nonfossil power source the U.S. can rely on for the rest of this century and probably for some time afterward. Let us now examine the objections that have been raised against this source of power.

Some concern has been expressed over the fact that nuclear reactors in routine operation release radioactivity through outflowing liquids. According to the standards originally set by the Atomic Energy Commission and now administered by the Nuclear Regulatory Commission, these releases must be kept "as low as practicable," and under no circumstances must the additional radiation exposure of a person living permanently near the fence of the power plant be greater than five millirem per year. Most modern fission power plants release far less than this limit. For the purposes of comparison an average person in the U.S. receives 100 millirem per year in natural radiation (from cosmic rays, radioactivity in the earth and in buildings and radioactive substances inside his body) and an average of about 70 millirem per year from diagnostic medical X rays. It has been estimated that in the year 2000 a person living in the U.S. would on the average receive an additional tenth of a millirem from nuclear reactors if 1,000 of them are deployed. Chemical plants for reprocessing the nuclear fuel may add a couple of tenths of a millirem, but the Nuclear Regulatory

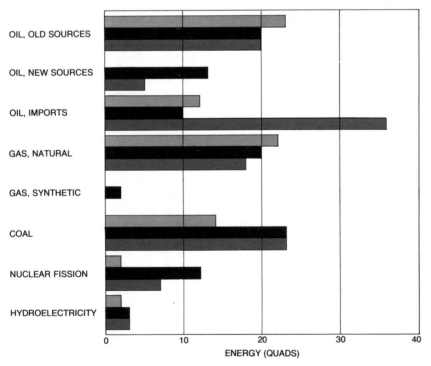

TWO ESTIMATES of the contribution of various fuels to the energy budget of the U.S. in 1985 are presented here along with the actual energy budget in 1973 (*light gray bars*). The chart is based on President Ford's State of the Union Message for 1975, in which he compared the expected impact of his administration's energy program (*dark gray bars*) with the expected situation if no action is taken (*colored bars*). The total U.S. energy consumption in 1973 was 75 quads; the total for 1985 in the absence of any major new programs is projected here to be about 112 quads, including some 36 quads of imported oil; Ford program, which includes a major energy-conservation effort aimed at saving about nine quads per year by 1985, envisions a total U.S. energy consumption of 103 quads for that year.

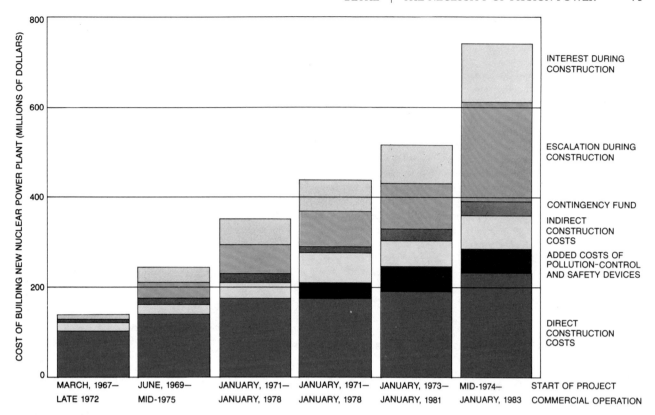

CONSTRUCTION COST of a 1,000-megawatt nuclear power plant of the light-water-reactor type has risen substantially for plants planned between 1967 and 1974 and expected to become operational between 1972 and 1983. As this ERDA bar chart indicates, however, a large fraction of the cost increase is due to inflation and interest during construction. Cost of building a coal-burning power plant has risen at a comparable rate during this period. (The bar for nuclear plants started in 1971 was revised upward in 1973.)

Commission is tightening the regulations further. In view of these very small numbers the controversy over the routine release of radioactivity, which was strong in the 1960's, has pretty much died down.

A more popular fear at present is that a reactor accident would release catastrophic amounts of radioactivity. Here it must be said first of all that a reactor is not a bomb. In particular, light-water reactors, which make up the bulk of U.S. reactors at present, use uranium fuel with a readily fissionable uranium-235 content of only 3 percent. Such material, no matter how large the amount, can never explode under any circumstances. (For breeder reactors, which can only come into operation about 20 years from now, the argument is slightly more complicated.)

It is, however, conceivable that a reactor could lose its cooling water, melt and release the radioactive fission products. Such an event is extremely unlikely, and one has never happened. There are at least three barriers to such a release. The radioactive fission products are enclosed in fuel pellets, and those pellets have to melt before any radioactivity is released. No such "meltdown"

has occurred in nearly 2,000 reactor-years of operation involving commercial and military light-water reactors in the U.S. Moreover, even if there were to be a meltdown, the release of radioactivity would be retarded by the very strong reactor vessel, which typically has walls six to 12 inches thick. Finally, once this reactor vessel melts through, the radioactive material would still be inside the containment building, which is equipped with many devices to precipitate the volatile radioactive elements (mainly iodine, cesium and strontium) and prevent them from escaping to the outside. Only if very high pressure were to build up inside the reactor building could the building vent and release major amounts of radioactivity. The chance of that happening is extremely small, even in the event of a meltdown.

One may nonetheless ask: Exactly how likely is such a reactor accident? Obviously it is very difficult to estimate the probability of an event that has never happened. Fortunately most of the conceivable failures in a reactor do not lead to an accident. Reactors are designed so that in case of any single failure, even of a major part of the reactor, the reactor can still be safely shut down. Only when

two or more essential elements in the reactor fail simultaneously can an accident occur. This makes a probabilistic study possible; an estimate is made of the probability of failure of one important reactor element, and it is then assumed that failures of two different elements are independent, so that the probability of simultaneous failure of the two is the product of the individual probabilities. This, however, is not always true. There can be "common mode" failures where one event triggers two or three failures of essential elements of the reactor; in that case the probability is the same as that of the triggering event, and one does not get any benefit from the multiplication of small probability numbers. The probability of such common-mode failures is of course the most difficult to estimate.

Working on the basis of these principles, a Reactor Safety Study commissioned three years ago by the AEC estimated the probability of various types of reactor accident. The results were published in draft form in August, 1974, in a document that has come to be known as the Rasmussen report, named for the chairman of the study group, Norman C. Rasmussen of the Massachusetts Institute of Technology. The final report was published last October.

The methods applied in the Rasmussen report have been used for several years in Britain to predict the probability of industrial accidents. Experience has shown that the predictions usually give a frequency of accidents somewhat higher than the actual frequency. Several groups, including the Environmental Protection Agency and a committee set up by the American Physical Society, have since studied various aspects of the problem and have come out with somewhat different results. Those differences have been taken into account in the final Rasmussen report; the most important of them will be discussed here.

The basic prediction of the Rasmussen report is that the probability of a major release of radioactivity is about once in 100,000 reactor-years. (Common-mode failures were found to contribute comparatively little to the total probability.) Such an accident would involve the release of about half of the volatile fission products contained in the reactor. A release of that scale would have to be preceded by a meltdown of the fuel in the reactor, an event for which the report gives a probability of once in 17,000 reactor-years. Finally, the report predicts that the water coolant from a reactor will be lost once in 2,000 reactor-years, but that in most cases a meltdown will be prevented by the emergency core-coolant system.

There is at least some check on those estimates from experience. For one thing, there has never been a loss of coolant in 300 reactor-years of commercial light-water-reactor operation. Furthermore, there has never been a fuel meltdown in nearly 2,000 reactor-years of commercial and naval light-water-reactor operation. If Rasmussen's estimate were wrong by a factor of 20 (in other words, if the probability of a meltdown were once in 850 reactor-years), at least one meltdown should have occurred by now.

What would be the consequences in the extremely improbable event of a major release of radioactivity? The immediate effects depend primarily on the population density near the reactor and on the wind direction and other features of the weather.

For a fairly serious accident (one that might take place in a million reactor-years) Rasmussen estimates less than one early fatality but 300 cases of early radiation sickness. He also predicts that there could be 170 fatalities per year from latent cancers, a death rate that might continue for 30 years, giving a total of some 5,000 cancer fatalities. In addition there might be 25 genetic changes per year; counting the propagation of such changes through later generations, the total number of genetic changes could be about 3,000.

The number of latent cancers in the final version of the Rasmussen report is about 10 times as high as it was in the original draft report; that change was largely suggested by the study of the American Physical Society, as modified by a very careful study made by the Rasmussen group. A major release of radioactivity under average weather and population conditions (probably one in 100,000 reactor-years) would cause about 1,000 latent cancers, but it would not result in any cases of early radiation sickness.

It is obvious that 5,000 cancer deaths would be a tragic toll. To put it in perspective, however, one should remember that in the U.S. there are more than 300,000 deaths every year from cancers due to other causes. A reactor accident clearly would not be the end of the world, as many opponents of nuclear power try to picture it. It is less serious than most minor wars, and these are unfortunately quite frequent. Some possible industrial accidents can be more serious, such as explosions and fires in large arrays of gasoline storage tanks or chemical explosions. The danger from dam breaks is probably even greater.

The probability of a serious reactor accident was predicted in the Rasmussen report to be once in 10,000 years when there are 100 reactors, which is about the number expected for the U.S. in the year 1980. What if the number of reac-

ROUTINE EMISSION OF RADIATION (2)   NUCLEAR REACTOR ACCIDENTS (2)

ACCIDENT RISKS estimated for the entire U.S. population as the result of the operation of 100 nuclear power plants are compared here with the risks from several leading causes of accidents in terms of the average number of deaths per year attributable to each cause. (The averages for the latter categories are rounded to the nearest 1,000 fatalities.) The figure for the risk of death from nuclear accidents is based on the conservative assumption that there is likely to be one major release of radioactivity in the U.S. every 1,000 years, resulting in about 1,000 eventual deaths from cancer, and that once in 10,000 years there could be a more serious accident resulting in approximately 5,000 eventual deaths. The average risk from nuclear reactors is obviously extremely small compared with other risks that society accepts. It must be noted, however, that the nuclear-power risk can only be predicted, whereas the other risks are actuarial, that is, derived from statistics of actual events.

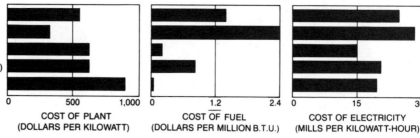

COAL
OIL
URANIUM (LIGHT-WATER REACTOR)
EXPENSIVE URANIUM (LIGHT-WATER REACTOR)
EXPENSIVE URANIUM (BREEDER REACTOR)

COST OF PLANT
(DOLLARS PER KILOWATT)

COST OF FUEL
(DOLLARS PER MILLION B.T.U.)

COST OF ELECTRICITY
(MILLS PER KILOWATT-HOUR)

COST ESTIMATES summarized in this bar chart were made in late 1973 by the Philadelphia Electric Company for prospective electric-power plants consuming three different types of fuel: coal, oil and uranium. The first three entries in each set of bars were made on the assumption that plant construction would start in 1974. (The bars representing fuel costs for coal and oil have been updated to 1975.) The bottom two entries refer to an indefinite date in the future when uranium is expected to become much more expensive. According to this study, electricity from nuclear fuel will continue to be substantially cheaper than that from fossil fuel.

tors increases to 1,000, as many people predict for the year 2000 or 2010? The answer is that reactor safety is not static but is a developing art. The U.S. is now spending about $70 million per year on improving reactor safety, and some of the best scientists in the national laboratories are engaged in the task. I feel confident that in 10 years these efforts will improve both the safety of reactors and the confidence we can have in that safety. I should think that by the year 2000 the probability of a major release of radioactivity will be at most once in 10 million reactor-years, so that even if there are 1,000 reactors by that time, the overall chance of such an accident will still be no more than once in 10,000 years.

Taking into account all types of reactor accidents, the average risk for the entire U.S. population is only two fatalities per year from latent cancer and one genetic change per year. Compared with other accident risks that our society accepts, the risk from nuclear reactors is very small [see *illustration on opposite page*].

A special feature of possible reactor accidents is that most of the cancers would appear years after the accident. The acute fatalities and illnesses would be rather few compared with the 5,000 estimated fatalities from latent cancers in the foregoing example. The problem is that many more than the 5,000 victims will think they got cancer from the radiation, and it will be essentially impossible to ascertain whether radiation was really the cause. The average probability that the exposed population will get fatal cancer from the released radioactivity is only about .1 percent, compared with the 15 percent probability that the average American will contract fatal cancer from other causes. Will the affected people in the case of a reactor accident be rational enough to appreciate this calculation? Or would an accident, if it occurs, have a psychological effect much more devastating than the real one?

The problem of nuclear energy that is considered most serious by many critics is the disposal of nuclear wastes. Will such wastes poison the atmosphere and the ground forever, as has been charged? It is true that the level of radioactivity in a standard 1,000-megawatt reactor is very high: about 10 billion curies half an hour after the reactor is shut down. The radioactivity then decays quite quickly, however, and so does the resulting heat.

When the spent nuclear fuel is unloaded from a reactor, it goes through a number of stages. First the highly radioactive material, still in its original form, is dropped into a tank of water, where it is left for a period ranging from a few months to more than a year. The water absorbs the heat from the radioactive decay and at the same time shields the surroundings from the radiation.

After the cooling period the fuel will in the future be shipped in specially protected trucks or railcars to a chemical-reprocessing plant. (No such plant is currently in operation, but a large one is being built in South Carolina and could go into operation next year.) In the chemical plant the fuel rods will be cut open (still under water) and the fuel pellets will be dissolved. The uranium and the plutonium will be separated from each other and from the radioactive fission products. The uranium and plutonium can be reused as reactor fuel and hence will be refabricated into fuel elements. The remaining fission products are the wastes. *nuclear waste is*

These substances are first stored in a water solution for an additional period to allow the radioactivity to decay further. Special tanks with double walls are now being used for that purpose in order to ensure against leakage of the solution.

After five years the wastes will be converted into solids, and after another five years they will be shipped to a national repository. Three different methods have been developed for solidifying wastes; one method now operates routinely at ERDA's reactor test station in Idaho to solidify the wastes from Government-owned reactors. The solid wastes can then be fused with borosilicate glass and fabricated into solid rods, perhaps 10 feet long and one foot in diameter. (Approximately 10 such rods will be produced by a standard 1,000-megawatt reactor in a year.) The rods are then placed in sturdy steel cylinders closed at both ends. It is difficult to see how any of the radioactive material could get out into the environment after such treatment, provided that the material is adequately cooled to prevent melting.

There are two possibilities for the national repository. One, for interim storage, would be in an aboveground desert area; the steel cylinders would be enclosed in a heavy concrete shield to protect the external world from the radiation. Cooling would be provided by air entering at the bottom of the concrete shield, rising through the space between the steel and the concrete and escaping at the top after having been heated by about 20 degrees C. Natural air circulation would be sufficient; no fans are required. The proposal for such a national repository has been studied and approved by a committee of the National Academy of Sciences.

The area required for such an interim-storage repository is not large. A standard reactor produces about two cubic meters of solid waste a year. The National Academy of Sciences committee estimated that all the wastes produced by U.S. reactors by 2010 could be stored on a tract of 100 acres. The cost is estimated at $1.5 billion, a small fraction of the probable cost of the reactors.

The second possibility for the national

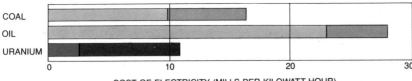

COAL

OIL

URANIUM

COST OF ELECTRICITY (MILLS PER KILOWATT-HOUR)

ACTUAL POWER COSTS for the first quarter of 1975 were obtained by averaging the total costs, including fuel (*color*), for 22 utilities that operate nuclear reactors as well as other plants. These figures, compiled by the Atomic Industrial Forum, are in accord with those in illustration on preceding page. (The capital investment in these currently operating plants was of course much lower than that estimated for the future plants in that example.)

repository is permanent storage deep underground. The preferred storage medium here is bedded salt, which presents several advantages. First, the existence of a salt bed indicates that no water has penetrated the region for a long time; otherwise the salt would have been dissolved. Water trickling through the storage site should be avoided, lest it leach the deposited wastes and bring them back up to the ground, an extremely slow process at best but still better avoided altogether. Second, salt beds represent geologically very quiet regions; they have generally been undisturbed for many millions of years, which is good assurance that they will also remain undisturbed for as long as is required. Third, salt flows plastically under pressure, so that any cracks that may be formed by mechanical or thermal stress will automatically close again.

The first attempt by the AEC to find a storage site in a salt mine in Kansas was unfortunately undertaken in a hurry without enough research. (Drill holes in the neighborhood might have allowed water to penetrate to the salt bed and the waste.) Now ERDA is carefully examining other sites. A promising location has been found in southeastern New Mexico. There are roughly 50,000 square miles of salt beds in the U.S.; only three square miles are needed for disposal of all the projected wastes up to the year 2010.

The method of disposal is this: In a horizontal tunnel of a newly dug mine in the salt bed, holes would be drilled in the wall just big enough to accommodate one of the steel cylinders containing waste. It has been calculated that the cylinders could be inserted into the salt 10 years after the waste comes out of the reactor. The residual heat in the waste, five kilowatts from one cylinder, is then low enough for the salt not to crack. (The high heat conductivity of salt helps here.) If the calculation is confirmed by experiment in the actual mines, the wastes could go directly from the chemical-processing plant into per-

manent disposal and interim storage would be unnecessary. Otherwise the wastes would be placed for some years in the interim repository and then be shifted from there to permanent storage underground.

It seems to me virtually certain that a suitable permanent storage site will be found. It is regrettable that ERDA is so slow making a decision and announcing it, but after the difficulties with the Kansas site it is understandable that ERDA wants to make absolutely sure the next time.

Most of the fission products have short half-lives, from a few seconds to a few years. The longest-lived of the common products are cesium 137 and strontium 90, with half-lives of about 30 years. The problem is that the wastes also contain actinides: elements heavier than uranium. In the present chemical process .5 percent of plutonium and most of the other actinides go with the wastes. Plutonium 239 has a half-life of nearly 25,000 years, and 10 half-lives are required to cut the radioactivity by a factor of 1,000. Thus the buried wastes must be kept out of the biosphere for 250,000 years.

Scientists at the Oak Ridge National Laboratory have studied the possible natural events that might disturb radioactive-waste deposits and have found none that are likely. Similarly, it is almost impossible that man-made interference, either deliberate or inadvertent, could bring any sizable amount of radioactivity back to the surface.

The remaining worry is the possibility that the wastes could diffuse back to the surface. The rate of diffusion of solids in solids is notoriously slow, and experiments at Oak Ridge have shown that the rate holds also for the diffusion of most fission products in salt. Ultimately this observation will have to be confirmed in the permanent storage site by implanting small quantities of fission products and observing their migration.

In the meantime one can draw further confidence from a beautiful "experiment" conducted by the earth itself. It has been discovered that in the part of Africa now called the Gabon Republic there existed some 1.8 billion years ago a natural nuclear reactor. A metal ore in that area is extremely rich in uranium, ranging from 10 to 60 percent. Whereas the present concentration of uranium 235 in natural uranium is .72 percent, the concentration 1.8 billion years ago was about the same as it is in present-day light-water reactors (3 percent). The ore also contained about 15 percent water. Therefore conditions were similar to those in a light-water reactor (except for the cooling mechanism). In the natural nuclear reactor plutonium 239 was formed, which subsequently decayed by emitting alpha radiation to form uranium 235. The interesting point is that the plutonium did not move as much as a millimeter during its 25,000-year lifetime. Moreover, the fission products, except the volatile ones, have stayed close to the uranium, even after nearly two billion years.

Assuming that plutonium is made in appreciable amounts, it must be kept from anyone who might put it to destructive use. Contrary to a widespread fear, however, there is little danger that plutonium could be stolen from a working nuclear reactor. The reactor fuel is extremely radioactive, and even if an unauthorized person were to succeed in unloading some fuel elements (a difficult and lengthy operation), he could not carry them away without dying in the attempt. The same is true of the used fuel cooling in storage tanks. The places

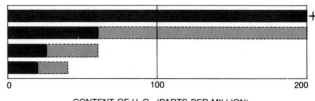

VARIOUS ORES

CHATTANOOGA SHALE

CHATTANOOGA SHALE

CONWAY GRANITE

CONTENT OF U_3O_8 (PARTS PER MILLION)

URANIUM RESOURCES OF THE U.S. are listed in this bar chart on the basis of ERDA estimates. The top row of bars refers to high-grade uranium ores. The prices throughout

from which plutonium might in principle be stolen are the chemical reprocessing plant (after the radioactive fission products have been removed), the fuel-fabrication plant or the transportation system between the plants and the reactor where the refabricated fuel elements are to be installed.

Transportation seems to be the most vulnerable link. Therefore it is probably desirable to establish the chemical plant and the fuel-fabrication plant close together, leaving only the problem of transportation from there to the reactor. Actually the problem of secure and safe transportation is essentially solved, at least in the U.S. The sophisticated safeguards now in force for nuclear weapons can be easily adapted for the transportation of nuclear materials. The protection of plants against theft is also being worked on and does not appear to present insuperable problems. For example, people leaving a plant (including employees) can be checked for possession of plutonium, even in small amounts, by means of automatic detectors, without requiring a body search. These direct measures for safeguarding plutonium are necessary and cannot be replaced by simple inventory-accounting procedures, which would be far too inaccurate. By ensuring that no plutonium (or fissionable uranium) has been diverted from U.S. plants one can be reasonably confident that no terrorists in this country can make an atomic bomb (which, by the way, is not as easy as some books and television programs have pictured it).

It has been asserted that the proposed measures for safeguarding plutonium and similar measures for protecting nuclear power plants from sabotage will interfere with everyone's civil liberties. I do not see why this should be so. The workers in the nuclear plants, the guards, the drivers of trucks transporting nuclear material and a few others will be subject to security clearance (just as people working on nuclear weapons are now). I estimate their number at less than 20,000, or less than 1 percent of our

present armed forces. The remaining 200 million Americans need suffer no abridgement of their civil liberties.

Plutonium has been called the most toxic substance known. The term toxicity can be misleading in this context, because it implies that the danger lies in some chemical action of plutonium. Experiments with animals have shown that it is the level of radioactivity of the plutonium that counts, not the quantity inhaled, as is the case with a chemical poison. Nonetheless, the radioactive hazard is indeed great once plutonium is actually absorbed in the body: .6 microgram of plutonium 239 has been established by medical authorities as the maximum permissible dose over a lifetime, and an amount approximately 500 times greater is believed to lead to lethal cancer.

Plutonium can be effectively absorbed in the body if microscopic particles of it are inhaled. About 15 percent of the particles are likely to be retained in the lung, where they may cause cancer. Fortunately there is little danger if plutonium is ingested in food or drink; in that case it passes unchanged through the digestive tract, and only about one part in 30,000 enters the bloodstream. Therefore effective plutonium poisoning of the water supply or agricultural land is virtually impossible.

Some opponents of nuclear power have maintained that because of the very low maximum permissible dose even small amounts of plutonium in the hands of terrorists could cause great damage. This point has been put in perspective by Bernard Cohen, who has investigated in theory the effect of a deliberate air dispersal of plutonium oxide over a city. He finds that on the average there would be one cancer death for every 15 grams of plutonium dispersed, because only a small fraction of the oxide would find its way into people's lungs. Other, soluble compounds of plutonium would be even less effective than an insoluble oxide. A terrorist who manages to steal six kilograms or more of plutonium could probably do more damage

by fashioning a crude bomb from it than by dispersing it in the air of a city.

Will the spread of nuclear reactors encourage the proliferation of nuclear weapons? That in my opinion is the only really serious objection to nuclear power. The availability of fissionable material is obviously a prerequisite for making nuclear weapons. Even after the material is available, however, the manufacture of a nuclear bomb is still a massive undertaking: in each of the six countries that have so far conducted nuclear explosions, thousands of scientists and technicians have worked on the development of the weapon. Nonetheless, a number of additional countries would be capable of this effort if they wanted to make it, and if they had the material.

Many countries in need of nuclear power will soon be in the market for the purchase of nuclear power plants from any country willing to sell them. Nuclear power plants sold in international trade are usually put under the inspection system of the International Atomic Energy Agency (IAEA) in order to ensure that no fissionable material is diverted for military purposes. The IAEA needs strengthening and more money for its force of inspectors. An important additional safeguard would be to prevent the proliferation of nuclear chemical-processing plants, since it is from those plants rather than from the reactors that fissionable material could be diverted. A good proposal is that the chemical processing be centralized in plants for an entire region rather than dispersed among plants for each nation. Another approach would be to have the country supplying the reactor lease the fuel to the customer country with the requirement that the used fuel be returned.

The original fuel for a light-water reactor is mostly uranium 238 enriched with about 3 percent of readily fissionable uranium 235. If an explosive were to be made from this fuel, the two isotopes would have to be separated, a procedure that requires a high level of technology. The used fuel contains in

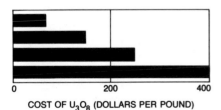

COST OF U_3O_8 (DOLLARS PER POUND)

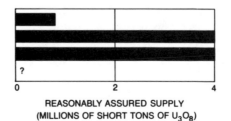

REASONABLY ASSURED SUPPLY
(MILLIONS OF SHORT TONS OF U_3O_8)

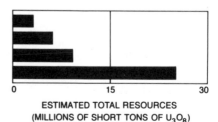

ESTIMATED TOTAL RESOURCES
(MILLIONS OF SHORT TONS OF U_3O_8)

are in 1975 dollars and include rehabilitation of the land in the case of low-grade materials (*bottom three rows of bars*). Other estimates of total resources, measured in millions of short tons of uranium oxide (U_3O_8), range up to three times those given here.

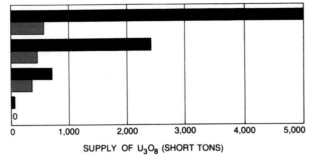

LIGHT-WATER REACTOR

HIGH-TEMPERATURE
GAS-COOLED REACTOR

CANDU REACTOR WITH
THORIUM BREEDING

BREEDER REACTOR

SUPPLY OF U₃O₈ (SHORT TONS)

URANIUM REQUIRED for the initial fueling (*colored bars*) and 40-year operation (*gray bars*) of a 1,000-megawatt nuclear power plant is indicated (again in millions of short tons of U_3O_8) for the four principal types of reactor system. For the breeder reactor the plutonium required for initial fueling is expected to be available as a by-product of previously operational light-water reactors. Candu is the Canadian name for their heavy-water natural-uranium reactor, which can be modified to convert thorium into the isotope uranium 233.

addition some plutonium, which can be separated from the uranium by chemical procedure, a less difficult task. The resulting plutonium has a high concentration of plutonium 240 (with respect to plutonium 239), which could be used to make rather crude bombs by a country just beginning in nuclear-weapons technology. Breeder reactors contain more plutonium per unit of power, with a smaller percentage of plutonium 240. I personally would therefore recommend that breeder reactors not be sold in international trade.

Proliferation would not be prevented if the U.S. were to stop building nuclear reactors for domestic use or if it were to stop selling them abroad. Western Europe and Japan not only need nuclear power even more than the U.S. does but also have the technology to acquire it. Moreover, they need foreign currency to pay for their oil imports and so they will want to sell their reactors abroad. The participation of the U.S. in the reactor trade may enable us to set standards on safeguards, such as frequent IAEA inspection, that would be more difficult if we left the trade entirely to others.

It has been alleged that nuclear power is unreliable. The best measure of reliability is the percentage of the time a plant is available for power production when the power is demanded. This "availability factor" is regularly reported for nuclear plants and runs on the average about 70 percent. There are fewer good data on the availability of large coal-fired plants, but where the numbers exist they are about the same as those for nuclear plants.

The "capacity factor" is the ratio of the amount of power actually produced to the amount that could have been produced if the plant had run constantly at full power. That percentage is usually lower than the availability factor for two

reasons: (1) some nuclear power plants are required for reasons of safety to operate below their full capacity, and (2) demand fluctuates during each 24-hour period. The second factor is mitigated by the operation of nuclear reactors as base-load plants, that is, plants that are called on to operate as much of the time as possible, because the investment cost is high and the fuel cost is low. A reasonable average capacity factor for nuclear power plants is 60 percent. One utility has estimated that at a capacity factor of 40 percent nuclear and coal-fired plants generating the same amount of electricity would cost about the same; operation at 60 percent therefore gives the nuclear plant a substantial edge.

But are not nuclear power plants expensive to build? An examination of the construction cost of such plants planned between 1967 and 1974, and expected to become operational between 1972 and 1983, shows that the cost of a 1,000-megawatt power plant of the light-water-reactor type has risen from $135 million to $730 million in this period [*see illustration on page 71*]. Closer inspection reveals, however, that a large fraction of the cost increase is due to inflation and to a rise in interest rates during construction; without those factors the 1974 cost is $385 per kilowatt of generating capacity. This figure represents a cost increase of about 300 percent, which is more than the general inflation from 1967 to 1974. The main cause must be looked for in the steep rise of certain construction costs, particularly labor costs, which rose about 15 percent per year, or 270 percent in seven years.

The cost of building coal-fired plants has risen at a comparable rate. A major factor here has been the requirement of "scrubbers" to remove most of the sulfur oxides that normally result from the burning of coal. Coal plants equipped

with scrubbers may still be about 15 percent cheaper to build than nuclear plants. Any massive increase in coal production would, however, call for substantial investment not only in the opening and equipping of new mines but also in the provision of additional railroad cars and possibly tracks, particularly in the case of Western mines. If this "hidden" investment is included, the capital cost of coal-burning power plants is not very different from that for nuclear plants. Even disregarding this factor the overall cost of generating electricity from nuclear fuel is already much less than it is for generating electricity from fossil fuel, and recent studies indicate that nuclear power will continue to be cheaper by a wide margin [*see illustration on page 73*].

There is some truth in the charge that "nuclear power does not pay its own way," since the Government has spent several billion dollars on research on nuclear power and several more billions will undoubtedly have to be spent in the future. On the other hand, the Government is also spending about $1 billion a year as compensation to coal miners who have contracted black-lung disease.

It has also been said that uranium will run out soon. It is true that the proved reserves of high-grade uranium ore are not very large, and the existing light-water reactors do require a lot of uranium. If all reactors were of this type, and if the U.S. were to set aside all the uranium needed for 40 years of reactor operation, then the total uranium-ore resources of the U.S. would only be enough to start up 600 reactors, a number that might be reached by the year 2000. Beyond that date it will be important to install reactors that consume uranium more efficiently. The most satisfactory alternative to emerge so far is the breeder reactor, which may be ready for industrial operation by 1990. The breeder in effect extracts the energy not only from the rare isotope uranium 235 but also from other isotopes of uranium, thereby increasing the supply of uranium about sixtyfold. Even more important, with the breeder the mining of low-grade uranium ore can be justified both economically and environmentally. With these added resources there is enough uranium in the U.S. to supply 1,000 reactors for 40,000 years [*see bottom illustration on preceding two pages*].

As interim alternatives two other types of reactor are attractive: the high-temperature, gas-cooled, graphite-moderated reactor and the Canadian natural-uranium reactor ("Candu"), which is moderated and cooled by heavy water.

The Candu reactor can be modified to convert thorium by neutron capture into the fissionable isotope uranium 233.

In weighing the overall health hazard presented by nuclear reactors it is appropriate to compare nuclear plants with coal-burning power plants. Recent findings indicate that even if scrubbers or some other technology could reduce the estimated health effects from coal burning by a factor of 10 (which hardly seems attainable at present), the hazard from coal would still exceed that from nuclear fuel by an order of magnitude [*see illustration at right*]. This comparison is not meant as an argument against coal. The U.S. clearly needs to burn more coal in its power plants, and even with coal the hazard is not great. The comparison does point up, however, the relative safety of nuclear reactors.

In sum, nuclear power does involve certain risks, notably the risk of a reactor accident and the risk of facilitating the proliferation of nuclear weapons. Over the latter problem the U.S. has only limited control. The remaining risks of nuclear power are statistically small compared with other risks that our society accepts. It is important not to consider nuclear power in isolation. Objections can be raised to any attainable source of power. This country needs power to keep its economy going. Too little power means unemployment and recession, if not worse.

MINING AND MILLING

TRANSPORTATION, MANUFACTURING AND OPERATION

POLLUTION

ACCIDENTS

OVERALL HEALTH HAZARDS presented by a nuclear power plant and a coal-burning power plant, both capable of generating 1,000 megawatts of electricity, are compared here in terms of the estimated number of deaths resulting from one year of operation. (Injuries and other health effects have been translated into equivalent deaths by a suitable formula.) The data were gathered primarily by C. L. Comar of the Electric Power Research Institute and L. A. Sagan of Stanford University. The coal-mining figure refers to underground mining; surface mining is much less dangerous. The figure for pollution from coal includes sulfur dioxide pollution only. As the second set of bars shows, even if sulfur dioxide "scrubbers" were to succeed in reducing the estimated hazard from coal-burning by a factor of 10, adverse health effects from coal power would still be greater than those from nuclear power.

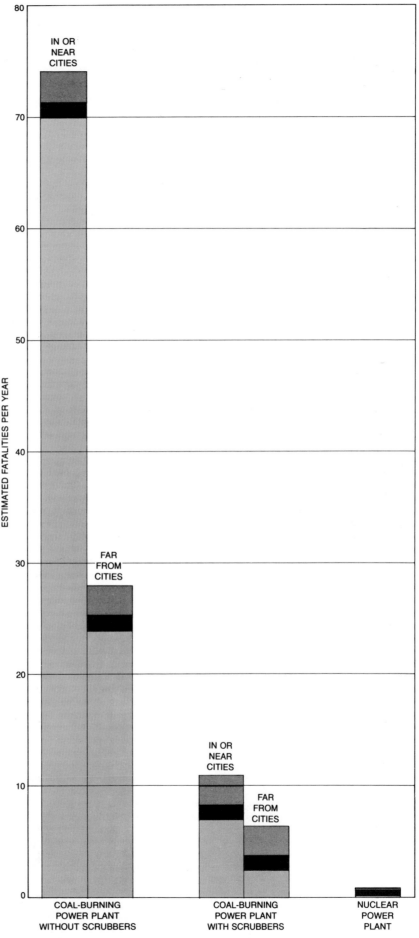

LETTERS TO THE EDITOR

April 1976

Sirs:

H. A. Bethe's article "The Necessity of Fission Power" [SCIENTIFIC AMERICAN, January] is an admirably clear and concise statement of the technical aspects of nuclear energy as it is related to the U.S. energy program. Unfortunately the article lacks perspective and may be misleading to the readers of *Scientific American.*

Perhaps the major omission in Professor Bethe's article is the failure to consider the *net* energy contribution of the proposed ambitious U.S. fission-energy program to the energy "account" of the economy. One approach to nuclear energy is to calculate the coal or oil equivalent of a pound of uranium 235. The ratio on a weight basis is more than a million to one and is the source of the popular view that fission energy is "free."

An alternative and more complete approach is to estimate the total energy consumed in the construction of the power plant and in the extraction, shipping and processing of the fuel. The net energy delivered to the society is obviously negative during the construction period and then becomes positive as the power plant functions as expected. . . .

It is particularly important to apply such an "energy analysis" to the nuclear option, since nuclear energy is so often considered free. Consider a nuclear plant with an electric-generating capacity of 1,000 megawatts. Estimates made by Peter Chapman (*New Scientist,* December 19, 1974) and W. Kenneth Davis (address given to the Atomic Industrial Forum, March 3, 1975) agree that the initial fueling requires 1×10^9 kilowatt-hours of electricity per year and that refueling requires $.4 \times 10^9$ kilowatt-hours per year. To the energy associated with plant construction Chapman assigns 1.8×10^9 kilowatt-hours of electricity and Davis assigns $.7 \times 10^9$ kilowatt-hours. The disagreement in this last item is well within the uncertainty of such figures. If we assume a capacity factor for the plant of .62 and properly incorporate the energy costs of refueling and other energy losses, we discover, utilizing Chapman's figures, that the effective output of the 1,000-megawatt plant is 523 megawatts.

Suppose we express one year's output of the plant not in electrical energy but in the thermal-energy equivalents. (Using the standard rate of 10,000 B.t.u.'s per kilowatt-hour of electricity, this leads to a multiplication of all electrical-energy units by a factor of 2.8.) This view of the energy balance examines the fossil-fuel account of the economy and the nuclear impact on it. . . .

In these terms we derive from Chapman's data that in 25 years a nuclear plant produces about 27 times as much "thermal equivalent" energy (kilowatt-hours of electricity $\times$ 2.8) as went into its construction and initial fueling. At that level the nuclear-reactor programs appear to be as advertised: a short- and medium-range solution to U.S. energy shortages and a solution to the hazardous dependence on foreign imports.

This picture breaks down drastically when one includes the 10-year delay between conception and birth for nuclear plants, and the required growth in plant starts in order to achieve the hoped-for additional capacity in the year 2000 of 430×10^3 megawatts. The present nuclear capacity is 31×10^3 megawatts, with 59×10^3 megawatts under construction. Estimates by the Energy Research and Development Administration (ERDA) and by others suggest 520×10^3 megawatts as the nuclear capacity in 2000.

A simple analysis, based on six new starts in the first year, 18 percent growth in starts up to 1990 (no starts after 1990) and the thermal-equivalent return ratio of 27, then shows that the net energy returned to the society would still be negative (-3.3×10^{15} B.t.u.'s) by 1990, and that by 1995 only 11.3×10^{15} B.t.u.'s (net thermal equivalent) would have been derived from such a massive nuclear program. Indeed, by the year 2000, 520 gigawatts of electric capacity would be at hand, but over the preceding 25 years the net thermal-equivalent benefit to the society would be only about 70×10^{15} B.t.u.'s. . . .

There are other growth scenarios. Half of the additional capacity could be started now and half 10 years from now. With this growth pattern about 10×10^{15} B.t.u.'s (thermal equivalent) would be lost to the society in the first 10 years and a net of 90×10^{15} B.t.u.'s would be gained in the next 10 years (1995). The overall result of such a program would be about 7×10^{15} B.t.u.'s per year over a 25-year period. . . .

These growth scenarios assume an increase in electrical-energy demand by about a factor of two in the year 2000, with the additional generating capacity being produced by nuclear fission. If we ignore the energy deficits and the long payback period, it is interesting to touch a little on the economics. If we ignore fission-fuel costs and assume a capital cost for nuclear plants of about $1,000 per kilowatt of electricity, we can estimate the cost of nuclear energy on the basis of the above two scenarios over the next 25 years. An addition of 430×10^3

megawatts costs about $430 billion. Under the first scenario (18 percent growth in starts, 71×10^{15} B.t.u.'s net in 2000) the energy cost is about $6 per 10^6 B.t.u.'s (thermal equivalent). Under the second scenario (half of the capacity started now, half 10 years from now and 180×10^{10} B.t.u.'s net in 2000) the cost is $2.40 per 10^6 B.t.u.'s. The comparison should be oil at about $2 per 10^6 B.t.u.'s ($12 per barrel) and coal at about $1 per 10^6 B.t.u.'s ($25 per ton).

There may be arguments for or against the nuclear option, and reasonable men may differ. However, it does seem to me that it cannot be argued that fission power can make a net major contribution to U.S. sources of energy before the turn of the century. There is only one short-term solution to the U.S. energy problem and that is conservation. Coal, oil and gas will continue to make the most positive net contributions.

Solar power, geothermal power, wind power, bioconversion power, fusion power and even fission power fit into relatively long-term solutions. For fission the technology has been largely developed by heavy Government investment. The point of this letter is to emphasize that it is not the energy panacea, and its future priorities have to be carefully evaluated. The other technologies have only recently been getting similar attention. I personally believe that these alternative techniques are not as inherently impractical as Professor Bethe suggests. . . .

GEORGE D. CODY

Princeton, N.J.

Sirs:

There may indeed be a necessity for fission power, as Professor Bethe says, but that does not mean the fission power is necessarily going to be there. In fact, there is good reason to believe that light-water-reactor technology is so inefficient and the fuel supply is so limited that any reactor completed after 1976 will not have a lifetime supply of fuel.

The most recent figures on the availability of uranium show that the reserves total approximately 600,000 tons. Probable resources (that is, undiscovered material thought by some to be available if enough exploration effort is expended) are estimated, in an arcane way, to amount to possibly another 1.2 or 1.3 million tons. A very substantial exploration effort has been under way in the U.S. for the past four years to discover this postulated material. Additions to reserves have averaged about 12,000 tons per year, whereas the fuel used has equaled, or has slightly exceeded, that rate. The record shows that it is easy to total up postulated resources based on geological analogies but that it is difficult to convert these undiscovered resources to fuel you can count on. If you are optimistic and assume that in the next 30 years we shall add an average of

20,000 tons per year to the reserves, you will then have a total of 1.2 million tons as the total stock available.

Then the question is: How much uranium do you use during the 30-year (not 40-year) lifetime of a reactor? The published estimate is approximately 230 tons of U_3O_8 (yellowcake) per year, excluding initial fueling, for a 1,000-megawatt plant with a capacity factor of 75 percent. This means that you can generate 32×10^6 kilowatt-hours per short ton of U_3O_8. This figure implies that in its 30-year lifetime a plant will use about 6,900 tons of yellowcake (without uranium or plutonium recycling) and that if you go on the basis of existing reserves, you need only build 100 plants to preempt supplies, and that if exploration is successful, fewer than 200 nuclear plants will preempt the total supply.

However, it is not that simple. If you look into the operation of existing plants, you will find very little data that show how much uranium was burned and how much power was produced. Excluding such inefficient plants as Dresden 2 and Yankee Rowe, the average power production by light-water-reactor technology is about 18×10^6 kilowatt-hours per short ton of U_3O_8. This increases the fuel requirement to about 410 tons per year and makes the needed 30-year lifetime supply almost 12,500 tons. Hence 57 1,000-megawatt plants will preempt the supply, and the planned capacity of units now in operation or under construction and due to be in service by the end of 1976 may be slightly in excess of this figure. The additional fuel needed for the initial core in each reactor has not been included and increases the total required for each unit. But that is close enough.

After this year (1976) it is doubtful that a lifetime supply will be demonstrably available for any reactor just coming on line. And a finding rate of 20,000 tons per year implies that not more than two new 1,000-megawatt plants can come on line during any 18-month period and have a lifetime supply in view—and this means about twice the success in converting resources to reserves as has been the average experience over the past four years. If you agree with Professor Bethe that a reactor lifetime of 40 years is reasonable, you can reduce the preceding figures proportionally.

As for the breeder, the first commercial breeder, under the most optimistic estimates, such as one made by John Patterson of ERDA in May, 1975, will not be on line until 1993 at the earliest. And if it has the same doubling time as the much praised French Phénix (30 to 60 years), it will have little significance in the energy picture. And Professor Bethe says that without the breeder there is no justification for the mining of low-grade ores (less than 1,000 parts per million). Thus the figure of 600,000 tons of reserves becomes very significant.

The only conclusion that can be reached, based on the available information, is that any reactor that will not be on line by the end of 1976 (1977 at the latest) should be redesigned for the use of coal. Hence the thesis of Professor Bethe's article, that there is a nuclear source of energy that could be brought in were it not for the critics (safety, economy, danger of sabotage or nuclear blackmail, etc.), is unsound because the resource base is absent....

Nuclear fission, therefore, is not a solution to the energy shortage. It is part of the problem: it diverts men, materials and money from the only long-term supply of fuel we have—coal.

RAPHAEL G. KAZMANN

Professor of Civil Engineering

JOEL SELBIN

Professor of Chemistry
Louisiana State University
Baton Rouge

Sirs:

It is surprising that so knowledgeable an authority as Professor Bethe should ignore the very significant contribution to our energy resources that could come from the direct conversion of sunlight into electricity. The feasibility and reliability of photovoltaic conversion has been conclusively established by its successful use as the power source in many space vehicles. The solar cells used for space vehicles are much too expensive for terrestrial use, but current research indicates that, with mass-production techniques being developed, the cost of manufacture will drop to the extent that this inexhaustible and pollution-free source of electrical energy will become competitive with nuclear fission.

Current estimates suggest that the cost of a photovoltaic generating plant per kilowatt peak output will be about $500. As was noted by Professor Bethe, an average output (24 hours per day, 365 days per year) of one kilowatt would require an installation providing a peak output of four kilowatts, which would cost about $2,000, if the projections mentioned above are realized.

The cost of a fission generating station per installed kilowatt-hour is quoted by Professor Bethe as being about $750, and this would seem to indicate that the photovoltaic generation of electricity will never be economically attractive. Three considerations suggest otherwise. In the first place, the cost of the installation for nuclear or fossil-fuel generating stations is only part of the cost; fuel, maintenance and obsolescence must also be paid for. A photovoltaic system uses no fuel, and the absence of moving parts minimizes maintenance costs and maximizes life expectancy. Secondly, it is misleading to compare costs on the basis of the "round the clock, round the calendar" operation of a nuclear installation. On the average, electric-generating plants in the U.S. generate only 50 percent of their rated output, averaged over the entire year.

Thirdly, it should be remembered that in those parts of the country that have the most sunshine the heaviest demand for electricity comes when the sun is shining, because the air-conditioning load is superimposed on the industrial and commercial demand. During those periods the output of a solar generating plant would be close to four kilowatts for every $2,000 worth of installation. Since that peak load must be provided by one means or another, the photovoltaic option may be the most economical. Finally, it is not unlikely that combined photovoltaic and heating units will become available for home installation. A unit giving an average output of one kilowatt of electrical energy would produce about 8,700 kilowatt-hours per year, which is about the amount consumed by the "average family." The cost of such an installation, including battery storage, would be less than $3,000. If a lifetime of 15 years is assumed, the annual cost, including interest at 8 percent and repayment, would be about $360 per year. The cost per kilowatt-hour would then be just over four cents. This calculation is for the most highly insolated part of the country (the Southwest), but even in the Northeast, where there is only six-tenths as much sunshine, the cost would be seven cents, which is comparable to the cost of the electricity delivered by the utility company. It should be added that the cost of heating would be very substantially reduced, because the solar heat would provide at least half of the energy required for space heating and hot water.

BRUCE CHALMERS

Division of Engineering
   and Applied Physics
Harvard University
Cambridge, Mass.

Sirs:

The problem of the energy needed to build a nuclear reactor has been much discussed, and I certainly agree with Dr. Cody that it is essential for a judgment of the usefulness of nuclear reactors. My appraisal of this problem differs from Dr. Cody's on three points.

1. Reactor building is a steady process. Many reactors are under construction now and will come on line in the next five or 10 years. It is therefore unrealistic to assume that construction starts now and then stops at some later time. Instead I shall assume that the number of reactors being committed, and also the number being completed, increases exponentially with time.

At present we have about 31 gigawatts of nuclear reactors. I shall take the same numbers as Dr. Cody, that is, 520×10^3 megawatts, as the capacity in 2000. This corresponds to a yearly growth rate of 11.2 percent. Alternatively I shall take President Ford's aim of 225 reactors in 1985; this corresponds to a growth rate of 19.8 percent. To bracket these two numbers I shall use alternative growth rates of 10 percent and 20 percent.

2. The energy input is not uniform during construction. The fuel needs to be provided only shortly before the reactor goes into operation, let us say one year. The first two or three years of reactor building are occupied by paperwork, obtaining a license and so forth. I shall assume that the energy needed in construction is spent in equal amounts from two to seven years before the reactor is completed.

3. Of the energy assigned by Chapman to plant construction, 1.8×10^9 kilowatt-hours of electricity, about a third is required to obtain the heavy water that British reactors use. U.S. reactors do not use heavy water, so that Chapman's figure is reduced to 1.2×10^9 kilowatt-hours. I shall take the average between this and Davis' figure, namely 1.0×10^9 kilowatt-hours of electricity.

I then calculate the ratio of the energy spent in new reactor construction and fueling to the net energy produced by the reactors in operation, making the same assumptions on capacity factor and so forth that Chapman made. The result is that the new construction absorbs 5.8 percent of the produced energy if the growth rate is 10 percent per year and 16.5 percent of the produced energy if the growth rate is 20 percent per year. Neither of these figures is alarming. In an economy with steady construction of nuclear reactors the energy consumed in construction is less than 20 percent of the energy produced by the reactors.

Concerning the letter from Professors Kazmann and Selbin, the problem of uranium availability is clearly very important. Realizing this, ERDA established a Fuel Cycle Task Force early in 1975 that made a detailed assessment of uranium-ore prospects and published it in ERDA-33.

The important point is that one should consider not only proved reserves but also total probable resources. (If we had taken only proved reserves, oil would have run out many decades ago.) The proved reserves of uranium oxide are indeed about 600,000 short tons, as Kazmann and Selbin state. However, the probable resources are 2.4 million short tons, including ores of concentration down to 200 parts per million U_3O_8 (yellowcake). (The resources include copper leach residues and phosphates.)

ERDA-33 states the following: "Based on statements from experts in the field, it seems reasonable to assume that, without going into such low-grade deposits as shales, lignites or seawater, and with reasonable assurance of a market for this product, the uranium mining industry could, with proper incentives, carry out an exploration, mining and milling program which, while taxing the money market and the ability of the equipment suppliers, could produce on the order of 2,400,000 tons of U_3O_8 equivalent." As is clear in the letter from Professors Kazmann and Selbin, the rate of exploration for uranium must increase.

Again quoting ERDA-33: "A comprehensive Government program (National Uranium Resource Evaluation—NURE) is currently under way to evaluate domestic uranium resources and to identify areas favorable for uranium exploration. A preliminary evaluation is to be completed by the end of 1976 and a complete evaluation around 1980."

At present we are unfortunately in a vicious circle: public acceptance of nuclear power is uncertain, therefore there is not enough drilling for yellowcake, hence the proved reserve of uranium is insufficient, and this again diminishes public acceptance.

Professors Kazmann and Selbin make a rather low estimate of the energy that can be extracted from uranium and accordingly state that in its 30-year lifetime a 1,000-megawatt plant will use about 6,900 tons of yellowcake. According to the AEC evidence in the report WASH-1139, you need only 5,900 tons without recycling, or 5,200 tons with plutonium recycling. The difference is small but not negligible.

Professors Kazmann and Selbin are also concerned because actual plants have extracted only about 55 percent of the available energy from the fuel. However, as ERDA-33 states, "the causes of these difficulties have been identified, and both manufacturers now maintain they are confident their fuel will meet the guarantees of about 25,000 MWd/t for Boiling Water Reactor fuel and approximately 33,000 MWd/t for Pressurized Water Reactor fuel."

Moreover, even if energy extraction is inadequate, the remainder of uranium 235 in the fuel can be recovered in chemical processing plants. Although such reprocessing is expensive, it will become economically advantageous when the price of uranium ore increases. With reprocessing we may also regain the uranium 235 that remains in fully used fuel, which will reduce the ore requirement by another 10 or 15 percent.

Using the probable resource of 2.4 million short tons, we can supply 350 reactors with their lifetime fuel if we adopt the Kazmann-Selbin figure, 405 if we adopt WASH-1139 and 460 if we also include the recycling of plutonium. This is about the number of reactors now expected to be built by 1995. Moreover, there is an independent assessment by Milton F. Searl of the Electric Power Research Institute, who estimates that the total available yellowcake in the U.S. is about 10 million tons.

Nevertheless, I wholeheartedly agree with Professors Kazmann and Selbin that a nuclear plant making a better use of the available uranium is urgently needed. In my article I mentioned the high-temperature gas-cooled reactor and a modification of the CANDU reactor using uranium 233. If this version of the CANDU reactor were adopted generally, our uranium resources could be stretched by a factor of about five.

Professors Kazmann and Selbin mention the breeder, which I also discussed. If breeders come on line in the 1990's, they will be just in time to relieve the uranium-ore situation. Accordingly no time should be lost in developing the breeder. A breeder with a doubling time of 30 to 60 years, like the French Phénix, would already be very useful, because after the first fuel investment no further fuel would be needed for the lifetime of the breeder. Indeed, such a breeder would provide for a gradual expansion of the nuclear-power establishment. On the other hand, most people in the business are confident a breeder can be built that will have a doubling time of about 20 years. This should be sufficient to allow expansion of the nuclear-power industry at a reasonable rate after the year 2000.

The letter from Professor Chalmers is good news, namely that photovoltaic generation of electricity from the sun's energy may be possible at a very reasonable price: $500 per kilowatt peak output. It is to be hoped that industrial realization of the scheme will keep the price at this low level, and that this development is receiving full support from ERDA.

Professor Chalmers is also correct in his economic analysis. Solar electricity will indeed be available just at the time of the heaviest demand for electricity and is therefore doubly welcome. However, the use of this peaking-power installation evidently presupposes that there are also generating plants for the base load of electricity. Nuclear plants are particularly suitable for this, so that a combination of nuclear and solar installations would be most advantageous economically.

H. A. BETHE

Cornell University
Ithaca, N.Y.

August 1976

Sirs:

In your April issue there is a letter from George D. Cody commenting on H. A. Bethe's article, "The Necessity of Fission Power," in the January issue that contains references to one of my papers dealing with "net energy." Cody's calculations are at variance with the facts, and although Professor Bethe has dealt with this in his reply appearing in the same issue, I should like to discuss the matter further because of the naïve and unrealistic types of calculation along the lines suggested by Cody that are appearing in various journals.

Cody's calculations seem to be based on starting to build six generating units of 1,175 megawatts (MW) each in 1976 (a total of 7,050 megawatts) and increasing the capacity 18 percent each year to 72 units of 1,175 megawatts each in 1990 (84,400 megawatts), then stopping. That would lead to an increase of 430,000 megawatts, or a total of 520,000 megawatts, of nuclear-plant generating capacity in the year 2000. This is based on a capacity of 31,000 megawatts in operation now, 59,000 megawatts under construction and a 10-year lead time between the initiation of a plant and its going into operation.

These numbers are incorrect. The mathematical approach disregards the way in which capacity would be added, and is being added, on a realistic basis. At the end of 1975 nuclear plants with a capacity of 39,000 megawatts were licensed for operation and plants with a capacity of another 71,000 megawatts had received construction permits. The time required for design, manufacturing and construction, the time during which there would be significant dollar inputs and energy, is seven years or less. Three years or more are now required for environmental and safety reports, hearings and so on. A reasonable, practical program can be outlined for achieving the much too modest goal of 520,000 megawatts at the end of the year 2000 that would lead to the following nuclear capacities and amounts of power needed for the design, manufacturing and construction of the nuclear plants.

End of year	Nuclear capacity (MW)	Annual rate of additions (MW)
1976	39,000	
1980	82,000	12,500
1982	110,000	15,200
1985	161,000	18,700
1990	266,000	23,100
1995	388,000	25,700
2000	520,000	26,600

This program would match the present plant-building capacity and predicted expansion to provide about 15,800 megawatts for design, manufacturing and construction in 1976. That figure would rise to 26,600 megawatts in 1993, the year in which actual work would start on the plants for the year 2000. These figures are average annual ones, since historically the number of plants ordered and put into production each year has varied with circumstances.

The program outlined in my table would represent an annual growth rate in "starts" of a little more than 3 percent, which is quite different from the 18 percent cited by Cody. The actual plant-

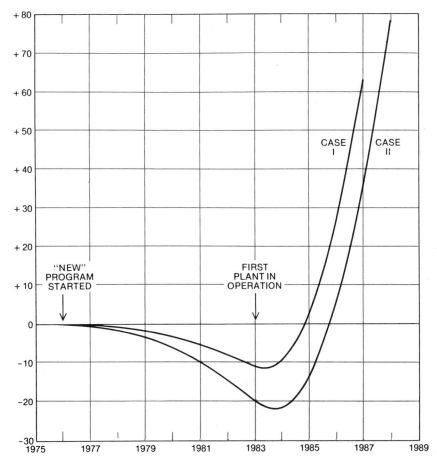

The net energy balance of a nuclear-power program

building capacity is substantially higher, and growth rates could also be higher. If the capacity were fully utilized, it would give more than the total of 520,000 megawatts assumed for the year 2000.

It is easy to demonstrate (as I did in the paper mentioned by Cody) that the current nuclear-energy output is greatly in excess of the energy required to build new nuclear plants. If, however, one asks the question in the form he did, unrealistic as it is, it can still be shown that the energy required for a "new" nuclear program is paid back rapidly once the plants start producing power. (It is difficult to repay it sooner, although that seems to be one of the principal complaints!)

Assume that the "new" program is that nuclear plants will go into the process of design, manufacturing and construction starting in 1976 (and into operation starting in 1983) on a scale that will match the capacity figures cited in my table. These are the plants on which we might still have a reasonable basis for a "go" or "no-go" decision. Assume that the plants operate at an average capacity of 70 percent and that 5 percent of their output goes to the nuclear-fuel cycle (mostly for uranium enrichment). The net production factor would then be 66.5 percent for base-load operation, which should hold until the year 2000.

My construction energy estimates, which are based on actual plant figures and energy ratios rather than on dollars (as the references cited by Cody are), lead to a requirement of 1,580 kilowatt-hours per kilowatt of capacity to build the plant and provide the first load of fuel.

How long does it take to return all the energy invested under these assumptions? From the curve for Case I in the accompanying chart it can be seen that the period is about 1.8 years from the time (the beginning of 1983) that the reactors from the "new" program begin operation. Thus by the end of 1984 there would be a net energy profit from this program.

If we were to reduce the assumed load factor to 65 percent and increase the fuel-cycle losses to 6 percent, and if we were also to assume that the Chapman-Price estimates for energy used in plant construction are correct (which I dispute), then we would get the energy balance shown by the curve for Case II in the chart. In this most unfavorable case it takes 2.7 years for the net balance to become positive, that is, it takes until the fall of 1985. In both cases the reactors scheduled for completion after 1990 are included in the construction energy requirement, which Cody does not do. These payout periods would look most inviting in any financial analysis. Indeed, they would be as good as one could reasonably expect from any program of energy production.

Cody also makes the statement that "the net thermal-equivalent benefit to the society would be only about 70×10^{15} B.t.u.'s" up to the year 2000. This insignificant amount of energy is equal to the entire amount consumed in the U.S. in one year today. Here again, however, the figure is in error. The figures in my calculations for the period from 1983 to the year 2000 are 189 quads (10^{15} B.t.u.'s) for Case I and 167 quads for Case II. Since the maximum debit shown in the chart is from 11 to 22 gigawatt-years, or from one to two quads, a net positive return of this scale by the year 2000 seems to be a good reason for pursuing nuclear power rather than one for opposing it.

Nuclear power cannot provide all the solutions to our growing energy problems. Along with coal and our best efforts at conservation, however, it will have to play a major role for the next generation or more. It is irresponsible to suggest that other energy sources will be developed and put into large-scale use before that time. It is equally irresponsible to claim too much for conservation, which so far has accomplished very little, even though a great deal more can and should be done. Professor Bethe has in my opinion done a public service with his factual and well-reasoned article in *Scientific American,* and I commend it to the attention of all thoughtful people.

W. KENNETH DAVIS
Vice-President
Bechtel Power Corporation
San Francisco

7

The Disposal
of Radioactive Wastes
from Fission Reactors

by Bernard L. Cohen
June 1977

*A substantial body of evidence indicates that the
high-level radioactive wastes generated by U.S.
nuclear power plants can be stored satisfactorily in
deep geological formations*

The task of disposing of the radioactive wastes produced by nuclear power plants is often cited as one of the principal drawbacks to the continued expansion of this country's capacity to generate electricity by means of the nuclear-fission process. Actually the task is not nearly as difficult or as uncertain as many people seem to think it is. Since 1957, when a committee of the National Academy of Sciences first proposed the burial of such wastes in deep, geologically stable rock formations, a substantial body of evidence has accumulated pointing to the technical feasibility, economic practicality and comparative safety of this approach. In recent years a number of alternative schemes—some of them involving undersea burial—have also been put forward, but deep underground burial remains the best understood and most widely favored solution to the problem of nuclear-waste disposal.

In what follows I shall describe the nature of the wastes produced by nuclear power reactors, evaluate their potential impact on public health and the environment and outline current plans to dispose of them in secure underground repositories.

What are the special characteristics of nuclear-plant wastes, and how do they differ from the wastes produced by the combustion of other fuels to generate electricity? For the sake of comparison it might be helpful to consider first the wastes resulting from the operation of a large (1,000-megawatt) coal-burning power plant. Here the principal waste is carbon dioxide, which is emitted from the plant's exhaust stacks at a rate of about 600 pounds per second. Carbon dioxide is not in itself a dangerous gas, but there is growing concern that the vast amounts of it being released into the atmosphere by the combustion of fossil fuels may have deleterious long-term effects on the world's climate. The most harmful pollutant released by a coal-burning power plant is sulfur dioxide, which is typically emitted at a rate of about 10 pounds per second. According to a recent study conducted under the auspices of the National Academy of Sciences, sulfur dioxide in the stack effluents of a single coal-fired plant causes annually about 25 fatalities, 60,000 cases of respiratory disease and $12 million in property damage. Among the other poisonous gases discharged by coal-burning power plants are nitrogen oxides, the principal pollutants in automobile exhausts (a large coal-fired plant releases as much of these as 200,000 automobiles do), and benzpyrene, the main cancer-causing agent in cigarettes. Solid wastes are also produced, partly in the form of tiny particles. In the U.S. today such "fine particulate" material is considered second in importance only to sulfur dioxide as an air-pollution hazard; approximately a sixth of all man-made fine-particulate pollution comes from coal-burning power plants. Finally there is the residue of ashes, which for a 1,000-megawatt coal-fired plant accumulate at a rate of about 30 pounds per second.

The wastes from a nuclear power plant of equivalent size differ from the by-products of coal combustion in two important ways. First, their total quantity is millions of times smaller: when the wastes are prepared for disposal, the total volume produced annually by a 1,000-megawatt nuclear reactor is about two cubic meters, an amount that would fit comfortably under a dining-room table. The comparatively small quantities of radioactive materials involved here make it practical to use highly sophisticated waste-management procedures, whose cost must be viewed in relation to the price of the electricity generated. For a 1,000-megawatt plant that price is roughly $200 million per year.

The second distinguishing characteristic of nuclear wastes is that their potential as a health hazard arises not from their chemical properties but from the radiation they emit. There appears to be a widespread misapprehension that this factor introduces a considerable degree of uncertainty into the evaluation of the potential health hazards associated with nuclear wastes, but the truth is quite the opposite. The effects of radiation on the human body are far better understood than the effects of chemicals such as air pollutants, food additives and pesticides. Radiation is easy to measure accurately with inexpensive but highly sensitive instruments; indeed, that is why radioactive isotopes are used so widely in biomedical research. Moreover, a large body of information has been compiled over the years from human exposure to intense radiation, including the atomic-bomb attacks on Japan, medical treatment with different forms of radiation and the inhalation of radon gas by miners. The available data have been analyzed intensively by national and international groups, including the National Academy of Sciences Committee on the Biological Effects of

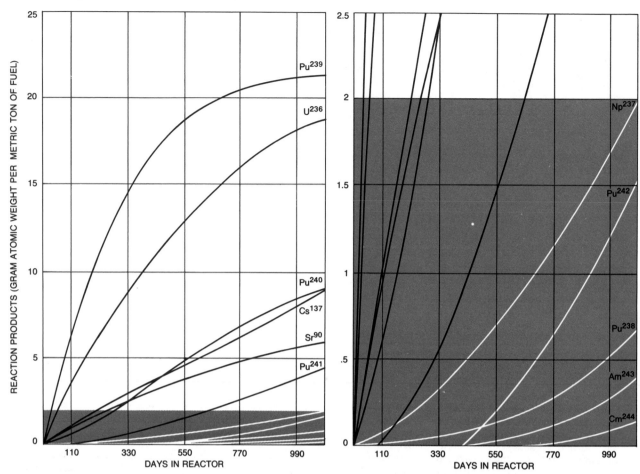

BUILDUP OF REACTION PRODUCTS per metric ton (1,000 kilograms) of uranium fuel in the active core of a typical U.S. power reactor of the light-water type is plotted here on two different vertical scales as a function of time over the three-year period the fuel customarily resides in the core. The hundreds of products resulting from the fission of uranium-235 nuclei in the fuel are represented by two characteristic fission fragments, strontium 90 and cesium 137, which together constitute about 5 percent of the total. All the other isotopes shown result from nuclear reactions in which uranium nuclei in the initial fuel are transmuted by neutron-capture reactions, followed in some cases by radioactive decay. Leveling off of the curve for fissionable plutonium 239 means that near the end of the effective life of the fuel this isotope is being consumed by fission reactions and neutron-capture reactions almost as fast as it is being created.

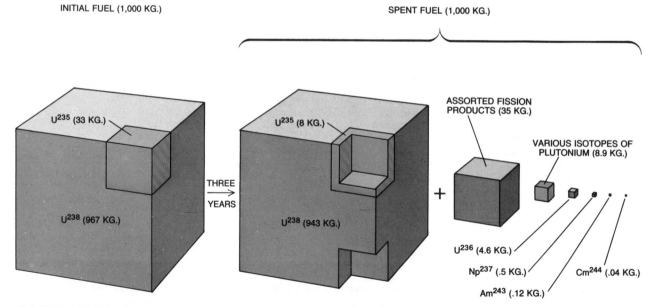

BLOCK DIAGRAM provides another graphic view of the transformation that takes place in the composition of the nuclear fuel in a light-water reactor over a three-year period. For every 1,000 kilograms of uranium in the initial fuel load (*left*) 24 kilograms of uranium 238 and 25 kilograms of uranium 235 are consumed (*center*), reducing the "enrichment" of uranium 235 from 3.3 percent to .8 percent. Uranium that is consumed is converted into 35 kilograms of assorted fission products, 8.9 kilograms of various isotopes of plutonium, 4.6 kilograms of uranium 236, .5 kilogram of neptunium 237, .12 kilogram of americium 243 and .04 kilogram of curium 244 (*right*).

Ionizing Radiation and the United Nations Scientific Committee on the Effects of Atomic Radiation. The result is a fairly reliable set of estimates of the maximum effects of various levels of radiation on the human body.

What are the radioactive substances in the waste products of a nuclear reactor, and how are they formed? In a light-water reactor (the type of nuclear plant now in general service for generating electricity in this country) the fuel consists initially of a mixture of two isotopes of uranium: the rare, readily fissionable isotope uranium 235 ("enriched" to 3.3 percent) and the abundant, ordinarily nonfissionable isotope uranium 238 (96.7 percent). The fuel mixture is fabricated in the form of ceramic pellets of uranium dioxide (UO_2), which are sealed inside tubes of stainless steel or a zirconium alloy. In the course of the reactor's operation neutrons produced initially by the fission of some of the uranium-235 nuclei strike other uranium nuclei, either splitting them in two (and thereby continuing the chain reaction) or being absorbed (and thereby increasing the atomic weight of the struck nucleus by one unit). These two types of reaction result in a variety of nuclear products, which can be plotted as a function of the time the fuel is in the reactor, usually about three years [see top illustration on opposite page].

The most important reaction in a light-water reactor is the fission of uranium 235, which creates hundreds of different products, of which strontium 90 and cesium 137, two characteristic fission fragments, constitute about 5 percent of the total. Another important reaction is the capture of neutrons by uranium-238 nuclei, which gives rise to plutonium 239. (Actually the neutron-capture reaction first yields uranium 239, which then decays radioactively in two steps to plutonium 239.) The plutonium 239 does not continue to build up linearly with time, because it may also participate in nuclear reactions. For example, a nucleus of plutonium 239 may fission when it is struck by a neutron, or it may absorb the neutron to become a nucleus of plutonium 240. The leveling off of the plutonium-239 curve means that near the end of the effective life of the fuel load this isotope is being destroyed by such processes at nearly the same rate as the rate at which it is being created.

Plutonium 240 can also capture a neutron and become plutonium 241, which can in turn either fission or capture another neutron and become plutonium 242. Plutonium 242 can be converted by the capture of still another neutron into americium 243 (after an intermediate radioactive decay from plutonium 243), and there is even an appreciable amount of curium 244 created by an additional neutron capture followed by a radioactive decay. By the same token successive neutron captures beginning with uranium 235 can respectively give rise to uranium 236, neptunium 237 and plutonium 238.

For every metric ton (1,000 kilograms) of uranium in the initial fuel load 24 kilograms of uranium 238 and 25 kilograms of uranium 235 are consumed in the three-year period, reducing the enrichment of the uranium 235 from 3.3 percent to .8 percent. In the process 800 million kilowatt-hours of electrical energy can be generated, and the uranium that is consumed is converted into 35 kilograms of assorted fission products, 8.9 kilograms of various isotopes of plutonium, 4.6 kilograms of uranium 236, .5 kilogram of neptunium 237, .12 kilogram of americium 243 and .04 kilogram of curium 244. Since only 25 kilograms of uranium 235 are consumed and a fifth of that amount is converted into uranium 236 and neptunium 237, one can easily calculate that only 60 percent of the energy-releasing fission reactions actually take place in uranium 235. Thirty-one percent occur in plutonium 239, 4 percent occur in plutonium 241 and 5 percent are induced by high-energy neutrons in uranium 238. (These figures are averages over the three years the fuel customarily is in the reactor. Near the end of that period only 30 percent of the fission reactions take place in uranium 235, with 54 percent occurring in plutonium 239, 10 percent in plutonium 241 and 5 percent in uranium 238. In view of the current public controversy over the projected future recycling of plutonium in nuclear reactors, it is inter-

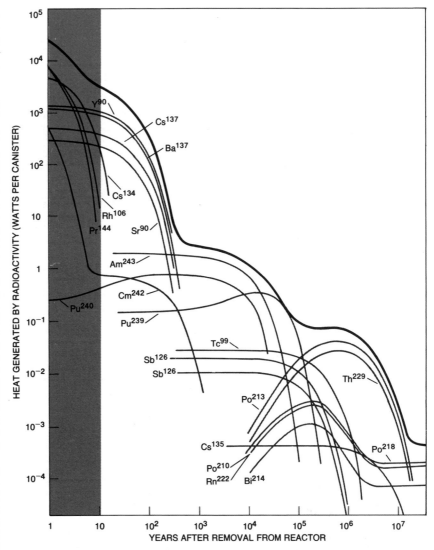

HEAT GENERATED by the various radioactive isotopes in the spent fuel from a nuclear power plant must be allowed to dissipate safely, which means that in any long-term storage plan the canisters containing the high-level wastes must be spread out over a fairly large area. The problem can be substantially alleviated by resorting to an interim-storage period of about 10 years (colored panel at left), after which the heat generated by each canister will have fallen off to about 3.4 kilowatts. The gray curves trace the contributions of the more important radioactive isotopes to the overall heating effect, which in turn is indicated by the black curve.

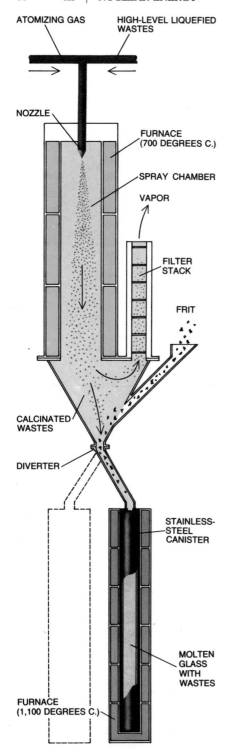

ATOMIZING GAS

HIGH-LEVEL LIQUEFIED WASTES

NOZZLE

FURNACE (700 DEGREES C.)

SPRAY CHAMBER

VAPOR

FILTER STACK

FRIT

CALCINATED WASTES

DIVERTER

STAINLESS-STEEL CANISTER

MOLTEN GLASS WITH WASTES

FURNACE (1,100 DEGREES C.)

CURRENT PLAN for handling high-level radioactive wastes calls for their incorporation into glass cylinders about 300 centimeters long and 30 centimeters in diameter. In the single-step solidification process depicted here the liquid high-level waste is first converted into a fine powder inside a calcining chamber (*top*), then mixed with glassmaking frit (*middle*) and finally melted into a block of glass within the thick stainless-steel canister in which it will eventually be stored (*bottom*). When canister is full, flow is switched by a diverter valve into a new canister (*broken outline*); hence the process is continuous.

esting to note that plutonium is already in intensive use as a nuclear fuel.)

After the spent fuel is removed from the reactor it is stored for several months in order to allow the isotopes with a short radioactive half-life to decay. (This temporary storage is particularly important with respect to an isotope such as iodine 131, one of the most dangerous fission products, which has a half-life of only eight days.) Thereafter one of the options would be to send the spent fuel to a chemical-reprocessing plant, where the fuel pins would be cut into short lengths, dissolved in acid and put through a series of chemical-separation processes to remove the uranium and plutonium, which would then be available to make new fuel. Everything else (except for certain gases, which would be discharged separately, and the pieces of the metal fuel pins that do not dissolve in the acid) is referred to as "high level" waste. In addition to all the fission products, which are responsible for the bulk of the radioactivity, the high-level wastes would in this case include the isotopes of neptunium, americium and curium, along with the small amounts of uranium and plutonium that would not be removed in reprocessing, owing to inefficiencies in the chemical separations.

The simplest and most obvious way to dispose of the remaining high-level wastes (once an economically sufficient quantity of them began to accumulate) would be to bury them permanently deep underground. On the face of it such an approach appears to be reasonably safe, since all rocks contain traces of naturally radioactive substances such as uranium, thorium, potassium and rubidium, and the total amount of this natural radioactivity in the ground under the U.S. down to the proposed nuclear-waste burial depth of 600 meters is enormously greater than the radioactivity in the wastes that would be produced if the country were to generate all its electric power by means of nuclear fission. Of course, the radioactivity of the nuclear wastes is more concentrated, but in principle that does not make any difference; the biological effects of radiation are generally assumed to have a linear relation to dosage, so that distributing a given total dosage among more people would not change the number of adverse health effects. (If this "linearity hypothesis" were to be abandoned, current estimates of the potential health hazards from nuclear wastes and all other aspects of the nuclear power industry would have to be drastically reduced.)

The detailed procedures for handling the high-level wastes are not yet definite, but present indications are that the wastes will be incorporated into a borosilicate glass (similar to Pyrex), which will be fabricated in the form of cylinders about 300 centimeters long and 30

centimeters in diameter. Each glass cylinder will in turn be sealed inside a thick stainless-steel casing. These waste canisters will then be shipped to a Federally operated repository for burial. One year's wastes from a single 1,000-megawatt nuclear power plant will go into 10 such canisters, and the canisters will be buried about 10 meters apart; hence each canister will occupy an area of 100 square meters, and all 10 canisters will take up 1,000 square meters. It has been estimated that an all-nuclear U.S. electric-power system would require roughly 400 1,000-megawatt plants, capable of generating 400,000 megawatts at full capacity, compared with the present average electric-power usage of about 230,000 megawatts. Accordingly the total high-level wastes generated annually by an all-nuclear U.S. electric-power system would occupy an area of less than half a square kilometer.

The main reason for spreading the canisters over such a large area is to dissipate the heat generated by their radioactivity. The problem of dealing with this heat can be substantially alleviated by waiting for 10 years after the reprocessing operation, at which time the heat generated by each canister will have fallen off to about 3.4 kilowatts. The advantage of delayed burial is seen more clearly when the heating effect is translated into the estimated rise in temperature that would result at the surface of a canister buried alone in rock of average thermal conductivity [*see top illustration on page 90*]. It is evident that burial after a wait of a year would lead to a temperature rise of 1,900 degrees Celsius, whereas waiting for 10 years would reduce the rise to 250 degrees C. The difference is critical, since glass devitrifies (crystallizes and becomes brittle) at temperatures higher than 700 degrees. In rock of average thermal conductivity the maximum average temperature of the rock just above and below the burial depth would be reached 40 years after burial, when the average temperature at the burial depth would be increased by 140 degrees [*see bottom illustration on page 90*]. If the canister were to be buried in salt, which has a much greater thermal conductivity, the rise in temperature at the burial depth after 40 years would be less: 85 degrees.

In salt an additional effect must be taken into account, since the heat will cause the migration of water toward the waste canister. Typical salt formations contain about .5 percent water trapped in tiny pockets. The solubility of salt in water increases with temperature, so that if the temperature on one side of the pocket is raised, more salt will go into solution on that side. This raises the salt content of the water above the saturation point for the temperature on the opposite side of the pocket, however, causing the salt to precipitate out of solution on that side. The net effect is a migration of the water pocket in the direction of the higher temperature, which is of course the direction of the buried waste canister. The rate of the migration depends on how rapidly the temperature increases with distance, and on how rapidly the temperature gradient, as I have explained, falls off with time.

This process is expected to lead to the collection of water around each canister at an initial rate of two or three liters per year; within 25 years a total of 25 liters will have collected, with very little further collection expected thereafter. Since the temperature at the surface of the canister would be higher than the boiling point of water, the water arriving at the canister would be converted into steam and would be drawn off by the ventilation system (assuming that the repository is not sealed). Small amounts of water would continue to migrate toward the canisters after 25 years, carrying corrosive substances such as hydrochloric acid arising from chemical reactions induced in the salt by the radiation from the canister. It is therefore usually assumed that the stainless-steel casings will corrode away, leaving the waste-containing glass cylinders in contact with the salt.

How can one evaluate the health hazards presented by such radioactive waste materials? The most direct hazard is from the gamma radiation emitted by the decaying nuclei. Gamma rays behave much like X rays except that they are even more penetrating. The effect of gamma rays (or any other form of ionizing radiation) on the human body is measured in the units called rem, each of which is equal to the amount of radiation that is required to produce the same biological effect as one roentgen of X radiation. ("Rem" stands for "roentgen equivalent man.") In analyzing the impact of radioactive wastes on public health the only significant radiation effects that need to be considered are those that cause cancer and those that induce genetic defects in progeny. According to the best available estimates, for whole-body radiation such as would be delivered by a source of gamma rays outside the body the risk of incurring a radiation-induced fatal cancer is approximately 1.8 chances in 10,000 per rem of radiation exposure. The estimated risk for total eventual genetic defects in progeny is about 1.5 chances in 10,000 per rem of radiation delivered to the gonads (with the effects spread out over about five generations). In the discussion that follows I shall be referring only to cancers, but it should be kept in mind that there are in addition a comparable (but generally smaller) number of genetic defects caused by exposure to gamma radiation.

The biological damage done by a gamma ray is in most situations roughly proportional to the ray's energy, so that

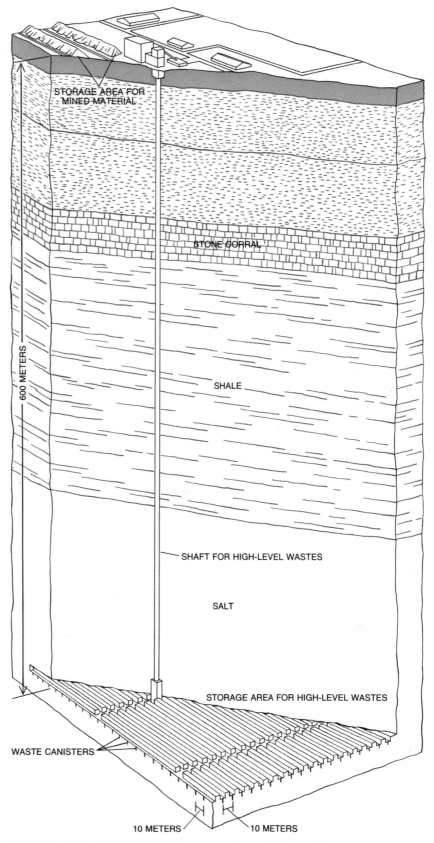

DEEP UNDERGROUND BURIAL is at present the method favored by most nuclear power experts in the U.S. for the long-term storage of high-level radioactive wastes. In this idealized diagram of a proposed Federally operated repository in southeastern New Mexico the waste canisters are shown emplaced at a depth of 600 meters in a geologically stable salt formation. In order to dissipate the heat from the canisters they would be buried about 10 meters apart; thus each canister would occupy an area of about 100 square meters. On this basis the total high-level wastes generated annually by an all-nuclear U.S. electric-power system (assuming roughly 400 1,000-megawatt plants) would occupy an area of less than half a square kilometer.

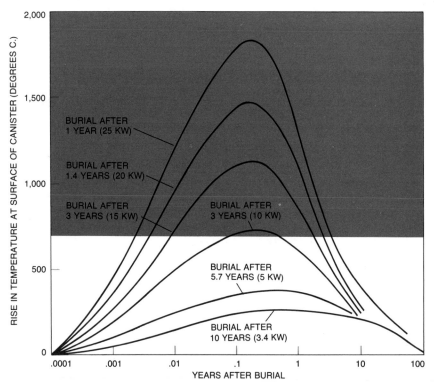

ADVANTAGE OF DELAYED BURIAL is evident in this graph, in which the heating effect of a single waste canister is translated into the estimated rise in temperature that would result at the surface of the canister if it were buried alone in rock of average thermal conductivity. The numbers labeling each curve indicate the heat generated by the canister (in kilowatts) after a given interim-storage period (in years). Thus burial after one year (*top curve*) would cause a temperature rise of 1,900 degrees Celsius, whereas waiting for 10 years (*bottom curve*) would reduce the increment to 250 degrees C. Colored area at top symbolizes critical fact that glass devitrifies (crystallizes and becomes brittle) at temperatures higher than 700 degrees C.

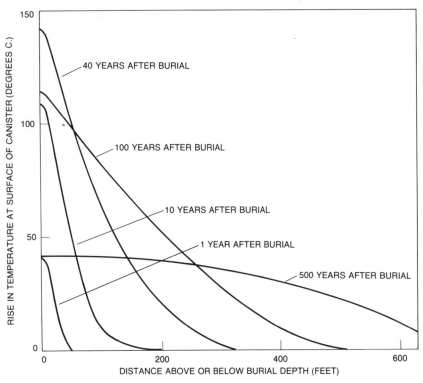

MAXIMUM AVERAGE TEMPERATURE of the rock just above and below the burial depth of the waste canister would be reached 40 years after burial, when the average temperature at the burial depth would be increased by about 140 degrees C. If the waste canister were to be buried in salt, the corresponding temperature increments would be considerably reduced.

one first plots the gamma-ray energy emitted per second (in watts) by the wastes resulting from one full year of a U.S. energy budget based on all-nuclear generation of electric power [*see bottom illustration on opposite page*]. From such a graph one can see that for the period between eight and 400 years after reprocessing the dominant contribution to the total gamma-ray emission is made by cesium 137 and its immediate decay product barium 137. During this four-century period the total gamma-ray hazard falls by more than four orders of magnitude.

One way to grasp the potential hazard presented by this amount of gamma radiation is to consider what would happen if the source of radiation were to be distributed over the entire land surface of the U.S. The number of fatal cancers per year induced in that case could be as high as many millions. Clearly the material that gives rise to the radiation must be confined and handled with great care. On the other hand, gamma rays are attenuated by about a factor of 10 per foot in passing through rock or soil, so that there would be no danger of this type from wastes that remain buried deep underground.

A measure of the care that must be taken in handling the waste canisters is indicated by the fact that a dose of 500 rem (which has a 50 percent chance of being fatal) would be received in 10 minutes by a human being standing 10 meters away from an unshielded new waste canister. There is no great technical difficulty, however, in providing shielding adequate for safe and effective remote handling of the waste canisters.

If any of the radioactive wastes were to enter the human body, their biological effects would be enhanced, since the radiation they would emit would strike human tissue in all directions and since the exposure would continue for some time. Accordingly one must consider the two major possible entry routes: ingestion and inhalation. The ingestion hazard can be evaluated in terms of the number of cancer-causing doses in the wastes produced by one year of all-nuclear electric power in the U.S. [*see illustration on page 92*]. In this graph the value of 10^6 at 10^4 years, for example, means that if all the wastes, after aging for 10,000 years, were to be converted into digestible form and fed to people, one could expect a million fatal cancers to ensue. This "worst case" scenario assumes, of course, that many millions of people are involved, but in view of the linear relation between dose and effect generally assumed for calculating such radiation risks it does not matter how many millions there are. The derivation of such a graph is rather complex, involving for each radioactive species the probability of transfer across the intestinal wall into the bloodstream; the probability of transfer from the blood into each body organ; the time the radioactive substance spends in each organ; the energy of the radiation emitted by the substance and the fraction of the energy absorbed by the organ; the mass of the organ; the relative biological effects of the different kinds of radiation emitted, and finally the cancer risk per unit of radiation absorbed (in rem).

Feeding all this radioactive material to people is hardly a realistic scenario, however, so that one might consider instead the consequences if the wastes were to be dumped in soluble form at random into rivers throughout the U.S. For this scenario, which comes close to assuming the most careless credible handling of the disposal problem, the graph shows that a million fatalities could result. It is unlikely anyone would suggest such dumping, but in any event it is clearly not an acceptable method of disposal.

In evaluating the inhalation hazard by far the most important effect that must be taken into account is the induction of lung cancers [see illustration on page 93]. Here again the graph shows the consequences of a situation approximating the most careless credible handling of the wastes: spreading them as a fine powder randomly over the ground throughout the U.S. and allowing them to be blown about by the wind.

Much attention is given in public statements to the potential hazards represented by the scales in such graphs that show the number of cancers expected if all the radioactive materials involved were to be ingested or inhaled by people. One often hears, for example, that there is enough radioactivity in nuclear wastes to kill billions of people. To put such statements in perspective it is helpful to compare the known hazards of nuclear wastes with those of other poisonous substances used in large quantities in the U.S. [see illustration on page 94]. Such a comparison shows that there is nothing uniquely dangerous about nuclear wastes. Nevertheless, it is often emphasized that radioactive wastes remain hazardous for a long time. Nonradioactive barium and arsenic, on the other hand, remain poisonous forever. It might also be argued that the other hazardous substances are already in existence, whereas nuclear wastes are a newly created hazard. Roughly half of the U.S. supply of barium and arsenic, however, is currently imported, and hence these hazards are also being introduced "artificially" into our national environment. One other important difference often goes unnoted, and that is that the chemical poisons are not carefully buried deep underground as is the plan for the nuclear wastes; indeed, much of the arsenic is used as a herbicide and hence is routinely scattered around on the ground in regions where food is grown.

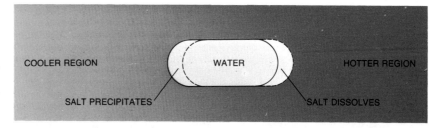

IN SALT the heat from the waste canister would cause the migration of tiny pockets of water in the direction of the higher temperature, since the salt would tend to go into solution on the hotter side of the pocket (right) and to precipitate out of solution on the cooler side (left).

Actually such quantitative representations of potential hazards are virtually meaningless unless one also takes into account the possible pathways the hazardous agents can take to reach man. Therefore I shall now turn to that subject. It is generally agreed the most important health hazard presented by nuclear wastes arises from the possibility that ground water will come in contact with the buried wastes, leach them into solution, carry them through the overlying rock and soil and ultimately into food and water supplies. Human exposure would then be through ingestion. From the analysis of the ingestion route outlined above one can deduce that the hazard from ingested radioactive mate-

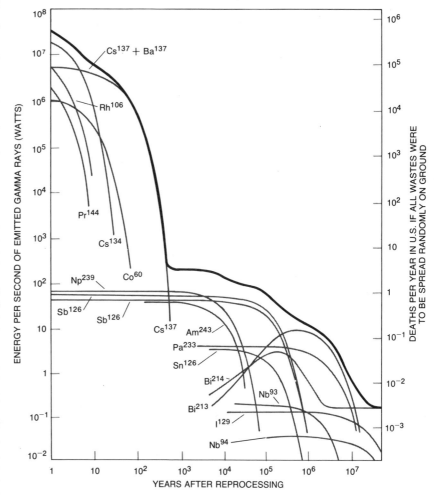

MOST DIRECT HEALTH HAZARD presented by radioactive wastes arises from the gamma radiation emitted by the decaying nuclei. The biological damage done by a gamma ray is in most situations roughly proportional to its energy; hence in this graph the gamma-ray energies emitted per second by various radioactive isotopes in the wastes resulting from one full year of an all-nuclear U.S. electric-power system (again assuming 400 1,000-megawatt plants) are plotted according to the scale at left. The black curve shows that between eight and 400 years after reprocessing the total gamma-ray hazard falls by more than four orders of magnitude. Scale at right indicates the total number of fatal cancers expected per year if the source of this amount of gamma radiation were to be spread at random over entire land surface of the U.S.

rial is high at first but much less after a few hundred years. In fact, one can calculate that after 600 years a person would have to ingest approximately half a pound of the buried waste to incur a 50 percent chance of suffering a lethal cancer. It is reasonable to conclude that it is very important the wastes be isolated from human contact for the initial few hundred years. I shall first take up that problem but shall return to the longer-term one.

When people first learn that nuclear wastes must be isolated for hundreds of years, their immediate response is often to say this is virtually impossible: man's social institutions and political systems and the structures he builds rarely last that long. This response, however, is based on experience in the environment encountered on the surface of the earth. What one is actually dealing with are rock formations 600 meters below the surface. In this quite different environment the characteristic time intervals re-

quired for any substantial change are on the order of millions of years.

In addition to the general security of the deep underground environment a great deal of extra protection is provided for the critical first few hundred years by the various time delays intrinsic to any conceivable release process. The most important of these additional safeguards has to do with the selection of a storage site, which is determined by geological study to be not only free of circulating ground water now but also likely to remain free of it for a very long time to come. In geological terms a few hundred years is a short time, so that predictions of this kind can be highly reliable. Since the patterns in which groundwater flows can be changed by earthquakes, only tectonically stable areas would be chosen. Salt formations offer additional security in this regard, because when salt is subjected to pressure, it flows plastically. Thus it is capable of sealing cracks that develop from tecton-

ic activity. This property of salt also removes the scars of the burial operations, leaving the canisters sealed deep inside a gigantic crystalline mass.

Suppose, however, water does somehow manage to get into cracks in the rock formation in which the waste is buried. What happens then? The rock would of course be chosen to be impervious to water, so that there would be a second delay while the rock was being leached away before the waste glass was exposed to water. It would seem that there would not be much delay in salt because it is so soluble in water, but in fact the quantities of water deep underground are not large and the mass of salt is huge. For example, if all the ground water now flowing in the region of the proposed Federal waste-repository site in New Mexico were somehow diverted to flow through the salt, it would take 50,000 years for the salt enclosing one year's deposit of nuclear wastes to be dissolved away.

A third delay arises from the time it would take to leach away the waste glass itself. There is some uncertainty on this point, and the matter is complicated by the fact that leaching rates increase rapidly with temperature, but it seems fairly certain that the low rate at which the glass can be leached away will offer considerable protection for at least a few hundred years. If new leaching-rate studies indicate otherwise, it would not be too difficult or expensive to switch to ceramics or other more resistant materials for incorporating the wastes.

A fourth delay arises from the length of time it ordinarily takes water to reach the surface. Typical flow rates are less than 30 centimeters per day, and typical distances that must be covered are tens or hundreds of kilometers. For anything to travel 100 kilometers at 30 centimeters per day takes about 1,000 years.

The radioactive wastes would not, however, move with the velocity of the ground water even if they went into solution. They would tend to be filtered out by ion-exchange processes. For example, an ion of radioactive strontium in the wastes would often exchange with an ion of calcium in the rock, with the result that the strontium ion would remain fixed while the calcium ion would move on with the water. The strontium ion would eventually get back into solution, but because of continual holdups of this type the radioactive strontium would move 100 times slower than the water, thus taking perhaps 100,000 years to reach the surface. For the other important waste components the holdup is even longer.

IF ALL WASTES WERE TO BE INGESTED, the biological effects on the human population of the U.S. would be considerable. As this graph shows, the number of cancer-causing doses in the wastes produced by one year of all-nuclear electric power in the U.S. is such that if all the wastes, after aging for 10,000 years, were to be converted into digestible form and fed to people, one would expect a million fatal cancers to ensue (*scale at left*). If instead the wastes were to be converted into soluble form and immediately after reprocessing dumped at random into rivers throughout the U.S., the result could again be a million fatalities (*scale at right*).

As a result of all these delays there is an extremely high assurance that very little of the wastes will escape through the ground-water route during the first few hundred years when they are most dangerous. Indeed, the time delays offer substantial protection for hundreds of thousands of years. I shall give no credit for this factor, however, in the following discussion of the potential longer-term hazard.

As we have seen, the "50 percent lethal" dose of nuclear wastes ingested after 600 years would be half a pound. This is hardly a potent poison, and its dangers seem particularly remote when one considers that the material is carefully buried in low-leachability form isolated from ground water a third of a mile below the earth's surface. Many more potent poisons are routinely kept in the home. It is true, however, that nuclear wastes remain poisonous for a very long time, so that they could conceivably present a hazard.

To evaluate this long-term risk one must develop an estimate of the probability that the wastes will escape into the environment. How can this be done? One way is to make a comparison between an atom of nuclear waste buried at a depth of 600 meters and a typical atom of radium somewhere in the rock or soil above the waste canister, assuming that the waste atom is no more likely than the radium atom to escape and find its way into a human being. This would seem to be a conservative assumption, since "the rock or soil above the waste canister" includes the material near the surface, where the erosive forces of wind, surface runoff, freeze-thaw cycles, vegetation and so on are active.

It is difficult to calculate the escape probability for an atom of radium in a particular area, but the average escape probability over the entire continental U.S. can be estimated. To make such a comparison meaningful one can assume that the wastes are buried in a uniform distribution over the entire country, but for calculating averages it is equivalent to assume that they are buried at random locations across the country and always at the same depth. When the assumption is stated this way, it is clearly conservative; one would think that by making use of all the information available from geology, hydrology and lithology one could choose a burial site that would be much securer than a randomly chosen one.

Having made these two basic assumptions—random burial and an equal escape probability for atoms of waste and radium—one need only estimate the average probability that an atom of radium in the top 600 meters of the U.S. will escape. One approach has two steps: calculating the probability that a radium atom will escape from the soil into rivers and multiplying this number by the probability that a given sample of water will be ingested by a human being. The average concentration of radium in rivers (two grams per 10 trillion liters) and the total annual water flow in U.S. rivers (1.5 quadrillion liters) are known quantities; the annual transfer of radium

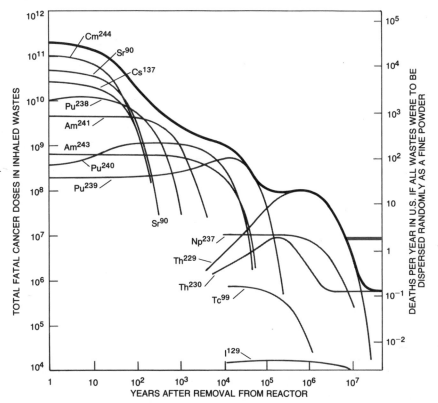

IF ALL WASTES WERE TO BE INHALED, the most important health hazard would be the induction of lung cancers. In this graph again the scale at left shows the total number of cancer-causing doses in the wastes produced by one year of all-nuclear electric power in the U.S. The scale at right shows the number of deaths expected by the inhalation route if all these wastes were to be spread as a fine powder randomly over the ground throughout the U.S. In both this graph and the one on the opposite page the short colored line at the lower right indicates the corresponding long-term health hazard represented by the natural radioactivity in the uranium ore that would be consumed by such an all-nuclear electric-power system in the U.S.

from the soil into rivers is the product of these two numbers, or 300 grams. Since radium is a product of the radioactive decay of uranium, from the average concentration of uranium in rock (2.7 parts per million) one can readily estimate the amount of radium in the top 600 meters of the U.S. as being 12 billion grams. The annual transfer probability is the ratio of the annual transfer to the total quantity, or .000000025 per year. The inverse of this number, 40 million years, is then the average lifetime of rock in the top 600 meters of the U.S. Therefore the assumption is that each atom of buried nuclear waste has less than one chance in 40 million of escaping each year. About one part in 10,000 of river flow in the U.S. is ingested by human beings, but owing to various purification processes the fraction of the radium in river flow that is ingested is closer to 1.5 part in 100,000. Multiplying this number by the annual probability for escape into rivers (.000000025), one finally obtains the total annual transfer probability of a radium atom from the rock into a human being. It is roughly four chances in 10 trillion.

There are at least two flaws in this calculation. It ignores transfer through food, a factor that reduces the transfer

probability, and it assumes that all the radium ingested is taken up by the body, a factor that increases the transfer probability. These problems can be avoided and the calculation can be simplified by estimating the number of human cancers induced annually by ingested radium (12) and dividing that number by the number of cancer-causing doses of radium in the top 600 meters of the U.S. (30 trillion). The first quantity is obtained from actual measurements of the amount of radium in cadavers combined with generally accepted estimates of the risk of a person's getting cancer from the radium. The result for the annual transfer probability obtained by this method is in close agreement with the figure derived by the preceding method. It therefore is reasonable to multiply the dosage scale in the ingestion graph on the opposite page by .0000000000004 (four chances in 10 trillion) to obtain the number of fatalities expected annually from the nuclear wastes produced annually by an all-nuclear U.S. electric-power system.

What all of this means is that after the first few hundred years of storage, during which we would be protected by the time delays discussed above, one could expect about .000001 fatality per year or less attributable to the buried waste. When this toll is added up, it comes to .4 fatality for the first million years plus an additional four fatalities over the next 100 million years.

If one is to consider the public-health effects of radioactivity over such long periods, one should also take into account the fact that nuclear power burns up uranium, the principal source of radiation exposure for human beings today. For example, the uranium in the ground under the U.S. is the source of the radium that causes 12 fatal cancers in the U.S. per year. If it is assumed that the original uranium was buried as securely as the waste would presumably be, its eventual health effects would be greater than those of the buried wastes. In other words, after a million years or so more lives would be saved by uranium consumption per year than would be lost to radioactive waste per year.

The fact is, however, that the uranium now being mined comes not from an average depth of 600 meters but from quite near the surface. There it is a source of radon, a highly radioactive gaseous product of the decay of radium that can escape into the atmosphere. Radon gas is the most serious source of radiation in the environment, claiming thousands of lives in the U.S. per year according to the methods of calculation used here. When this additional factor is taken into account, burning up uranium in reactors turns out to save about 50 lives per million years for each year of all-nuclear electric power in the U.S., more than 100 times more than the .4

life that might be lost to buried radioactive wastes.

Thus on any long time scale nuclear power must be viewed as a means of cleansing the earth of radioactivity. This fact becomes intuitively clear when one considers that every atom of uranium is destined eventually to decay with the emission of eight alpha particles (helium nuclei), four of them rapidly following the formation of radon gas. Through the breathing process nature has provided an easy pathway for radon to gain entry into the human body. In nuclear reactors the uranium atom is converted into two fission-product atoms, which decay only by the emission of a beta ray (an electron) and in some cases a gamma ray. Roughly 87 percent of these emission processes take place before the material even leaves the reactor; moreover, beta rays and gamma rays are typically 100 times less damaging than alpha-particle emissions, because their energies are lower (typically by a factor of 10) and they deposit their energy in tissue in less concentrated form, making their biological effectiveness 10 times lower. The long-term effect of burning uranium in reactors is hence a reduction in the health hazards attributable to radioactivity.

In this connection it is interesting to note that coal contains an average of about 1.5 parts per million of uranium, which is released into the environment when the coal is burned. The radon gas from the uranium released by one year of an all-coal-powered U.S. electric-generating system would cause about 1,000 fatalities per million years, a rate

three orders of magnitude greater than the result obtained above for the wastes from an all-nuclear-powered system.

If the risk of ingesting radioactive waste materials with food or water is so low, what about the risk of inhaling them as airborne particulate matter? The potential hazards from inhaling such materials are much greater and longer-lasting than the hazards from ingesting them. It is difficult, however, to imagine how buried nuclear wastes could be released as airborne particulates. The largest nuclear bombs yet considered would not disturb material at a depth of 600 meters. Meteorites of sufficient size to do so are extremely rare, so that their average expected effect would be a million times lower than that from ingestion. Volcanic eruptions in tectonically quiet regions are also extremely rare; moreover, they disturb comparatively small areas, so that their effects would be still smaller.

Release through ground water could lead to a small fraction of the radioactivity being dispersed at the surface in suspendable form, but calculation indicates that for this pathway to be as hazardous as ingestion all the wastes would have to be dispersed through it. Wastes dispersed at the surface would also constitute an external-radiation hazard through their emission of gamma rays, but another calculation demonstrates that this hazard too is less than that of ingestion.

None of the estimates I have given so far takes into account the possible release of nuclear wastes through human intrusion. Let us therefore consider that

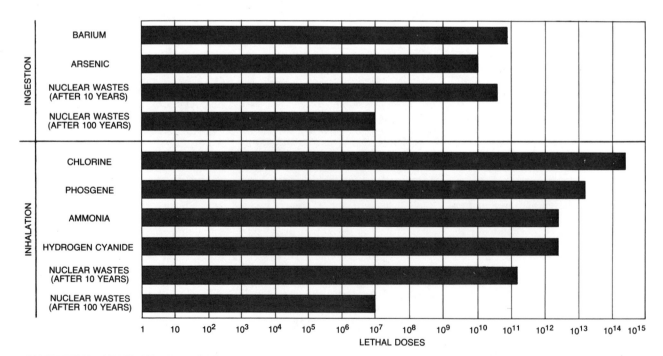

COMPARISON OF HEALTH HAZARDS presented by high-level radioactive wastes from nuclear reactors with those of other poisonous substances routinely used in large quantities in the U.S. demonstrates that there is nothing uniquely dangerous about the nuclear wastes. Moreover, the author notes, "chemical poisons are not carefully buried deep underground as is the plan for the nuclear wastes; indeed, much of the arsenic is used as a herbicide and hence is routinely scattered around on the ground in regions where food is grown."

possibility. Buried waste would not be an attractive target for saboteurs because of the great amount of time, effort, equipment and personal danger that would be needed to remove it. Only release through inadvertent human intrusion, such as drilling or mining, needs to be considered. The current plan is to retain Government ownership of repository sites and to maintain surveillance and long-lasting warning signs, so that this problem would exist only if there were a total collapse of civilization. One of the criteria for the choice of a repository site is that there be a lack of valuable minerals and the prospect of discovering them. (Indeed, the principal factor delaying the development of the proposed New Mexico site is the possibility that it may hold potash deposits.) Nevertheless, if there were random exploratory drilling in the area at the rate of the current average "wildcat" drilling for oil in the U.S., the effects would still be much less than those of release in ground water. If there were mining in the area (presumably for minerals not now regarded as valuable), the operations would have to be on a scale approaching that of the entire current U.S. coal-mining enterprise before their effects would equal those of ground-water release.

Wastes buried in salt might seem to be a poor risk against the possibility of intrusion by mining, since salt is widely mined. The quantity of salt underground, however, is so huge that on a random basis any given area would not be mined for tens of millions of years. Again the probability of release through this pathway is comparable to that through ground water, except that here the wastes are in insoluble form and, if ingested, much less likely to be taken up by the body. A pathway would seem to exist through the use of salt in food, but only 1 percent of the salt mined in the U.S. is so used, and it is purified by allowing insoluble components to settle out. Thus exposure through this pathway would be reduced roughly to that through the use of salt in industrial processes. All in all, then, the probability of the release of stored nuclear wastes through human intrusion is less than that of their release through ground water.

It is often said that by producing radioactive wastes our generation places an unjustifiable burden on future generations in requiring them to guard against their release. Here it should first be recognized that the estimate of the health effects of nuclear wastes I have given—an eventual .4 fatality for each year of all-nuclear power—was based on no guarding at all. The estimate was derived from a comparison with radium, and no one is watching this country's radium deposits to prevent them from getting into rivers through various earth-

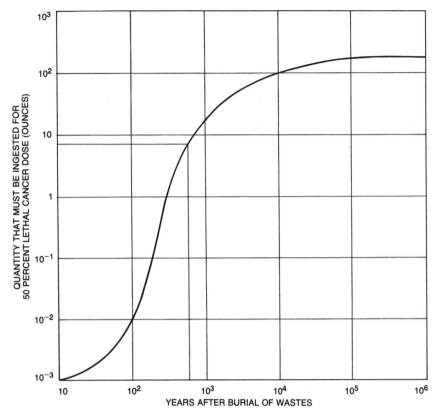

DANGER FROM INGESTED WASTES can be shown to be very great at first but much less after a few hundred years. As this graph shows, after 600 years a person would have to ingest approximately half a pound of the buried wastes to incur a 50 percent chance of contracting a fatal cancer. Such a calculation suggests that although it is obviously very important to isolate such wastes from human contact for a few hundred years, it is less imperative thereafter.

moving operations. Therefore guarding buried nuclear wastes would only serve to reduce that already small toll.

Even if guarding should be considered advisable, it would not be very expensive or difficult. Once the repository is sealed the guarding would consist only in making periodic inspections of the surface area—about 10 miles square for the wastes from 1,000 years of all-nuclear power—to make sure that the warning signs are in good order and to see that no one has unexpectedly undertaken mining or deep drilling. In addition occasional water samples might be drawn from nearby rivers and wells to check for increased radioactivity. Hence keeping watch on the wastes accumulated over 1,000 years of all-nuclear electric power in the U.S. would provide a job for only one person at a time.

Perhaps the best way to put into perspective the burden we are placing on our descendants by storing nuclear wastes is to compare that burden with others we are placing on them. Probably the worst will be the burden resulting from our consumption of the earth's high-grade mineral resources. Within a few generations we shall have used up all the world's economically recoverable copper, tin, zinc, mercury, lead and dozens of other elements, leaving fewer

options for our descendants to exploit for materials. Moreover, we are burning hydrocarbons—coal, oil and gas—at the rate of millions of tons each per day, depriving our descendants not only of fuels-but also of feedstocks for making plastics, organic chemicals, pharmaceuticals and other useful products. These burdens are surely far heavier than any conceivable burden resulting from the appropriate burial of nuclear wastes.

What makes this comparison particularly pertinent is that the only way we can compensate our descendants for the materials we are denying them is to leave them with a technology that will enable them to live in reasonable comfort without these materials. The key to such a technology must be cheap and abundant energy. With cheap and abundant energy and a reasonable degree of inventiveness man can find substitutes for nearly anything: virtually unlimited quantities of iron and aluminum for metals, hydrogen for fuels and so on. Without cheap and abundant energy the options are much narrower and must surely lead back to a quite primitive existence. It seems clear that we who are alive today owe our descendants a source of cheap and abundant energy. The only such source we can now guarantee is nuclear fission.

LETTER TO THE EDITOR

October 1977

Sirs:

Bernard L. Cohen's article "The Disposal of Radioactive Wastes from Fission Reactors" [SCIENTIFIC AMERICAN, June] contains serious errors and potentially misleading statements. Considerations of space limit my discussion to a few flaws that seem important.

Cohen's comment (pages 87 and 88) regarding the present extensive use of plutonium as a fuel in light-water reactors is true, but for the unwary reader it is probably deceptive. The controversy concerning plutonium recycling has little to do with plutonium that sits securely within an operating reactor core. It concerns only the fate of the plutonium after its removal from the reactor.

The parenthetical remarks on the "linearity hypothesis" (page 88) assume that if that hypothesis is abandoned, it will be replaced by a threshold model or some other reduced-toxicity model. If hypotheses such as those of Karl Z. Morgan ("Suggested Reduction of Permissible Exposure to Plutonium and Other Transuranium Elements," *American Industrial Hygiene Association Journal,* pages 567–575; August, 1975) concerning alpha-active isotopes become generally accepted, however, then abandonment of the linear hypothesis could lead to an increase in nuclear-related health-hazard estimates, not "drastic reductions."

Cohen must substantiate his claim (page 92) that "ceramics or other more resistant materials" will be able to maintain their integrity and low leachability over long periods when they are subjected to thermal heating, intense irradiation, large overburden pressures and surrounding brine.

Cohen's discussion of ion-exchange processes that tend to retard radioactive-waste transport through aquifers (page 92) oversimplifies the matter. It has previously been criticized by F. von Hippel, D. G. Jacobs and J. E. Turner in *Physics Today* (pages 68–69, August, 1976, and pages 15 and 86, November, 1976).

The calculated probability of waste release (page 93) is based on an assumed "equal escape probability for atoms of waste and radium." This assumption is unwarranted, because one can devise scenarios of waste-repository failure that are not applicable to radium in the earth. For example, (1) the possibility exists that waste-generated thermal ex-

pansion or post-backfill mine subsidence might fracture overlying shale formations (depicted on page 89), resulting in unanticipated ground-water intrusion (see W. Hambleton, "The Unsolved Problem of Nuclear Wastes," *Technology Review,* pages 15–19; March–April, 1972), and (2) radium in the earth does not have one or more "sealed" mine shafts leading to it, and little work has been done on the feasibility of creating long-lived water-impervious seals. Moreover, the knowledge that New Mexico is currently arid is useful in terms of the relatively short half-lives of strontium 90 and cesium 137 but is not terribly germane to the half-lives of iodine 129 or actinides (more than 24,000 years). Failure to address such issues renders Cohen's probability calculation meaningless.

The author's belief (page 94) that nuclear fission cleanses the earth of radioactivity is incomplete. He neglects to mention that uranium deposits which were sitting safely (more or less) underground are dug up in order to fission them. In the process radium, radon and their daughter products are left exposed to the elements in tailing piles at ground level. On the basis of the linear hypothesis and an assumed global population of four billion one can calculate that each reactor-year of operation eventually gives rise to more than 300 latent cancer fatalities from this cause alone.

The comparison of chemical and radiological hazards (page 94) should note that chemical toxins can, if it is desired, be neutralized by appropriate chemical transformations. Radiological toxins remain hazardous regardless of their chemical form, provided one can postulate some plausible mechanism for "administering" a dose. In addition, the absolute toxicity shown for nuclear waste after 100 years (10^7 lethal doses) is in contradiction with the information presented in the graphs on pages 92 and 93 (approximately 10^{10} lethal doses).

The discussion of the probability and nature of various future human activities (page 95) is entirely speculative. I might note, however, that "random exploratory drilling" may not be a helpful basis for prediction. Assuming some continuity to human nature, curiosity will long be with us. Thus the evidence of past drilling (which would be blatant if markers are left as warnings) might well increase the probability of future human activity at the same site....

Regarding the author's concluding statement, in the future no source of high-quality energy is likely to be "cheap."

Cohen also fails to assess the risks associated with predisposal transportation and handling operations.

Clearly we have radioactive waste, and we should get on the ball to put it someplace where it is unlikely to do us (or future generations) harm. My own preference would be to store high-level wastes or unreprocessed spent fuel in easily retrievable form in a man-made structure located deep within a salt mine or some other relatively stable geological formation, pending a more thorough evaluation of alternative disposal concepts than (barring some unforeseen conceptual breakthrough) can be obtained within the next couple of decades....

RICHARD SCLOVE

Department of Nuclear Engineering
Massachusetts Institute of Technology
Cambridge

Sirs:

My reply to the letter from Mr. Sclove will follow the order of his comments.

My article was based on a more detailed paper published in *Reviews of Modern Physics* (January, 1977), a reputable and thoroughly refereed journal (there were 14 referees on my paper). If the article contains "serious errors," Mr. Sclove should submit a paper to that journal about them.

I thought my one-sentence point about burning plutonium was interesting, and I do not see how it can be regarded as being deceptive. There are those who seem to believe that any use of plutonium, even its very existence, is sinful.

The position that the "linearity hypothesis" is much more likely to overestimate the harmful effects of small radiation doses than to underestimate them is endorsed by all official radiation-safety groups, including the National Academy of Sciences Committee on the Biological Effects of Ionizing Radiation (the BEIR Report), the U.S. National Commission on Radiation Protection and Measurements (NCRP Publication No. 43) and the United Nations Scientific Committee on the Effects of Atomic Radiation (*Ionizing Radiation: Levels and Effects*). The proposal that it may be otherwise, referred to by Mr. Sclove, was rejected by the International Commission on Radiation Protection in its decision not to lower permissible exposures. Under the circumstances I believe my one-sentence parenthetical remark was justified. In any case it did not form a part of my analysis.

My article made no statement about ceramics such as the one mentioned in Mr. Sclove's letter. My comment on ceramics is based on the work of G. McCarthy and Rustum Roy at Pennsylvania State University.

The criticism by von Hippel in *Physics Today* of my treatment of the effects of ion exchange on the transport of radioactive wastes was answered thoroughly in my reply to his letter (printed immediately following it). At the time von Hippel wrote that he was studying the problem further, and although he has written to me at least twice since then on other subjects, he has not mentioned that problem again. The Jacobs-Turner letter was essentially a more detailed treatment than mine rather than a criticism of mine, and there is little disagreement between us, as was explained in my reply printed with their letter.

The fracture of overlying shale by thermal or subsidence problems was studied by the Oak Ridge group and is not considered to be a serious problem. The shaft-sealing problem is being worked on and is not believed to be notably difficult. The shaft area is minuscule, it is quite distant from the buried waste in most designs and the salt is self-sealing. Moreover, there *are* holes in the ground (for example caves, mines and fissures) containing the radium to which the waste is compared. This is not to mention the comparison between the radium and the waste with respect to ground water, rivers, winds, animals, freeze-thaw cycles, vegetation working on surfaces and so on. My paper does *not* assume that the burial site is arid or will remain so. I do not agree that there is anything unfair in my comparing the escape probabilities of buried waste and the radium in the ground above it.

The discussion of tailings from uranium mining and milling is a rather separate issue that is discussed in other papers by me (for example *Bulletin of the Atomic Scientists*, February, 1976, page 61). That problem is being actively worked on and does not seem very difficult or costly to solve. Incidentally, the same calculation that Mr. Sclove cites as indicating 300 eventual cancer fatalities from that source yields a saving of 5,000 cancer fatalities from the burning up of the uranium.

The chemical toxicity of arsenic and barium is *not* essentially affected by ordinary changes in their chemical form. There was a mistake in the illustration on page 94 in that "(after 100 years)" should have read "(after 500 years)." The correct numbers appear in the text and in the illustrations on pages 92 and 93.

I do not understand or sympathize with Mr. Sclove's objection to my treatment of the drilling problem. Can he suggest a more appropriate treatment?

Nothing frightful can result from drilling in the far distant future.

I see no reason why nuclear energy should not become cheaper in the future, and remain so for millenniums.

Dangers to the public from waste handling and transport have been studied in other papers and found to be of far less consequence than those discussed here. Radioactivity releases from transport accidents are estimated to cause an average of much less than .01 fatality for each year of all-nuclear power.

It is difficult to quarrel with Mr. Sclove's call for more research, and the research effort is being accelerated. It is also difficult for me to understand, however, the intensity of his concern for the wastes from nuclear power, which have never harmed anyone and which my paper shows will probably never cause as much as one eventual fatality per year, when we are killing at least 10,000 Americans each year with the air-polluting wastes from coal burning, our only viable alternative source of electric power. Moreover, I cannot understand his insistence that the most conservative assumptions be made and the most improbable events be considered in evaluating the dangers from nuclear wastes, when the opposite philosophy is followed in evaluating dangers from coal-burning wastes and all other risks in our society. There are papers in the scientific literature that estimate annual fatalities from coal-burning air pollution at many times 10,000. Those papers have never been refuted. Moreover, radiation effects are far better understood than the effects of air pollution are, and the radiation levels under consideration here are a minute fraction of the natural background of radiation, whereas in the case of air pollution the levels are many times the natural background. Is it too much to ask that there be some balance and perspective in the consideration of our waste problems?

BERNARD L. COHEN

Department of Physics and Astronomy
University of Pittsburgh
Pittsburgh, Pa.

8

The Reprocessing of Nuclear Fuels

by William P. Bebbington
December 1976

The economics of fission power would be much improved if spent fuel were processed to remove fission products and plutonium and reclaim uranium. The industry needed for the task does not yet exist in the U.S.

Nineteen years after the first American nuclear power station went into service at Shippingport, Pa., the U.S. still has no commercial facility licensed to recover plutonium and unburned uranium 235 from the spent fuel of nuclear power reactors. Only one private plant was ever licensed to operate, and it was shut down in 1972 for modifications and enlargement. Its owners, Nuclear Fuel Services, Inc., of West Valley, N.Y., have since withdrawn their application for a license to reopen. Between 1966 and 1972 the plant reprocessed somewhat less than 650 tons of spent fuel. In Barnwell County, S.C., the separation facilities of a $500-million reprocessing plant with a capacity of 1,500 tons per year, owned by Allied-General Nuclear Services, were completed about a year ago. The owners are awaiting a Nuclear Regulatory Commission license, which in turn hinges on Government decisions on waste storage and on rules governing the utilization of recovered plutonium. Britain, France and several other countries reprocess spent fuel from nuclear power reactors in government facilities, but that is not the policy in the U.S. As a result the spent fuel from the nation's 62 operating fission power reactors has been piling up at repository sites. The current inventory is now about 2,500 metric tons.

Unlike coal, the fissionable fuel of a nuclear power reactor cannot be "burned" until all that is left is an essentially worthless and innocuous ash. The fresh fuel for American power reactors usually contains between 2.5 and 3.5 percent of the fissionable isotope uranium 235, having been enriched from the natural value of .7 percent uranium 235 by the gaseous-diffusion process. The remainder of the uranium in the fuel (and in the natural ore) is almost entirely the nonfissionable isotope uranium 238. When the nuclei of uranium 235 fission in the reactor, they give rise to a great variety of radioactive products, many of which act as fission "poisons" by absorbing the neutrons needed to keep the chain reaction going. By the time the uranium-235 content of the fuel has decreased to about 1 percent, the combined effects of depletion and by-product poisoning make it necessary to replace the fuel.

In addition to uranium 235 the spent fuel contains between .7 and 1 percent of plutonium 239, synthesized from uranium 238 by the absorption of a neutron. Plutonium 239 is even more fissionable than uranium 235, and the Federal Government is now deciding whether or not to approve the use of reactor fuel containing a mixture of the two nuclides. The fissionable material recovered from the spent fuel of three reactors is sufficient to fuel a fourth. The economics of the nuclear power industry will be strongly influenced by the decision that will allow or not allow the use of recycled, mixed fuels.

At the moment a reactor is shut down its spent fuel contains some 450 synthetically produced nuclides, including uranium 237 and neptunium 239, which decay into neptunium 237 and plutonium 239. The methods for chemically separating plutonium from uranium and its fission by-products were developed during World War II to provide highly purified plutonium for atomic bombs. Plutonium 239 was separated from the unenriched metallic uranium that served as the fuel of the Manhattan District reactors at Hanford, Wash. The heat from the reactors was discarded in the cooling water, and initially only the plutonium was recovered.

Uranium, neptunium and plutonium are members of the actinide series of elements, whose chemical properties are similar to those of the lanthanide series of rare-earth elements. Some months before Enrico Fermi and his coworkers demonstrated that plutonium could be made by a chain reaction in a uranium pile Glenn T. Seaborg and his colleagues had separated and purified several micrograms of pure plutonium metal that had been created by the bombardment of uranium in cyclotrons. The early studies revealed that plutonium had chemical properties that varied with its oxidation state and that could thus be exploited for separation processes. Those useful properties included the solubility of plutonium phosphates and fluorides in aqueous solutions (compared with the insolubility of the phosphates and fluorides of fission products) and the fact that certain plutonium ions could be extracted with organic solvents.

The fission products are isotopes of elements ranging in atomic number from 30 (zinc) to 66 (dysprosium). Most of them are radioactive, with half-lives that range from less than a second to thousands of years. The fission products are the chief source of the heat and radiation from spent fuel. Only a dozen or so combine intense radiation and long half-life with chemical and physical properties that are troublesome in reprocessing or in the ultimate disposal of wastes.

The first step in the treatment of spent nuclear fuel is to store it for several months in water-filled pools at the nuclear power station. During this period the radioactivity and the evolution of heat decrease by a factor of about 10,000. For example, the radioactivity of iodine 131, which has a half-life of 8.14 days, decreases by a factor of between 3,000 and 30,000. Indeed, iodine 131 is the chief determinant of how long the fuel is allowed to cool: the decay of the volatile element removes it as a problem in reprocessing.

The designers of the chemical separation plants at Hanford recognized that the technological innovations required for conducting chemical operations by remote control behind thick concrete walls were demanding enough without trying to achieve such niceties as the optimization of the process. They chose the simple batch operations that had been developed by Seaborg for working with microgram amounts of plutonium. Briefly, the uranium rods were first dissolved in acid, leaving an aqueous solution in which plutonium ions were extremely dilute. Bismuth and lanthanum were added as "carriers," so that when bismuth phosphate and lanthanum fluoride were subsequently precipitated out, they would carry with them plutonium phosphate and plutonium fluoride in quantities of precipitate large enough to separate. By repeated dissolutions and precipitations, with intervening changes in oxidation state, plutonium was separated from uranium and fission products. Simple tanks were used for the dissolutions and precipitations; centrifuges were used for separating the precipitates.

The processes worked well and safely, without any significant damage to the health of workers or to the environment. Removal of the fission products was efficient, and more than 95 percent of the plutonium was recovered. Operating ca-

pacity so far exceeded expectations that of four chemical-separation buildings planned only three were built and only two were operated. Uranium was not recovered, and the volume of waste was large because of the bismuth phosphate and lanthanum fluoride that had been added. Considering how long it takes to design and construct nuclear power facilities today it seems almost unbelievable that barely two and a half years elapsed between the initial demonstration of the chain-reacting pile on December 2, 1942, and the explosion of the first plutonium bomb on July 16, 1945.

The important legacy of Hanford to the nuclear-fuel-reprocessing industry was the concept of remote operation and maintenance, together with the innovations of engineering design that were needed to implement it. The buildings were long, thick-walled concrete

structures that enclosed the "canyons," or process spaces. The piping was embedded in the walls and ended in connectors precisely located at standard positions on the inside and near the top of the canyons. It was connected to the process equipment by accurately made jumpers that could be installed and removed by cranes that traveled the length of the canyons on rails. The crane operator, protected by heavy shielding and observing his tasks through a periscope, could remove and reinstall any of the equipment by using impact wrenches to manipulate the connectors at the ends of the jumpers. All liquids were transferred either by gravity or by steam-jet ejectors. Ingenious gang valves were developed to ensure that the steam lines were purged with air so that condensation could not suck radioactive solutions out of the shielded spaces.

After the war a major effort was

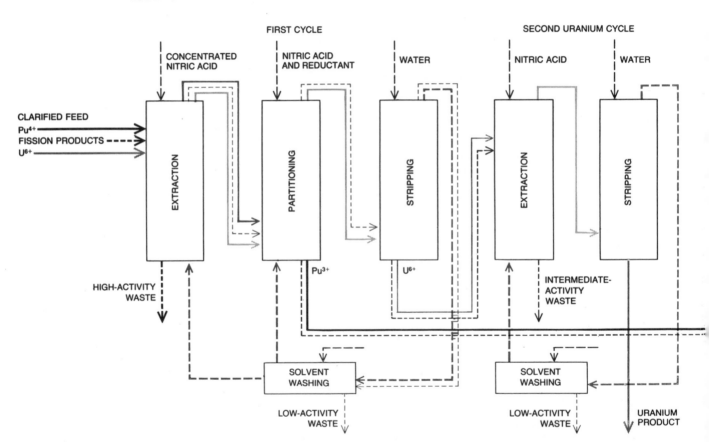

PUREX PROCESS for recovering uranium and plutonium from the spent fuel of power reactors employs TBP (tributyl phosphate) dissolved in a kerosenelike hydrocarbon as the separating agent. When uranium and plutonium ions are in a highly electron-deficient state, that is, are highly oxidized, they are more soluble in the TBP-hydrocarbon solution than they are in an aqueous solution. Under the same conditions the hundreds of radioactive by-products created when uranium-235 atoms fission in a reactor are more soluble in a strongly acid aqueous solution than in the organic one. This simplified diagram shows seven vertical columns in which organic and aqueous solutions are forced to travel countercurrently in intimate contact, so that substances more soluble in one solution than in the other can be efficiently separated. The feed mixture entering the first extraction column is the spent fuel in aqueous solution. In addition to the highly radioactive by-products it typically contains about 1 percent of unfis-

sioned uranium 235, more than 90 percent of the nonfissionable isotope uranium 238 and between .5 and 1 percent of mixed plutonium isotopes, primarily plutonium 239 and plutonium 240, the first produced from uranium 238 by the capture of a neutron and the second from plutonium 239 by the capture of another neutron. The uranium ions are in a highly oxidized state, deficient in six electrons (U^{6+}); the plutonium ions are deficient in four electrons (Pu^{4+}). The aqueous feed enters the first extraction column near the middle; the TBP solvent enters at the bottom. The uranium and plutonium are extracted by the upflowing solvent; the fission products are "scrubbed" out of the solvent by the downflowing aqueous stream of nitric acid and leave from the bottom of the column. The uranium-plutonium mixture passes to the second, or partitioning, column, where the plutonium is "stripped" out of the solvent by countercurrent contact with nitric acid that contains a reductant that reduces the plutonium to

launched to develop technically superior processes that could operate continuously rather than in batches and that could recover both uranium and plutonium with high yields. Solvent extraction received the most attention because it had previously been successfully applied to the purification of uranium. In solvent extraction aqueous and organic solutions flow in opposite directions (countercurrently) through a column or some other kind of mixing chamber that disperses one of the solutions in small droplets through the other. In the solvents that were used hexavalent uranyl ions, $(UO_2)^{++}$, and plutonyl ions, $(PuO_2)^{++}$, together with tetravalent plutonium ions, Pu^{4+} (plutonium atoms from which four electrons have been removed), are soluble, whereas trivalent plutonium ions, Pu^{3+}, and fission-product ions are not. Thus the solvent can extract the uranium and plutonium (in

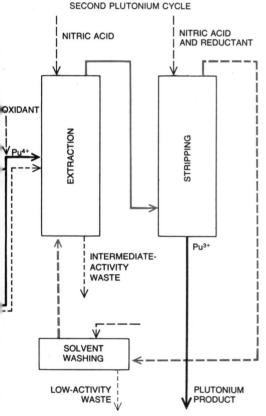

SECOND PLUTONIUM CYCLE

NITRIC ACID   NITRIC ACID AND REDUCTANT

OXIDANT

Pu⁴⁺

EXTRACTION

STRIPPING

Pu³⁺

INTERMEDIATE-ACTIVITY WASTE

SOLVENT WASHING

LOW-ACTIVITY WASTE   PLUTONIUM PRODUCT

the 3+ state (Pu³⁺), making it insoluble in the organic solvent. Simultaneously upflowing solvent scrubs the last traces of uranium from the aqueous solution of plutonium, which leaves from the bottom of the partitioning column. In the third, or stripping, column the uranium is removed from the organic solvent by dilute nitric acid. In the second uranium and plutonium cycles the extraction and stripping are repeated separately. In passing through the system, particularly in the first cycle, solvent is somewhat degraded by intense radiation and by chemical attack. Degradation products, along with traces of fission products, are removed from solvent with alkaline solutions.

its highly oxidized forms) from the aqueous feed solution, which retains most of the fission products. In separating the plutonium from the uranium the plutonium is reduced to the trivalent Pu^{3+}, making it insoluble in the solvent, which then contains all the uranium.

This play on oxidation states gave rise to the name Redox for the first solvent-extraction process to be applied on a large scale. The Redox process, with Hexone (methyl isobutyl ketone) as the organic solvent, was put into operation at Hanford in 1951. Later processes with other solvents exploited the same oxidation-reduction cycle. To force the highly oxidized ions of uranium and plutonium into the solvent high concentrations of nitrate ions are needed. In most chemical processing nitric acid is used to supply nitrate ions. Since the Redox solvent Hexone is decomposed by high concentrations of nitric acid, however, aluminum nitrate was used instead. This added greatly to the quantity of highly radioactive waste. Hexone also had the disadvantage of being highly volatile and flammable.

Shortly after the war the British built production reactors and a separation plant at Windscale in Cumbria. As an extraction solvent they chose Butex (β,β' dibutoxy diethyl ether). Butex is chemically stable in strong nitric acid, making it unnecessary to resort to aluminum nitrate. It is also denser and less volatile than Hexone, but it is more expensive. In the early 1950's, when the U.S. built a major new plant for producing plutonium and the hydrogen isotope tritium on the Savannah River near Aiken, S.C., tributyl phosphate (TBP) was selected for the solvent-extraction process. When TBP is dissolved in a kerosenelike solvent, it is chemically even stabler than Butex, is cheaper than Hexone and gives better separations than either. The TBP, or Purex, process is now used in all reprocessing plants.

The Purex process comprises three cycles of extraction with TBP. Extraction is preceded by a "head end" step in which the spent fuel is dissolved and the solution is clarified, a process that varies with the nature of the fuel and the cladding of the fuel rods. At the Savannah River Plant, where the reactors are fueled with natural uranium metal clad in aluminum, the cladding material is removed by dissolving it in an aqueous solution of sodium hydroxide and sodium nitrate. The uranium oxide elements that fuel all American power reactors are encased in long, slender tubes made either of stainless steel or of the zirconium alloy Zircaloy. Such rods are prepared for processing by chopping the tubes into short sections and dissolving out the oxide ("chop-leach"). (Chemical and electrochemical dissolution of the oxide fuel rods has been demonstrated, but it calls for process equipment made of alloys that are highly resistant to cor-

rosion and adds to the volume of liquid wastes.) The solutions from the head-end dissolvers are usually centrifuged to remove finely divided solids that would interfere with the solvent extraction. A substance such as manganese dioxide is sometimes precipitated to help clarify the solution and carry down some of the fission products.

When the separated uranium and plutonium streams emerge from the Purex process, they contain only about a millionth as much radioactivity due to fission products as the feed material did. At this low level of radioactivity the products in the two streams can be purified further and converted into the preferred final forms by fairly conventional chemical operations with relatively little radiation shielding. Evaporation, ion exchange, adsorption, precipitation and calcination have all been employed at one time or another. In the Government plants plutonium is reduced to the metallic form needed for weapons. If plutonium is ever used as fuel in nuclear power plants, plutonium oxide would be the preferred form, as is uranium oxide (U_3O_8). If the uranium is to be returned to the gaseous-diffusion plants for reenrichment, it is converted into uranium hexafluoride (UF_6), which is a gas at room temperature. Largely because of the differences between fuel forms two different practices have been adopted with regard to wastes. The wastes at Hanford and Savannah River are made strongly alkaline; this makes it possible to store them in tanks of carbon steel, which are placed in underground concrete vaults. Power-reactor wastes are concentrated in acid form and thus call for stainless-steel tanks.

The Purex process, with some modification, also lends itself to reprocessing the spent fuels from reactors using highly enriched uranium (such as the reactors of nuclear submarines) in which only traces of plutonium are formed. The chemical processing plant at the Idaho National Engineering Laboratory near Idaho Falls, Idaho, reprocesses the fuels from naval propulsion reactors and from research reactors of various kinds. The Idaho reprocessing plant differs markedly from the units at Hanford and Savannah River in being designed for direct (as opposed to remote) maintenance. The process equipment must be chemically decontaminated inside and out to allow men to enter the cells for repairs and replacements.

The efficiency of the Purex process depends heavily on the design of the solvent-extraction apparatus in which two immiscible liquids, one aqueous and one organic, are brought into intimate contact and then cleanly separated. This is done in an apparatus called a contactor. The simplest apparatus for countercurrent solvent extraction is the packed column, a vertical tube usually fitted with metal or ceramic rings that break up the liquid phases and direct them into tortuous paths through the column. The lighter organic solution flows up through the column as the heavier aqueous solution flows down. In such a column the mixing is not vigorous and the flow rates are low. As a result the column must be very tall to achieve a good separation, which complicates a plant that must be heavily shielded and must avoid the use of pumps. When packed columns were installed for the Butex process at Windscale, the result was a process building 20 stories high (with the fuel dissolver at the top) to allow the radioactive streams to flow downward by gravity.

The effectiveness of the extraction column can be greatly increased and its height can be reduced by "pulsing," so that the phases are drawn back and forth through perforated plates as they pass through the column. The pulsing can be done either by means of a piston or by applying air pressure to an external chamber. Pulsed columns were installed in the Purex plants at Hanford and at Idaho Falls.

One alternative to the extraction column is the "mixer-settler," in which the organic and the aqueous solutions are repeatedly mixed and separated in banks of from 12 to 24 horizontal stages, each consisting of a square mixing chamber at one end and a long settling chamber at the other. The mechanical agitator that mixes the solutions also propels the liquids from stage to stage. The chief drawback of mixer-settlers is the large volume of uranium and plutonium that is held up in liquid inventory. Because of the large holdup the solvent is subject to considerable damage from radiation and chemical activi-

ty. Among the advantages of the system are that the contactor can be readily adapted to remote maintenance, as it is at Savannah River, or be remote from the motors that drive the mixers, as it is at Windscale. Mixer-settlers are used in most of the European reprocessing plants.

An improvement on both the extraction column and the mixer-settler is a centrifugal contactor developed at Savannah River. The settling section that accounts for the large holdup of uranium and plutonium in the mixer-settler is replaced by a small centrifugal separator mounted on the same shaft as the mixing vanes. Typically arrayed in groups of six, the centrifugal units are more efficient than the mixer-settler, have only 2 percent of the volume, need only a small fraction of the time to come to a steady state of operation or to be flushed out and cause only about a fifth as much damage to the solvent. A few years ago an 18-stage centrifugal contactor replaced a 24-stage mixer-settler as the extraction contactor in the Savannah River Purex plant.

An axial-flow, multistage centrifugal contactor called the Robatel has recently been developed by a French company, Saint-Gobain Techniques Nouvelles. In this device eight stages are arrayed along the single vertical shaft of the centrifuge bowls. The apparatus has been selected as the extraction contactor for the first Purex cycle of the Barnwell plant of Allied-General Nuclear Services.

The large-scale use of nuclear energy for the generation of electricity got under way in Britain before it did in the U.S. To reprocess the magnesium-alloy-clad uranium-metal fuel from their first power reactors the British chose to modify and expand the plant at Windscale, which had originally been built to separate weapons-grade plutonium. They designed the plant on the principle that the equipment inside the shielded cells where the most highly radioactive material was handled would never be repaired or modified. This no-maintenance principle called for materials and equipment of the highest quality. Reliability was ensured by building two complete primary separation plants, one to serve as a spare. In addition each plant had a complete spare dissolver and a spare first solvent-extraction contactor. The spare primary plant was never needed as such, and it was later modified and increased in capacity. In 1957 it began reprocessing power-reactor fuel. Decontamination and modification of the first plant were then found to be feasible, with the result that the modified original plant served the British nuclear power program until 1964.

In the early 1960's the British designed and built their second-generation reprocessing plant, this time based on

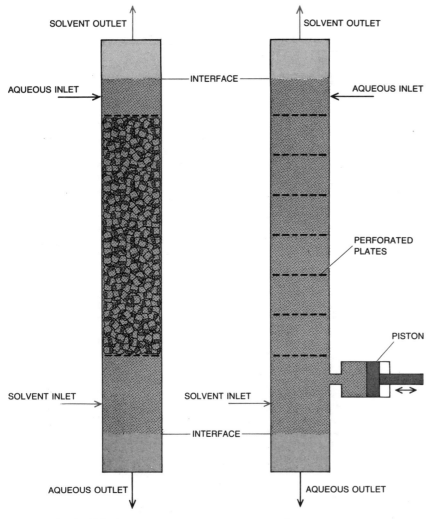

SIMPLE EXTRACTION COLUMNS were used to provide intimate contact between the solvent and aqueous solutions in the first spent-fuel-reprocessing plants. In the column at the left contact is provided by a packing of randomly oriented ceramic or metal objects, usually in the shape of rings or saddles. The aqueous solution in the column flows downward under the influence of gravity; the lighter-organic solution travels upward. A smaller, more efficient extractor (*right*) can be built if a piston or air pressure is employed to "pulse" the fluid in the column so that two solutions are repeatedly drawn back and forth through tiers of perforated plates.

WORLD'S LARGEST REPROCESSING PLANT for recovering uranium and plutonium from the spent fuels of nuclear power reactors is the Windscale plant at Seascale on the west coast of England. The facility was originally built to reprocess uranium from Britain's first plutonium-production reactors, which are barely visible in the distance. The multistory building with the tallest stack is the original plutonium-extraction facility. Its height was needed so that process streams of organic and aqueous solutions could flow countercurrently by gravity through extraction columns. The special railroad cars in the foreground are delivering spent nuclear fuel from power stations operated by the United Kingdom Central Electricity Generating Board. The Windscale plant, operated by British Nuclear Fuels Ltd., a government-owned corporation, can process 2,000 to 2,500 tons of fuel per year, and has processed fuels from several other nations.

OPERATING FLOOR AT WINDSCALE supports rows of motors that turn the agitators of "mixer-settlers," multistage horizontal chambers that perform the same function as extraction columns in bringing organic and aqueous solutions into intimate contact (*see illustration on* *page 104*). The plant operators are shielded by a thick concrete floor from the intense radiation in the mixer-settlers below. All equipment that may require repair or replacement is located above the floor. The mixer-settler cells themselves were designed never to be entered.

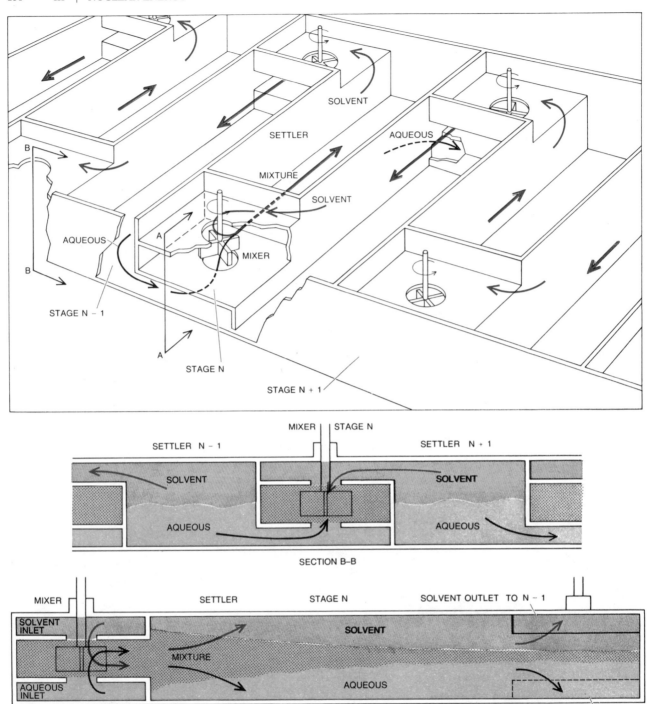

MIXER-SETTLERS, in which flow is maintained by paddles rather than by gravity, were first used at Savannah River Plant in the U.S. Aqueous and solvent solutions flow countercurrently through horizontal "stages." Each stage consists of mixing chamber and settling one.

Purex solvent instead of Butex. With a capacity to reprocess 2,000 to 2,500 metric tons of fuel per year, the new plant is the largest of its kind in the world. The no-maintenance principle was again followed. In 1969 a chop-leach dissolving facility was added as a head end to handle uranium oxide fuel (clad in either stainless steel or Zircaloy) that had been subjected to much longer "burnup" (exposure in the reactor) than earlier uranium-metal fuel rods and that as a result had a higher content of fission

products. Chop-leach dissolving is followed by one cycle of Butex extraction to bring the fission-product content of the oxide fuel into line with that of the uranium-metal feed to the main plant.

There is also a small reprocessing plant in northern Scotland for the spent, highly enriched uranium fuel from the adjacent Dounreay reactor and from materials-testing reactors. At Dounreay the fuel cycle is completed under conditions that anticipate the more stringent requirements of later generations of

commercial power reactors. Enriched-uranium fuel assemblies that have been irradiated to high burnup are reprocessed after short cooling times and are refabricated into new fuel elements.

The Windscale reprocessing plants (and the Dounreay operations as well) are now a part of British Nuclear Fuels Ltd., an independent (albeit government-owned) corporation that provides complete fuel-cycle services and has reprocessed spent nuclear fuels from other countries, including West Germany,

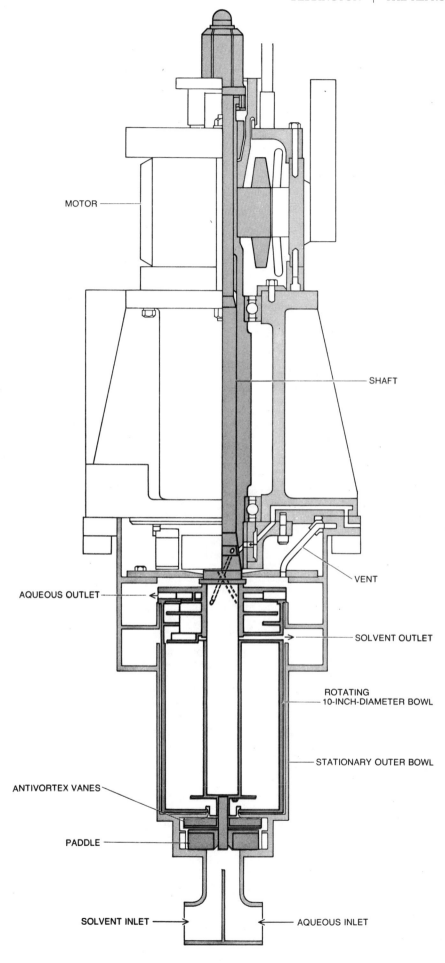

MOTOR

SHAFT

VENT

AQUEOUS OUTLET

SOLVENT OUTLET

ROTATING
10-INCH-DIAMETER BOWL

STATIONARY OUTER BOWL

ANTIVORTEX VANES

PADDLE

SOLVENT INLET

AQUEOUS INLET

Italy, Japan, Spain and Sweden. It is a member of United Reprocessors GmbH, a joint Anglo-French-German company set up to coordinate requirements and operations in Europe.

In the U.S. the Atomic Energy Commission, now merged into the Energy Research and Development Administration (ERDA), supported nuclear power by research and development in national laboratories, constructed demonstration reactors (such as the one at Shippingport, which was built in collaboration with public utilities) and stored (and in a few instances reprocessed) spent fuels from those reactors. The AEC did not, however, take the responsibility for either fuel fabrication or fuel reprocessing. The same policy continues under ERDA. Only the enrichment of uranium and the ultimate disposal of highly radioactive wastes are Government responsibilities.

Interest in the reprocessing of nuclear fuels developed among the suppliers of nuclear power equipment who felt the need to assure their customers of a closed fuel cycle, the chemical companies that had the necessary technological skill and background, and the oil companies that hoped to expand their operations into other energy sources. Several ventures into reprocessing emerged. The first, Nuclear Fuel Services (originally a subsidiary of W. R. Grace & Co. but now owned jointly by the Getty Oil Company and the Skelly Oil Company), designed and constructed a plant with a capacity of 300 tons of spent fuel per year on a site owned by the state of New York in West Valley, N.Y. After six years of operation the plant was shut down in 1972 for a planned expansion to 750 tons per year, for the correction of some deficiencies in the process, for the improvement of environmental-protection features and for the installation of waste facilities needed to meet new regulatory requirements. The plant used the Purex process in pulsed columns. Fuel was prepared for processing by chop-leach. The chop-leach equipment could be maintained or replaced remotely; the Purex-process cells were maintained directly. At last reports, however, the estimated cost of the modifications had risen from $15 million to $600 million, and Nuclear Fuel Services had withdrawn its application to the Nuclear Regulatory Commission for permission to reopen the plant.

CENTRIFUGAL CONTACTOR for mixing and separating solvent and aqueous phases was developed at the Savannah River laboratory of the Energy Research and Development Administration. Role of settling chamber in mixer-settler is taken over by a bowl on same shaft as mixing paddle. Drive motor, with its frame and bearings, is remotely replaceable.

The General Electric Company had meanwhile become convinced that relatively small reprocessing plants might be built to serve a group of power reactors within a short shipping radius. General Electric designed and built such a plant, the Midwest Fuel Recovery Plant at Morris, Ill., near the Dresden nuclear power station of the Commonwealth Edison Company. With a capacity of 300 tons per year, the Morris plant embodied major departures from the typical Purex-TBP process, with the aim of minimizing the contribution of reprocessing costs to the cost of nuclear power. The General Electric Aquafluor process involved TBP solvent extraction for the separation of uranium and plutonium from most of the fission products, ion exchange for separating uranium and plutonium from each other, and fluidized beds for the calcination of uranyl nitrate to the oxide (UO_3) and for the conversion of the oxide to the hexafluoride (UF_6).

Instead of the usual second solvent-extraction cycle for the uranium, General Electric incorporated a separation step that exploited differences in volatility for separating the fluorides of the fission products from the UF_6. This step reduced costs and eliminated some liquid waste, but it entailed the remote handling of radioactive powders. In the course of testing the plant equipment with nonradioactive feeds it was concluded that the problems of handling fine radioactive solids were far greater than had been anticipated and would preclude successful operation of the plant. It now seems that the plant cannot be modified economically to avoid such difficulties and meet the current requirements of the Nuclear Regulatory Commission.

In 1968 the Allied Chemical Corporation announced plans to build a 1,500-ton-per-year fuel-reprocessing plant on land in the Barnwell County industrial park, adjacent to (originally part of) the site of the Government's Savannah River Plant. Allied Chemical was joined by the General Atomic Company, jointly owned by the Gulf Oil Corporation and the Royal Dutch/Shell Group of Companies, as co-owner of Allied-General Nuclear Services, the operator of the Barnwell facility. Apart from its proximity, the Barnwell plant is independent of the Savannah River Plant. Construction at Barnwell was begun in 1971, and the originally planned facilities are now complete. They provide for receiving and storing fuel, chop-leach dissolving, Purex separations, storage of high-activity wastes and plutonium nitrate product, and the conversion of uranyl nitrate product into uranium hexafluoride.

Design and construction at Barnwell of "tail end" facilities for the solidification of the waste for shipment to a Fed-

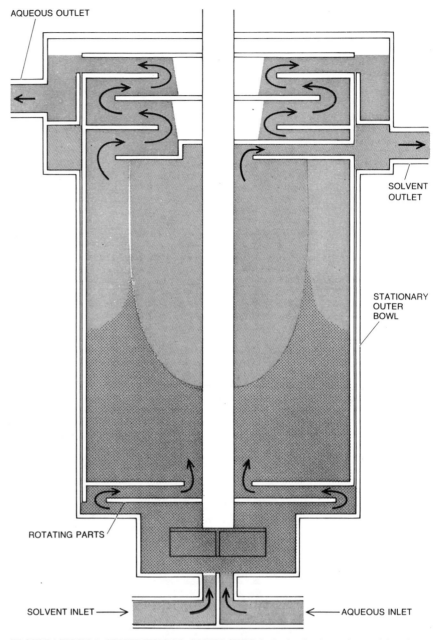

FLOW PATH IN A CENTRIFUGAL CONTACTOR is depicted schematically. After thorough mixing the organic and aqueous solutions travel upward into a rotating bowl where they are separated centrifugally. Heavier aqueous solution is thrown to outside of bowl; lighter organic solution is driven inward. Baffles at top of bowl direct two solutions to separate ports.

eral repository and for the conversion of plutonium nitrate to solid plutonium oxide await decisions by the Nuclear Regulatory Commission and ERDA on the specifications and destinations of those materials. So far Allied-General Nuclear Services has invested some $250 million in the Barnwell plant (more than three times the original expectation); the waste and plutonium facilities are expected to cost another $250 million. Half a billion dollars may seem like a large investment for a single reprocessing plant, but in the overall economics of nuclear power the outlay represents less than 1.5 percent of the value of the 50 to 60 nuclear power reactors whose spent fuel the Barnwell plant can reprocess.

In the Allied-General Nuclear Services separations facility the spent fuel will be chopped into short lengths by a shearing device that was conceived by its engineers and designed and built by Saint-Gobain Techniques Nouvelles. The uranium oxide pellets will be dissolved continuously from the cladding hulls in a series of vessels where fresh acid leaches the last traces of fuel from a batch of hulls. Solid particles are removed from the solution by centrifugation. There are two innovations in the Purex system: the first extraction contactor is the Saint-Gobain centrifugal unit I described above, and the separation of plutonium is achieved by reducing the plutonium electrolytically in an "electropulse" column, a development of the Allied-General Nuclear Services technical staff. The other contactors are also pulsed columns. Equipment that is subject to mechanical or electrical failure or to unusually corrosive conditions can be replaced remotely; the rest of the equipment is designed for direct maintenance.

Exxon Nuclear is a supplier of uranium and of reactor fuel assemblies and is actively interested in the rest of the fuel cycle, including enrichment and reprocessing. Earlier this year Exxon announced plans to build a 1,500-ton-per-year reprocessing plant on land that is now part of the ERDA site at Oak Ridge, Tenn. The company is awaiting a construction permit from the Nuclear Regulatory Commission. Although other companies have from time to time expressed interest in fuel reprocessing, no other commitments have been made.

One industry executive has summed up the current situation by saying: "At this moment the nuclear-fuel cycle does not exist." In the U.S., at least, this is true; even the design and construction of modified and new facilities are at a standstill pending the resolution of environmental and regulatory impasses.

Hearings on the recycling of plutonium as an oxide mixed with uranium are just getting under way, and waste handling is in limbo until the final disposal site and specifications are decided. Even receipt and storage of spent fuel at Barnwell awaits license hearings that are only now about to begin. The separations facility is ready and could be operated, with interim storage of waste and plutonium in solutions, but this too awaits completion of environmental and safety appraisals and subsequent license hearings. There is doubt about whether the plant will be cleared for start-up before the end of the decade.

The situation abroad is strikingly different. Both the British and the French have relied on military production facilities to process spent uranium from the first or second generation of power reactors, which used metallic fuel rather than oxide. The construction of additional facilities to reprocess uranium oxide fuel from the newer light-water (as opposed to gas-cooled) reactors has fallen behind schedule, but not seriously. The Windscale head-end facility for oxide fuel operated from 1970 to 1973,

MULTISTAGE CONTACTOR called the Robatel has been developed by the French firm Saint-Gobain Techniques Nouvelles. The rotating bowl of this centrifugal machine has a diameter of 80 centimeters, which is about three times diameter of the bowl in the Savannah River unit. The path through one stage of Robatel is shown on opposite page.

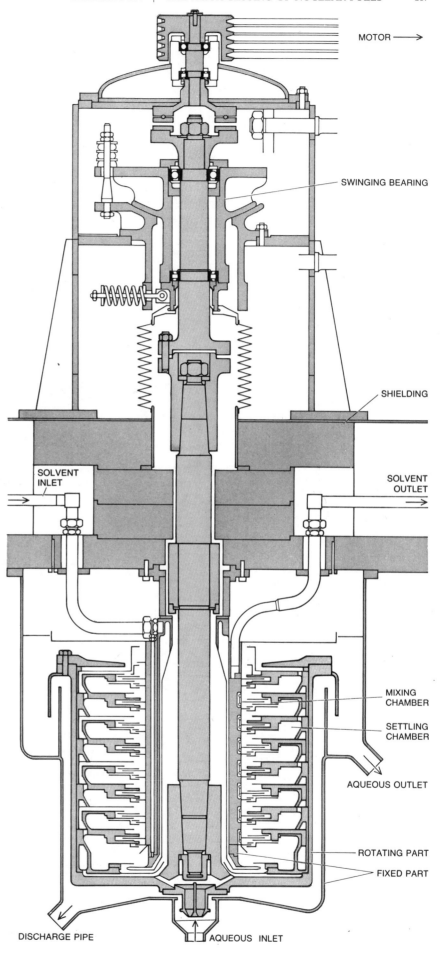

MOTOR →

SWINGING BEARING

SHIELDING

SOLVENT INLET

SOLVENT OUTLET

MIXING CHAMBER

SETTLING CHAMBER

AQUEOUS OUTLET

ROTATING PART

FIXED PART

DISCHARGE PIPE

AQUEOUS INLET

when it was shut down after a small release of radioactivity that led to a comprehensive review of the processing of highly irradiated oxide fuels. The British have now decided not to modify or rebuild the head-end facility and are planning to put up two more oxide-reprocessing plants with a capacity of 1,500 metric tons per year each, to be completed during the 1980's.

The original French reprocessing plant at Marcoule, with a capacity of processing 1,000 tons of uranium metal per year, has been running since 1958. A second plant of the same capacity went into operation at La Hague in 1967. A head-end facility able to handle about 800 tons of oxide fuel per year was recently added to the La Hague plant. The French are now proposing to build two more complete oxide-processing facilities at La Hague, each with a capacity of 800 tons per year, the first to be ready by 1984 and the second by 1986.

In West Germany a group of four chemical and nuclear engineering companies, which has been operating a small demonstration reprocessing plant, is now selecting a site for a plant with a 1,500-ton-per-year capacity that is expected to be operating by the late 1980's. The overall scheduling of European reprocessing facilities has been guided since 1971 by United Reprocessors, a consortium of British, French and West German enterprises. Its goal is to provide an integrated reprocessing capacity of about 20,000 tons per year by the early 1980's. There are small demonstration plants in Japan and India. Brazil and Pakistan have recently negotiated respectively with West Germany and France for the purchase of full-scale plants. Japan has also announced plans to build a large plant.

How can the U.S. nuclear power industry continue to operate without the reprocessing of its spent fuels? For the present there is enough uranium-enrichment capacity to allow once-through operation of existing nuclear power reactors, partly because the expansion of nuclear power facilities has been greatly retarded as a result of the economic slump, high construction costs and licensing delays. The volume of spent nuclear fuel accumulated per year is still manageable; in 1977 about 1,100 tons will be discharged, and its storage for long periods is simple and safe. The high-integrity, corrosion-resistant cladding is more than adequate to contain the fuel in the high-purity water of the storage basins. The capacity of the storage basins is being taxed, however, and modifications are being made to increase the size of some of them. The storage basin at Morris is in service, and the one at Barnwell has been completed.

Once-through operation of nuclear power reactors and the increasing investment in spent-fuel inventory add

FLOW SCHEME IN THE ROBATEL provides for eight stages of mixing and settling arranged one above the other. The schematic diagram shows the flow through a single stage. Briefly, the organic solution, traveling downward on the inside of the rotating bowl, is repeatedly mixed with the aqueous solution, which is conducted upward through a series of ports and baffles. Flow of organic solution is readily followed from diagram. At each stage aqueous solution leaves settler through ports labeled *A* and reappears in stage above through ports labeled *B*.

substantially to the cost of electric power from the nuclear plants, perhaps as much as 20 percent. These costs must ultimately be covered by the consumer. Even more important, the spent fuel constitutes a high-grade energy resource that must ultimately be "mined." The fuel cycle needs to be closed so that the technologies for reactor-fuel fabrication and for reprocessing can remain in step with reactor technology and meet their mutual requirements. It is particularly desirable that the commercial reprocessors have experienced staffs and demonstrated processes before they are called on to take up the more demanding task of reprocessing plutonium fuels

from breeder reactors and other advanced systems with high burnup rates and perhaps shorter cooling periods.

The U.S. has now had more than three decades of highly successful experience in reprocessing reactor fuels to extract hundreds of tons of plutonium, with no proof that these activities have done any significant harm to man or his environment. In spite of this experience critics of nuclear power point to such hazards as possible leaks from existing liquid-storage tanks containing highly radioactive wastes, the long life of highly radioactive wastes under any storage procedure and the harm that could be done by the routine discharge of effluents with

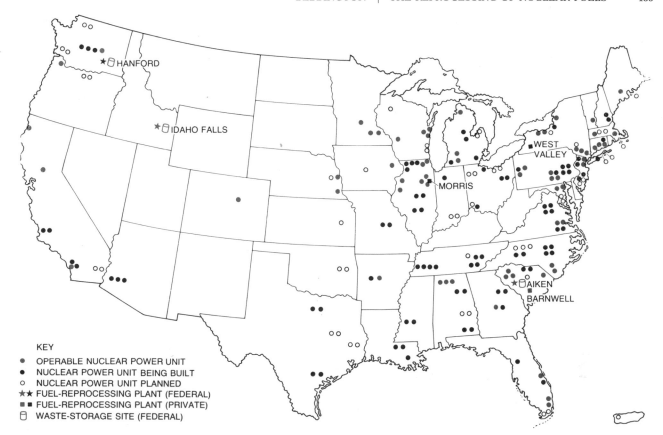

KEY

- ● OPERABLE NUCLEAR POWER UNIT
- ● NUCLEAR POWER UNIT BEING BUILT
- ○ NUCLEAR POWER UNIT PLANNED
- ★★ FUEL-REPROCESSING PLANT (FEDERAL)
- ■ ■ FUEL-REPROCESSING PLANT (PRIVATE)
- ⬚ WASTE-STORAGE SITE (FEDERAL)

SIXTY-TWO FISSION POWER REACTORS are now licensed to operate in the U.S. at 44 sites. Another 72 reactors are under construction and 61 more are in the planning stage. The operating plants have a total capacity of 44,650 megawatts, which represents just over 8 percent of the total U.S. electric-generating capacity. The 72 reactors under construction will add another 75,500 megawatts of capacity. Only one privately owned spent-fuel-reprocessing plant is substantially finished and awaiting a license to operate: the $500-million plant of Allied-General Nuclear Services in Barnwell, S.C. Two other reprocessing plants (*black squares*) have been constructed, one by the General Electric Company at Morris, Ill., the other by Nuclear Fuel Services, Inc., at West Valley, N.Y., but their owners have no present plans to put them into operation. The Federal Government has three major fuel-reprocessing facilities: one at Hanford, Wash. (now shut down), one on the Savannah River near Aiken, S.C., and one at Idaho Falls, Idaho. At these three sites the U.S. has also stored some 75 million gallons of radioactive wastes, the residues from more than 30 years of processing spent fuels, chiefly for extraction of plutonium. Fuel from power reactors is stored in pools adjacent to power plants and at reprocessing plants, including those at Morris and West Valley.

CHEMICAL REPROCESSING "CANYON," 800 feet long, at the Government's Savannah River Plant is one of two parallel remote-maintenance facilities. All piping and equipment can be removed and replaced by an operator riding in a heavily shielded crane that travels length of canyon on rails visible on facing walls. Pipe sections are balanced so that they hang level when they are lifted by crane. Connectors are operated by wrenches manipulated from crane. Reactors at plant were designed primarily to produce plutonium and tritium.

very low levels of radioactivity, by accidents in the reprocessing plants and by the theft of plutonium by terrorists. It is not possible in this article to answer all the questions that have been raised. Some of the known facts can nonetheless be discussed.

The principal fission products in aged high-radioactivity wastes are cesium 137 and strontium 90. Such wastes also contain traces of plutonium and uranium. Cesium 137 and strontium 90 have half-lives of about 30 years, so that several centuries of storage are needed for them to decay to negligible levels. (Twenty half-lives, or a factor of a million, are enough to leave most waste solutions innocuous.) The half-life of plutonium, however, is 24,000 years, but since it is a valuable product its loss in waste is kept as low as is practical. Typical wastes contain less than .1 percent of plutonium. At present all high-radioactivity waste is in solutions, sludges and masses of inorganic salts in underground tanks; none has been committed to ultimate disposal. The tanks are monitored, and the wastes can be (and have been) transferred to new tanks as the need arises.

There have been leaks in some of the older tanks at Hanford and Savannah River, but as could have been predicted from the characteristics of the wastes and the soils, the movement of the wastes away from the points of leakage has been slight. Strontium and plutonium are highly insoluble under the chemical conditions prevailing in the tanks, and cesium is strongly adsorbed by clay minerals in the soil. As successive groups of tanks have been built, standards of construction have been raised and the integrity of storage has been enhanced.

The alkaline Hanford and Savannah River wastes contain large concentrations of aluminum, iron and sodium salts, so that their volume is substantial (roughly 75 million gallons at present). On the other hand, only about 250 liters (65 gallons) of concentrated, acid, high-radioactivity waste remains from the reprocessing of a ton of power-reactor fuel. This amount can be reduced by demonstrated processes to as little as 75 liters of vitreous solid. The present policy of the Government is that the wastes shall be converted into solids by the reprocessors within 10 years of their production and that they shall then be delivered to Government sites for long-term storage.

The form of the wastes and the place for their ultimate disposal are technical problems that now require decisions and development rather than research. The places will almost certainly be geological formations; natural salt beds and other crystalline or sedimentary bedrock formations show promise. Salt deposits, and geological disposal in general, received a severe setback in public acceptance, however, when an abandoned salt mine in Kansas that had served for disposal tests was designated as the first site for actual demonstration disposals. It developed that this old mine was not well enough isolated from active salt workings nearby, and the project was abandoned.

Airborne and aqueous effluents from reprocessing plants normally carry the fission products krypton 85 and tritium, along with certain radioactive secondary by-products such as carbon 14, out into the environment. Although in total curies of radiation emitted the quantities of these substances are large, their radioactive, physical, chemical and biological characteristics are such that the radiation dose to man is a very small fraction of the dose he inescapably receives from the natural radioactivity in the environment. Some groups object to these routine releases on the grounds that any increase in the total dose of radiation, no matter how small, is harmful. This is the "linear hypothesis," based on the linear extrapolation of effects that can be observed at high doses to effects that cannot be observed at low doses.

Natural radioactivity exposes every person on the earth to an average annual dose of radiation amounting to about 100 millirems. In some populated areas the natural dose is several times higher. (For the purposes of comparison the official limit on the amount of radiation to which an individual worker in a nuclear plant may be exposed in the course of a year is five rems, or 50 times higher.) The amount of radiation added by nuclear operations is only a few millirems in the worst locations near the nuclear-plant fence. These small increases can be reduced, but only at great cost and in some cases with the substitution of other hazards. (An example is the storage of large quantities of krypton-85 gas under high pressure.)

Two general principles stated by the Committee on the Biological Effects of Ionizing Radiations of the National Academy of Sciences are pertinent. The first is: "No exposure to ionizing radiation should be permitted without the expectation of a commensurate benefit." The second is: "The public must be protected from radiation, but not to the extent that the degree of protection provided results in the substitution of a worse hazard for the radiation avoided. Additionally there should not be attempted the reduction of small risks even further at the cost of large sums of money that, spent otherwise, would clearly produce greater benefit."

Fears of accidents in the reprocessing plants have little foundation. The careful control required to maintain the efficiency of the process minimizes the probability of an accident. Moreover, the heavy shielding and sealed enclosures needed for routine protection from radiation confine an accidental release of radioactivity to a small area. The possibility that a critical mass of a fissionable isotope might accumulate is an understandable concern over an industry that grew out of the atomic bomb. The first effect of an accidental accumulation of a critical mass, however, is the almost instantaneous dispersal of the material, which immediately halts the chain reaction. The maximum energy release is small (equivalent to the combustion of between half a liter and five liters of gasoline), so that there is little damage to facilities and little dispersal of radioactivity. The burst of radiation, however, is serious, and a nearby worker could receive a grave dose. In the 30 years that the Government facilities have been operated there have been 12 accumulations of critical mass, five of them in chemical processing plants or laboratories. There was one fatality, which equaled the number from drowning and from shooting during the same period in those facilities. Chemical explosions and fires have had much more serious consequences in nuclear plants, but these too have had no significant off-site effects.

To many the greatest potential for disaster seems to be the possibility of terrorists' obtaining plutonium and making a bomb. The obstacles in the way of such a feat are great. Even a moderately effective bomb is a considerable technological achievement, and it is more difficult to make a bomb from power-reactor plutonium than from plutonium produced specifically for weapons. It is not generally appreciated that during the long exposure of fuel in a power reactor there is an accumulation of plutonium isotopes other than plutonium 239, particularly plutonium 240, which make it very much more difficult to assemble a supercritical mass of plutonium without an inefficient premature explosion. Weapons-grade plutonium is made with much shorter exposure in the reactor. Moreover, the radiation emitted by plutonium is easy to detect with sensitive instruments. (Such instruments can detect .25 gram in a volume of radioactive waste materials measuring several cubic feet.)

Diluting the plutonium oxide with uranium oxide before it is shipped from the reprocessing plant or "denaturing" it with a more radioactive material are possible deterrents to the hijacking of shipments. Such stratagems seem trivial, however, when one considers that there are tens of thousands of plutonium-containing weapons dispersed around the world and still more in weapons-fabrication plants. Perhaps our best hope is that someday plutonium will be more valuable for power-reactor fuel than for weapons, and that the nations will then beat their bombs into fuel rods.

9

Nuclear Power, Nuclear Weapons and International Stability

by David J. Rose and Richard K. Lester
April 1978

*Irresolution over domestic energy policy and the role
of nuclear power may act to undermine current U.S.
efforts to control the proliferation of nuclear weapons*

A year ago this month the Carter Administration put before Congress a comprehensive national energy plan that included as one of its key components a revision of this country's long-standing policy on the development of civilian nuclear power. The proposed change, which would have the effect of curtailing certain aspects of the U.S. nuclear-power program and of placing new restrictions on the export of nuclear materials, equipment and services, was based explicitly on the assumption that there is a positive correlation between the worldwide spread of nuclear-power plants and their associated technology on the one hand, and the proliferation of nuclear weapons and the risk of nuclear war on the other. This point of view has become the topic of a lively debate; at the periphery of opinion some see nuclear war lurking behind every reactor on foreign soil, whereas others argue that the connection between civilian nuclear power and nuclear-weapons proliferation is vanishingly small.

We shall advance here the heretical proposition that the supposed correlation may go the other way, and that the recent actions and statements of the U.S. Government have taken little account of this possibility. In brief, it seems to us that if the U.S. were to forgo the option of expanding its nuclear-energy supply, the global scarcity of usable energy resources would force other countries to opt even more vigorously for nuclear power, and moreover to do so in ways that would tend to be internationally destabilizing. Thus actions taken with the earnest intent of strengthening world security would ultimately

weaken it. We believe further that any policy that seeks to divide the world into nuclear "have" and "have not" nations by attempting to lock up the assets of nuclear technology will lead to neither a just nor a sustainable world society but to the inverse. In any event the technology itself probably cannot be effectively contained. We believe that the dangers of nuclear proliferation can be eliminated only by building a society that sees no advantage in having nuclear weapons in the first place. Accordingly we view the problem of the proliferation of nuclear weapons as an important issue not just in the context of nuclear power but in a larger context.

Fundamental tensions exist between the energy objectives and the nonproliferation objectives of U.S. policy, and on a different plane between the respective consequences of measures designed to achieve their primary effect either domestically or internationally. In what follows we shall analyze the complex set of interrelated issues that bear on the entire question of nuclear power and world security.

The most important of the new nuclear measures announced by the Administration last April were that the U.S. would defer indefinitely the reprocessing of spent nuclear fuel from domestic nuclear-power plants to recover and recycle plutonium and unused fissionable uranium, and that it would try to persuade other nations to follow its lead. Legislation submitted to Congress demonstrated the Administration's intention to restrict exports related to the nuclear-fuel cycle and to prevent the retransfer of exported U.S. nuclear tech-

nology to third parties. (A modified version of the Administration's nonproliferation bill has since then been approved by Congress.) Along with these restrictions the U.S. capacity to enrich uranium to standards suitable for use in conventional light-water (as opposed to heavy-water) reactors was to be increased to help meet the growing world demand for this service.

On the domestic front the regulatory requirements for installing light-water reactors were to be streamlined. The Administration also proposed a substantial slowdown of the U.S. program to develop a liquid-metal-cooled fast breeder reactor, a type of nuclear-power plant designed to create and consume fissionable plutonium out of the vast store of ordinary, nonfissionable uranium in the earth's crust. Specifically, the Clinch River breeder reactor, which was to have been built in Tennessee at a cost of some $2 billion, was scheduled for cancellation. The operation of breeder reactors requires reprocessing the nuclear fuel periodically, so that the retreat from fuel reprocessing and the deemphasis of breeder-reactor development complement each other. (Unfortunately the Clinch River reactor has become a focal point of the debate for both the critics and the proponents of nuclear power. The situation is doubly unfortunate because on the one hand that particular program was technologically and institutionally vulnerable and on the other the stopping of it has not helped resolve the deeper issues we discuss.)

The U.S. was also to redirect its nuclear research and development programs to place more emphasis on alternative fuel cycles and reactor designs that might offer reduced access to material suitable for use in nuclear weapons. This initiative has been carried into the world arena with the establishment of an international program to evaluate alternative technical and institutional strategies for the nuclear-fuel cycle.

In the months since the Administration's program was announced it has provoked much discussion within the U.S. and throughout the world. It has dissatisfied both critics and supporters of the U.S. nuclear-power program, and (partly because of the way it was presented) it has generated concern in many foreign capitals. Some of this country's partners in the development of nuclear power feel that they were not consulted adequately during the genesis of the new policy, and that policy communications, at least initially, have been clumsy and insensitive. Deeper-rooted anxieties underlie this irritation, however, since fuel enriched in the U.S. and reactors manufactured by U.S. companies still play significant roles in many national nuclear-power programs, and the effects of U.S. nuclear policies are widely felt.

What was the motivation behind the Administration's new nuclear policy and the related Congressional actions? Several possibilities come to mind. The first is simply that the Administration means what it says, namely that its goal is to increase international stability by taking actions thought to inhibit the proliferation of nuclear weapons. It would do so by reducing the availability of nuclear materials and technology helpful to a weapons program, even though the same materials

and technology had hitherto been commonly assumed to be a part of civilian power programs.

Another possibility is that both the Administration and Congress are undecided as to whether the collapse of the U.S. nuclear industry is desirable. This indecision contributes to the Government's apparent inability to formulate a coherent nuclear-energy strategy.

It is also possible that the Administration announced its policy in sympathetic response to the critics of nuclear power but expects that the policy will not work, and that after this demonstration of good faith a new, more pragmatic program will be unveiled at an appropriate time. One danger in such a tactic is that the Administration might delay its denouement too long. The reaction to an Administration generally perceived to be resigned to the demise of nuclear power in the U.S., or even actively to desire it, might develop its own irreversible momentum.

It may also be that the Administration, frustrated by the diplomatic rigidity of world discussions about international security and arms limitation, is casting about for some new approaches. Such approaches are to be found; we hope that this article is an example. These four general motivations are not mutually inconsistent, and the Government could shift its priorities among them as the consequences of its actions unfold.

Various circumstances created the context in which the Administration formulated its national energy program and in which its set of nuclear-energy and nuclear-nonproliferation

goals were to be pursued. The first circumstance was the urgent need to reduce the growing outflow of U.S. funds, currently running at more than $40 billion per year, to pay for imported petroleum. Such an expenditure, although not liable to bankrupt the country, was inconceivable as recently as 1973. Furthermore, the situation suggested excessive dependence on the policies of the principal oil-supplying nations. The increasing dependence on oil imports was (and is) seen as probably the major contributor to the U.S. energy problem.

The second circumstance was the presence of vast coal deposits in the U.S., amounting to perhaps 10 times the entire domestic oil and gas resource. Only 19 percent of the U.S. energy supply comes from the 650 million tons of coal currently mined each year. Estimates of the amount of coal recoverable at current prices with current technology range from 200 to 600 billion tons; thus even if coal production were to be tripled, the minimum estimate amounts to a 100-year supply. Moreover, the total amount of coal in the U.S. might be as much as 3,000 billion tons recoverable eventually at increased cost—a quarter of the earth's known reserves. Besides coal, an energy resource equivalent to perhaps 1,000 billion tons of coal resides in the oil-shale deposits in the vicinity of northwestern Colorado. Technological advances would surely make many of these resources available later, it was thought. In no absolute sense is the U.S. "running out of energy." Thus arose the goal of increasing coal production rapidly, to several times the present rate by the year 2000. In this

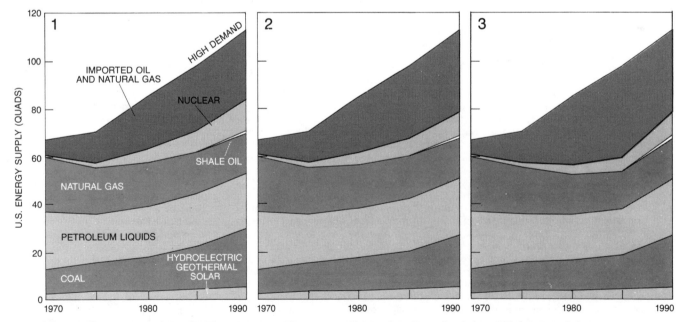

ALTERNATIVE U.S. ENERGY PROJECTIONS through 1990 are outlined in the set of graphs beginning on these two pages. The graphs represent 12 possible demand-and-supply scenarios constructed by the staff of the Congressional Research Service on the basis of different assumptions about economic growth rates, energy prices, the elas- ticity of energy demand and the constraints on various energy supplies. The tables from which the graphs were drawn were compiled in the course of a two-year study conducted at the request of several committees of Congress responsible for dealing with energy-related issues; the data appeared originally in *Project Interdependence: U.S.*

way the U.S. could reduce its dependence on imported petroleum, and perhaps also afford a more leisurely nuclear program.

The third circumstance was that energy conservation began to be taken seriously. Many studies under way between 1975 and 1977 showed not only that substantial increases in energy efficiency were possible but also that much energy was being wasted. Energy conservation had received significant recognition but little actual support during the previous Administrations, and Congress had not been overactive compared with what could have been done.

Several arguments had been marshaled against conservation, the main one being that economic activity and energy use were closely bound; hence restricting energy use would probably exacerbate a recession or cause one. In the short term energy and economic activity are indeed closely bound, because machines use energy, and they cannot be replaced overnight. By replacing more energy-intensive machines at the end of their life span with more energy-frugal ones, however, the energy demand could be cut in a matter of decades by 1 or 2 percent per year from what would have been otherwise forecast. With an economic growth rate of 3 or 4 percent per year, energy use might then grow at only half that rate; by the year 2000 the gross national product would have almost tripled, and the energy used per unit of economic output would decline to about 60 percent of its present value. Even so, domestic energy use would have increased by a factor of approximately 1.6, through the diligent exploitation of coal, solar power, light-water reactors and perhaps other technologies. (These numbers are meant only to indicate what many energy planners thought would be possible.)

The fourth major circumstance relates to several aspects of nuclear power itself. First, the U.S. industrial capacity to make light-water reactors is large—perhaps too large. A substantial part of this capacity would be needed to produce some of the base-load electric-generating plants, leaving coal for other electric plants and many other uses. Second, the nuclear-power industry, beleaguered by critics of many persuasions and by a Nuclear Regulatory Commission that it had come to regard as increasingly demanding, also needed some organizational relief. Thus arose the goal of simplifying procedures for fulfilling siting and other licensing requirements. The light-water-reactor industry was to be encouraged by these activities, and electric utilities would be encouraged to "go nuclear" by building light-water reactors wherever such plants were economically attractive.

The fifth circumstance relates to the uranium resources, particularly in the U.S., with which to fuel all those light-water reactors. Each reactor that produces 1,000 megawatts of electric power requires about 5,500 tons of uranium (in the form of uranium oxide) to operate during an expected 30-year life span. The Administration knew that the equivalent of about 680,000 tons of uranium oxide had been located in deposits in the U.S. with characteristics that would make economic recovery possible with current technology, together with an additional 140,000 tons that will be available between now and the end of the century as a by-product of other mineral-extraction operations. It also estimated that roughly another three million tons would be found when it was necessary and that this amount could be produced at a cost of $50 per pound or less. All this uranium would fuel about 700 reactors for their full life span or an even larger number if a full lifetime commitment of fuel were not made for each plant when it began operation.

Considering also that nuclear-power stations take 10 years or more to build and that orders would increase gradually, the Administration judged that adequate nuclear fuel would be available to last several decades into the next century. All this could be done without reprocessing spent fuel from the reactors. Besides, a number of studies had shown that the recycling of uranium and plutonium in the current generation of light-water reactors would at best be only marginally attractive economically and could in fact result in higher fuel-cycle costs than a "once through" fuel cycle.

This brings us to the sixth circumstance: the Administration's concern about the connection between the technology of nuclear-fuel reprocessing and the development of a nuclear-weapons capability. The past few years have brought increasing doubt about the ability of international safeguards to function satisfactorily in a "plutonium economy," that is, one in which large amounts of plutonium would be present at various stages of the fuel cycle in comparatively extractable form. The objective of international safeguards is

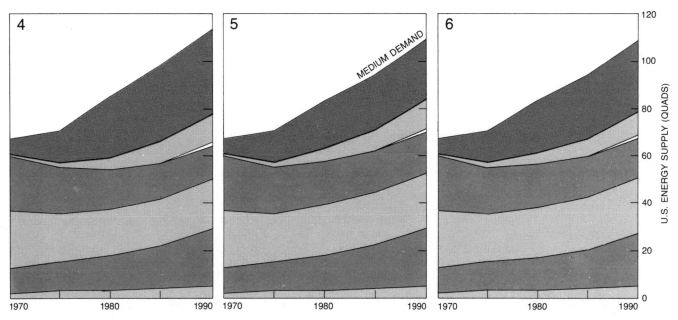

and World Energy Outlook Through 1990, a 939-page report published in November by the U.S. Government Printing Office. The first six scenarios in the set, depicted in the graphs on these two pages, are characterized as follows: high demand, high coal and nuclear supply (*1*); high demand, medium supply (*2*); high demand, low supply (*3*); high demand, low oil and gas supply, high coal and nuclear supply (*4*); medium demand, high coal and nuclear supply (*5*); medium demand, medium supply, also referred to in the study as the "base case" (*6*). All figures are given in "quads": quadrillions (10[15]) of British thermal units. The graphs are continued on the next two pages.

to detect the theft or diversion of nuclear material by nations early enough for diplomatic or other international countermeasures to achieve their objective before the material can be made into an explosive. It was argued, however, that plutonium that had been recovered by spent-fuel reprocessing and then recycled could be turned into an explosive so rapidly after national diversion of the recycled fuel that the ability of the safeguards system to work adequately would be fatally undermined, even if the loss of material were detected. International safeguards do not prevent diversion; they deter it through the threat of timely detection. If plutonium were adopted as a commercial fuel, the deterrence effect of safeguards would be lost. In addition, the presence of plutonium would increase the risks of nuclear terrorism, and there were unresolved questions about the effectiveness of the predominantly national safeguards that would be introduced to deal with this threat.

If, on the other hand, there were no reprocessing, there would be no plutonium either to fuel breeder reactors or to make plutonium-based nuclear weapons; a nation that did not reprocess the fuel from its nuclear-power reactors could not then imperceptibly slip into the position of having a nuclear-weapons capability, and it could not in some temporary passion easily pervert a civilian nuclear-power program. Either it would have to extract almost pure fissionable uranium 235 from natural uranium, an activity that is associated with nuclear weapons and not at all with conventional light-water or heavy-water re-

actors, or it would have to flagrantly set about reprocessing used reactor fuel to extract plutonium. Furthermore, both of these activities are widely thought to be beyond the present capability of any subnational group acting clandestinely.

If the U.S. was mainly worried about the international proliferation of nuclear weapons, why then did it stop the domestic reprocessing of nuclear fuel? The best answer seems to be that in order to argue its case persuasively, it would have, so to speak, to come to court with clean hands.

To be sure, breeder reactors would be delayed, perhaps indefinitely if some better prospect such as economic central-station solar power or controlled fusion came along. Clearly the latter options were to be encouraged. Meanwhile even the U.S. breeder program was to benefit because there would be time to explore a more varied set of technological options, both conceptually and through experiments on a modest scale. Far from being canceled, the breeder program would take on a needed diversity; perhaps more nearly proliferation-proof fuel cycles could be found.

Many doubts about the public acceptability of nuclear power had built up, and the new goals would surely be seen as being responsive to those doubts: no plutonium, no reprocessing, no breeder reactor in this century and so forth. Meanwhile little public concern had yet arisen over the social, environmental or health hazards of coal, the energy-supply option the Administration planned to promote most vigorously.

The Administration's perception of

this complex issue can be analyzed with the aid of a sequential logic diagram, which is useful in clarifying some of the main proliferation-related trains of thought and their impact on international security [see illustration on pages 120 and 121]. Two main paths appear in the diagram. First, there is a horizontal decision path that could be followed by a nation (let us call it Y) that does not now possess the nuclear technologies in question. The central questions affecting international stability are whether or not nation Y decides to develop a general capability with respect to nuclear weapons and, once it has decided to do so, how long it would take to acquire that capability. The second path consists of several vertically arranged inputs to Y's decisions. The U.S. sees the capability to reprocess nuclear fuel as a stimulant to weapons proliferation because Y would then have a source of weapons-grade fissionable material; such accessibility might be instrumental in a decision by Y to acquire nuclear weapons, and in any case once the decision had been made the reprocessing capability would help Y to move toward the right in the diagram: toward the weapons themselves.

The sequential arrangement of events depicted here is too simple. A decision to acquire nuclear weapons might be a prolonged process, and it could be made in parallel with (or could even follow) the acquisition of some of the technological components of a civilian nuclear-power program, which would also be useful for the development of a weapons system. For example, that would be the case with a commercial reprocessing

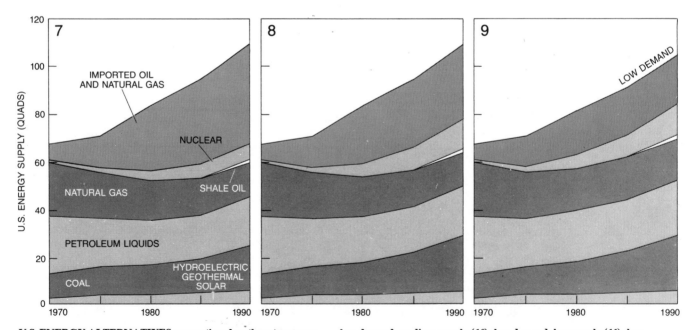

U.S. ENERGY ALTERNATIVES are continued on these two pages. The characterizations of the six remaining scenarios in the Congressional Research Service study are as follows: medium demand, low supply (7); medium demand, low oil and gas supply, high coal and nuclear supply (8); low demand, high coal and nuclear supply (9); low demand, medium supply (10); low demand, low supply (11); low demand, low oil and gas supply, high coal and nuclear supply (12). The authors of the Project Interdependence report point out that even with a low energy-demand projection coupled with high coal use, an expanded nuclear-power capacity, an increase of about 50

plant, assuming that in its normal operating procedure it produces plutonium separated partially or completely from the other constituents of irradiated fuel. Such a reprocessing plant would therefore be ambiguously perceived by observers, even if that were not the intention of a peaceful nation Y. The Administration saw that by blocking the connection at the top of the figure a substantial barrier would be erected against Y's either sidling consciously or sliding unconsciously into a technological competence applicable to the manufacture of weapons.

Of course, the Administration realized that other routes exist whereby Y could obtain fissionable material suitable for weapons. First, it might import the necessary technology from elsewhere, but for some time the U.S. has been actively attempting to close those routes by seeking to persuade the other major suppliers of nuclear materials and equipment (through bilateral channels and in the multilateral forum of the London Suppliers Group of nuclear exporting nations) to exercise restraint in the transfer of "sensitive" items that might offer increased access to weapons-grade material.

All the other routes involve a conscious decision by nation Y and a substantial effort on its part. It could develop its own civilian fuel-reprocessing technology, fully intending peaceful uses only, then have it subverted later after a change of attitude on the part of its government. Alternatively, it could attempt the production of weapons-grade plutonium in research reactors (as India did for its 1974 nuclear explosion)

or in a small clandestine reactor, in either case recovering the plutonium from the irradiated fuel in a small reprocessing plant built expressly for the purpose, a task much easier (and cheaper) than the development of the technology and the construction of the plants for commercial reprocessing. Still another alternative could involve diverting irradiated commercial fuel to a clandestine reprocessing plant from a temporary storage facility, where it might have been awaiting either commercial reprocessing or, in the case of a once-through fuel cycle, ultimate disposal.

Rather than work with plutonium from spent fuel Y might attempt to extract the fissionable isotope uranium 235 either from natural uranium (which contains less than 1 percent U-235) or from the low-enriched uranium used for power-reactor fuel (about 3 percent U-235), concentrating it to, say, 90 percent. For this approach there are several candidate technologies at various stages of development. For the past 25 years practically all enrichment, either for power-reactor purposes or for the production of weapons-grade uranium, has been carried out in the huge gaseous-diffusion plants of the U.S. and the U.S.S.R. Gaseous-diffusion plants do not need to be built on such a heroic scale, however; furthermore, other enrichment technologies now challenge the dominant position of gaseous diffusion. Ultracentrifuge enrichment is being actively developed in several countries and is on the verge of commercialization in some of them. It requires less power than gaseous diffusion and has other advantages as well. These factors,

together with the greater operational flexibility of centrifuge plants, suggest that this technology would offer a smoother path to weapons material, either with a plant built specifically for that purpose or through a facility built initially to fulfill civilian nuclear-power needs. It would be much easier to carry out the adjustments necessary to convert a gas-centrifuge plant from a low-enriched product to a high-enriched military-grade one than it would to similarly convert a gaseous-diffusion plant; alternatively a small string of centrifuges can be used over and over again progressively to enrich single batches of uranium.

Other enrichment technologies that are gaining in importance include the aerodynamic, or nozzle, approach, variations of which are being developed concurrently in West Germany and South Africa, and the laser technique for isotope separation, which is also being pursued in several countries. A pilot aerodynamic-enrichment plant in South Africa may have already been used to produce enough weapons-grade uranium for one or more explosive devices. Furthermore, although work on laser enrichment has so far been limited to laboratory research, enough information has been made available to suggest that the technique might ultimately provide a cheaper and more flexible route than any other enrichment process. All the known methods, at whatever stage of development, require sophisticated technology but nothing that is beyond the capability of many of the more advanced developing nations.

Turning to seemingly more bizarre ac-

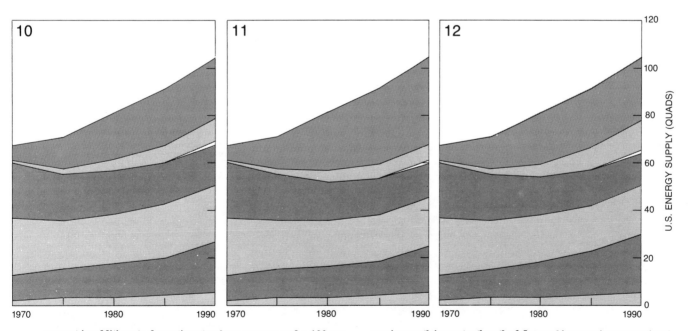

percent in additions to domestic natural-gas reserves and a 100 percent increase in additions to domestic oil reserves compared with the preceding decade (scenario No. 9) the U.S. would still have to import close to 20 quads of energy by 1985, equivalent to almost nine million barrels of crude oil per day. On the other hand, if the rate of eco- **nomic growth is greater than the 3.5 annual increase in gross national product projected in the base case and all areas of domestic energy supply turn out to be less productive than expected (scenario No. 3), the U.S. could be importing 17.7 million barrels of crude oil per day in 1985. The role of imported oil as a "swing" fuel (color) is evident.**

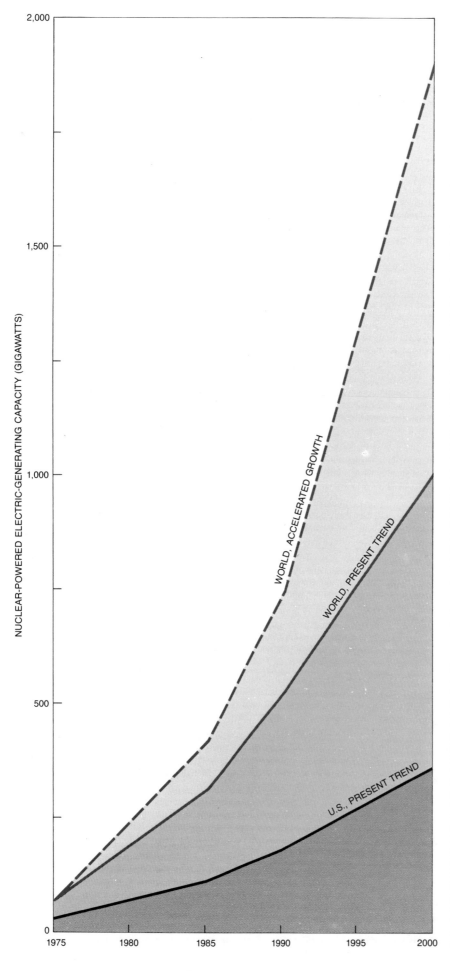

NUCLEAR-POWERED ELECTRIC-GENERATING CAPACITY (GIGAWATTS)

WORLD, ACCELERATED GROWTH

WORLD, PRESENT TREND

U.S., PRESENT TREND

tivities, country *Y* might employ agents to steal material from abroad or buy it in a black market, an open market or an intermediate "gray" market. It could receive such material as a gift or a loan from another government. It might steal an assembled weapon or even be given one. None of these activities can be excluded, and some may be more likely to occur than those we described above.

A really effective barrier to weapons proliferation would involve blocking all the lines marked with a black bar in the diagram. That is impossible, but the U.S. is only trying to make weapons proliferation substantially more difficult. Combining technological denial with an assortment of incentives related to the supply of enrichment services for light-water-reactor fuel (to be used by *Y* under tightly specified conditions) would, the Administration thought, significantly increase international stability.

Several other decision paths of a different nature exist, on which the U.S. Government has only an indirect influence; the principal ones are marked with gray bars. Why does *Y* decide to acquire a weapons capability in the first place? Why does it actually build nuclear bombs? Why might it use them? The answers to such questions depend on many things, and a long-term policy not to proceed at any one of these decision points makes all the technological elaborations irrelevant. The converse, however, is not necessarily true. As we implied when we were discussing reprocessing plants, a decision to proceed might be heavily influenced by the technological capability in place at that time.

When the U.S. policy described here is viewed from other countries or other domestic vantage points, it looks quite different. The main line of devel-

POTENTIAL TRENDS in the growth of nuclear power are shown through the end of the century for both the U.S. and the world (excluding China, the U.S.S.R. and the other countries of Eastern Europe). For the world the lower estimate is based on present trends in energy utilization and supply, including delays in the construction of new nuclear reactors, and assumes a continuation of these trends. The "accelerated growth" estimate assumes that the goals of ambitious nuclear programs will be met and that the world will return to higher rates of energy growth. The "present trend" estimate must be regarded as the more realistic of the two and may itself be too high. The data for these curves were obtained from a recent joint report by the nuclear-energy agency of the Organization for Economic Cooperation and Development (OECD) and the International Atomic Energy Agency (IAEA). U.S. projection is based on recent estimates by the Department of Energy. All such estimates must be viewed with caution, the authors point out, in view of the many uncertainties discussed in their article.

opment—in particular decisions by nation Y—can best be discussed in relation to a second diagram, which is similar to the first but starts with a U.S. decision to act restrictively and includes several additional logic paths. To understand its full significance we must start farther back, with the electric-power sector.

The U.S. electric-power industry abhors uncertainty about the future, for several reasons. One is the long time needed to construct new facilities (10 years is typical) and long expected life of these facilities (40 years, say). The industry also needs a stable fuel supply and is required by law to provide reliable service. In this regulated industry justifiable costs can be charged to the consumer. The present program of the Administration increases uncertainty about the future of nuclear power for several reasons. First, the decision against reprocessing spent fuel has raised fears, not yet completely alleviated by the Federal Government, that the electric utilities may in practice be left holding the spent fuel for a long time (for example by long-drawn-out court challenges to the Environmental Impact Statement for a Federal spent-fuel storage facility), a very unappealing prospect to them.

In addition, conflicting Federal opinions about the acceptability of nuclear power make the electric utilities both suspicious of Government motivations and better targets for anti-nuclear-power groups. Many electric utilities also fear that the Nuclear Regulatory Commission will order expensive, and in their view capricious, retroactive modifications to existing nuclear plants, in spite of current efforts to modify the Commission's legislative foundation. Many experts, concerned at the inherent uncertainty associated with the four-million-ton estimate for U.S. uranium resources that could ultimately be made available, feel that a "prudent planning estimate" for the purpose of setting nuclear-power policy should be appreciably lower. A National Academy of Sciences resource-evaluation group recently estimated that 1.8 million tons is all that is likely to be mined in the U.S. by the year 2010 at a cost of $30 per pound or less, even with a Government policy of maximum stimulation.

Furthermore, doubts have been expressed as to whether the U.S. uranium-supply industry, itself troubled by uncertainty about the size of the market for its product, will be prepared to invest in exploration, mining and ore-processing-plant construction at levels that will be sufficient to fuel a growing number of nuclear-power plants. Part of the uncertainty permeating the electric-utility sector stems from concern over the availability of nuclear-fuel supplies, so that the problem exhibits circular characteristics; it is also aggravated by the fact that the strength of the uranium supply industry's commitment to keeping power reactors adequately fueled is less than the utilities might find desirable. For example, it has been estimated that within a few years petroleum companies will own about 40 percent of all U.S. ore-processing capacity and as much as 50 percent of low-cost U.S. uranium resources. In short, the uranium suppliers do not constitute an industry "captive" to the electric-power sector. (Indeed, increasing corporate diversification in the various energy supply industries has led to the suggestion that the reverse might be true.) As a result uncertainty arising from the Administration's program may be compounded, unwittingly or otherwise. Similarly, each U.S. manufacturer of nuclear reactors has 75 percent or more of its business elsewhere (for example in other power systems), and the nuclear business is not essential to it. In a period of rising costs, large-scale cancellations of orders and excess production capacity, the business appears less than inviting.

The last point is worth further comment. The U.S. manufacturers of light-water reactors could turn out between 20 and 30 nuclear-power systems a year. The Administration has estimated that full implementation of its national energy policy would lead to an installed nuclear electric capacity of more than 300,000 megawatts by the year 2000. With 50,000 megawatts already installed, and a further 25,000 megawatts scheduled for completion by 1980, there would be, on the average, a dozen or so reactor systems completed each year for the last 20 years of the century, a situation that implies a large-scale restructuring of the reactor-manufacturing industry sooner or later.

Compounding these difficulties, the electric-utility sector suffers the additional one of raising enough capital. One cause is the host of uncertainties we have described. Another involves the general flight during recent inflationary times from long-term investment; the rate of economic return from the regulated utility industry has become unattractive, a circumstance that also affects other generating plants, particularly those that burn coal.

The upshot of all this is the paradoxical situation that although the existing nuclear reactors run pretty well and deliver economical electric power in many parts of the country, the nuclear industry in the U.S. may nonetheless be close to collapse. The proximate cause is a movement by the electric-utility industry and manufacturers away from nuclear power as they attempt to reduce their own institutional uncertainty, but deeper causes drive these changes and are coupled with the Administration's attitude toward nuclear power. This train of thought points toward the second possible motivation mentioned above: that the Administration, through internal indecision, is incapable of acting to prevent the nuclear industry from collapsing. The indications, however, are ambiguous.

What then is the U.S. electric-utility sector likely to do? The conventional option is coal, with the Administration's apparently enthusiastic backing. Oil and natural gas are expensive and in such uncertain supply that the Administration has submitted legislation prohibiting all new power plants from burning them, with only limited environmental and economic exceptions. Other legislative provisions would, through taxation and prohibitive clauses, encourage utilities not to burn oil and gas in existing facilities and to convert them to coal.

But what if the coal cannot be mined, transported and burned in time, and in socially accepted ways? Even before the Administration announced its national energy policy widespread doubts had grown about the wisdom, or even the possibility, of increasing coal production very quickly. In particular, can a goal of increasing coal production from its current level of 650 million tons per year to a projected total of 1.25 billion tons per year by 1985 actually be achieved? Industrial problems associated with such an expansion, land-use problems, states' policies and obstacles created by the Administration's environmental policy for coal have all been repeatedly raised as evidence to suggest that there will be hesitation on both the supply and demand sides of the coal industry. Even coal transportation, which currently accounts for about 30 percent of the U.S. rail tonnage, will be difficult for the disheveled railroad industry.

The environmental problems with coal appear to grow with time and increased understanding. The comparatively large amounts of disturbed land, the chemically and biologically active complex molecules present in coal and produced by the burning of it, and the ubiquitous nature of these effects create difficulties at local and national levels. On a global scale potentially the most serious long-range environmental impacts resulting from the large-scale burning of coal (or indeed of any fossil fuel) may arise from the effects of the increased concentration of carbon dioxide in the atmosphere. At this stage the problem is not well understood, and the potential contribution of planned U.S. coal-burning activities is therefore also shrouded with uncertainty. Nevertheless, in this problem area and others the prognosis looks more serious as more information accumulates. The U.S. electric-utility sector, generally aware of these difficulties, looks on coal with increasing anxiety.

Irresolution about nuclear power, increasingly apparent difficulty with coal, a partial ban against oil and a half-hearted attitude toward energy conservation make an impossible combination; something has to give. If the electric-utility industry waits a few years for public debate to resolve these issues, the concomitant pressure for rapid and comparatively pollution-free installations will drive it toward oil-burning plants. That would be doubly disastrous, because the conversion of some transportation, industrial, commercial and domestic systems from oil to efficiently used electric power, based on coal or nuclear fuels, is seen as a way of reducing oil imports. If the electric utilities are unable or unwilling to provide the means for this substitution, oil consumption will continue to exceed Administration targets.

Thus imported oil may once again fill the role the Administration has sought to prevent it from playing: the "swing" fuel that satisfies increased energy demands. Reinforcing this trend is the present concentration on increasing domestic production, which will have the effect of keeping up oil-based activities that must surely in the next decade or two be fed by imports.

The international importance of President Carter's attempts to reduce oil imports, and the dangers implicit in failing to do so, cannot be overemphasized. Today the U.S. imports nearly half of its immense consumption of 17 million barrels a day, an amount equal to almost a quarter of all internationally traded oil. A reduction to the Administration's stated import target of less than seven million barrels a day by 1985 from levels that would otherwise be reached if present trends are allowed to continue unabated would save annually an amount of crude greater than the current oil imports of Japan or half those of Western Europe. If the target is not met and the U.S. imports increase, the competitive pressures for oil, particularly Middle East oil, may reach dangerous proportions even without another politically motivated interruption in supplies. No other nation in the world (with the possible exception of Saudi Arabia, with its vast potential production capacity and its role of "swing" producer) can exert such an influence on the world's energy outlook through its domestic policies.

After this analysis, we can now enter the logic diagram on pages 122 and 123 at the point where the U.S. states its nuclear-policy position and then explore the international consequences.

All the foregoing trains of thought have been quite apparent to both developed and developing nations. Japan and most of the advanced industrial nations of Europe have meager coal or oil supplies themselves, relatively speaking; even North Sea oil harvested at the maximum planned capacity will supply only about 20 percent of Western Europe's needs. Thus all those countries, facing their own difficulties and the distinct possibility of a continuingly gluttonous

U.S., see an increased incentive to push ahead with their own nuclear programs, including reprocessing and breeder reactors. Further encouragement to do so seems to be arising from an unintended source. Along with the U.S., Australia (currently not a uranium producer but potentially one of the world's two or three major exporters of uranium within the next few years) and Canada (the world's biggest uranium exporter) have also recently imposed stringent proliferation-related controls on their exports. In all three cases these controls include the requirement that consent must be obtained from the exporting nation before any of the fuel can be reprocessed. In at least one case deliveries of fuel have been delayed pending agreement to these and other conditions by importing nations. Although the controls have been implemented with the intention of creating a more rigorous nonproliferation regime, for the uranium-deficient, fuel-supply-sensitive Japanese and Europeans such manifestations of external political involvement in their domestic fuel-cycle operations, together with the unsettling prospect of further mercurial behavior by their suppliers in the future, may ultimately act to increase the vigor with which these nations set about reducing their dependence on fuel imports by the use of plutonium and breeder reactors. According to statements of intent from many of them, that is already happening.

What will leaders in developing countries see? They will see a rich U.S. liable to forgo nuclear power, in spite of what it might say about preserving light-water reactors. They will see increased short-term pressure on the world's supply of oil, inevitably resulting in shortages and still higher prices. They will see pessimistic projections for world petroleum and gas resources that will allow neither continued profligate use by the industrialized nations nor any chance for their own countries to follow a development path anything like that followed by their predecessors. They will see an offer of nuclear-fuel supplies that binds them to the goodwill of the U.S. and other developed countries for even limited nuclear assistance. (The recent suggestion of an international nuclear-fuel bank is a partial response to this point.) They will see the growing appearance of a world divided into an oligopoly of developed states that turns into an oligarchy as nuclear power becomes more important throughout the world, and a coterie of less developed nations that must fall farther and farther behind. (This latter impression becomes reinforced by U.S. actions that treat certain of its industrialized trading partners as "exceptions" in view of their continued insistence on the need for reprocessing and breeder reactors in their countries without delay.) And they will see little promise of help from the U.S. to

become truly independent in terms of nuclear power.

The inevitable result will be increasing distrust of the U.S. and a growing sensation of unwanted dependence. Both developed and developing nations share these feelings of insecurity. Considering only uranium enrichment as an example, would a U.S. offer of more enrichment services suggest an extension of U.S. market control of enrichment supply, which in turn would suggest increased dependence on U.S. whims? Troubled in large part by the somewhat fickle nature of the decision-making process governing U.S. exports, the out-

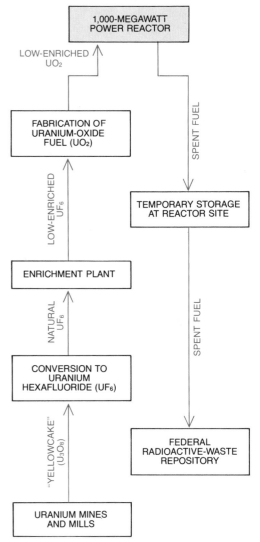

THREE NUCLEAR-FUEL CYCLES suitable with conventional light-water reactors are shown in these simplified diagrams, adapted from a recent report to the American Physical Society by its study group on nuclear-fuel cycles and waste management. In the prevailing "once through" approach (*left*) the spent-fuel rods, which still contain an appreciable amount of fissionable isotopes (principally "unburned" uranium 235 and plutonium 239 produced by the transmutation of the urani-

side world, including the less developed countries, sees more incentive to set up its own enrichment programs and to go nuclear with or without U.S. assistance.

Both developed and developing countries also share other reactions. Among both groups, for example, there are suspicions that the U.S. program is really designed to improve the sagging fortunes of U.S. nuclear exports: either to increase the attractiveness of U.S. light-water reactors or to curb the global movement toward plutonium and the breeder reactor until U.S. technology in these areas has caught up with capabilities in Western Europe. It has also been

pointed out that the exemplary nature of the U.S. decision to defer indefinitely the reprocessing of its own commercial power-reactor fuel was compromised from the outset by the fact that fuel reprocessing in connection with the U.S. weapons program would continue as before. Related comments address the entire network of U.S. weapons-manufacturing activities and deployed weapons, suggesting that the scale and wide distribution of this system presents a more attractive target to would-be proliferators than a commercial plutonium fuel cycle would. Furthermore, some observers have speculated that the sub-

sequent decision to use gas-centrifuge technology for the next increment of U.S. enrichment capacity rather than a more proliferation-resistant gaseous-diffusion plant compounds the already present inconsistencies.

To be sure, some of these reactions are mere rhetorical flourishes; nevertheless, they may still have wide-ranging international reverberations. Moreover, many of the reactions are contradictory. How can a rich U.S. liable to forgo nuclear power also be attempting to increase its share of nuclear exports? How can one explain the fact that some of the nations where complaints about the in-

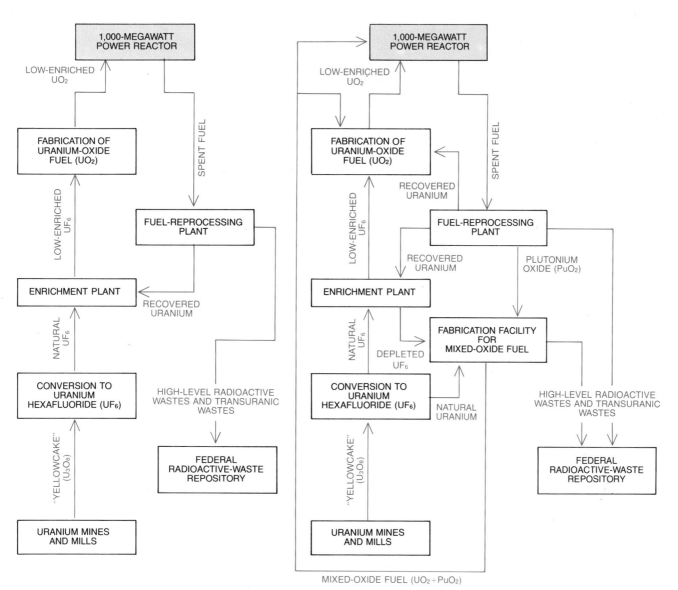

um-238 nuclei in the fuel are disposed of without reprocessing; disposition of the spent fuel can in principle be either temporary or permanent. In the uranium-recycle option (*center*) the spent fuel is reprocessed to recover only the residual uranium, which can then be enriched in the fissionable isotope U-235 or used as it is to replace some of the virgin natural uranium in the fabrication of new fuel assemblies. In the uranium-and-plutonium-recycle option (*right*) the spent fuel is reprocessed to separate both uranium and plutonium from the wastes. The recovered plutonium can then be combined with uranium having a very low concentration of U-235, in effect substituting the plutonium for some of the U-235 in the normally low-

enriched fuel. Useful mixed-oxide fuels can be made by combining plutonium with uranium derived from a number of different sources, including the normal low-enriched uranium product from an isotope-separation plant, the uranium recovered from spent fuel or the depleted "tails" from a uranium-enrichment plant. It has been estimated that with both uranium and plutonium recycling the industrial operations required to supply enriched uranium could be reduced by about 20 percent in the year 2000 compared with what they would be for either the uranium-recycle or no-recycle options. This saving would of course require the introduction of the costly and complicated fuel-reprocessing and mixed-oxide fuel-fabrication operations.

consistency of U.S. policies toward domestic and military reprocessing have been heard are also those that rely most heavily on the presence of the U.S. nuclear deterrent for their defense? Such contradictions, however, do no more than mirror the ambiguities and contradictions we have recognized in the U.S. policy, as it attempts to strengthen the barriers between peaceful and violent uses of nuclear energy and simultaneously wrestles with an immense and growing demand for energy, both domestic and international.

The Nonproliferation Treaty, to which more than 100 nations are now parties, embodies an internationally negotiated agreement on the framework in which the energy v. proliferation enigma should be resolved. In it the non-nu-

clear-weapons states party to the treaty undertake not to develop or otherwise acquire any form of nuclear explosive and to accept international safeguards on all peaceful nuclear activities. In return for this commitment the right of all parties to the treaty to develop and use nuclear energy for peaceful purposes is affirmed, as is the right to participate in exchanges of equipment, materials and technology for the peaceful use of nuclear energy.

The restrictive export policies of the U.S. (and of other major nuclear suppliers) are viewed in many parts of the world as extending the inequalities that have always been inherent in the Nonproliferation Treaty between nuclear-weapons states and non-nuclear-weapons states. The new expression of these

inequalities is the attempt to influence criteria for the international distribution of certain "sensitive" peaceful technologies, particularly reprocessing. Implied in this policy is a redrawing of the line separating peaceful uses of nuclear energy from violent ones, and therefore a redefinition of proliferation. Traditionally the latter had been defined as the acquisition of nuclear weapons. Now, however, the new U.S. position is being interpreted as an attempt to redefine proliferation as the capability of acquiring nuclear weapons. Had this always been the case, it is argued, negotiating the Nonproliferation Treaty would have been impossible in the first place.

We make no attempt to determine whether in fact the U.S. would be failing to comply with its international legal

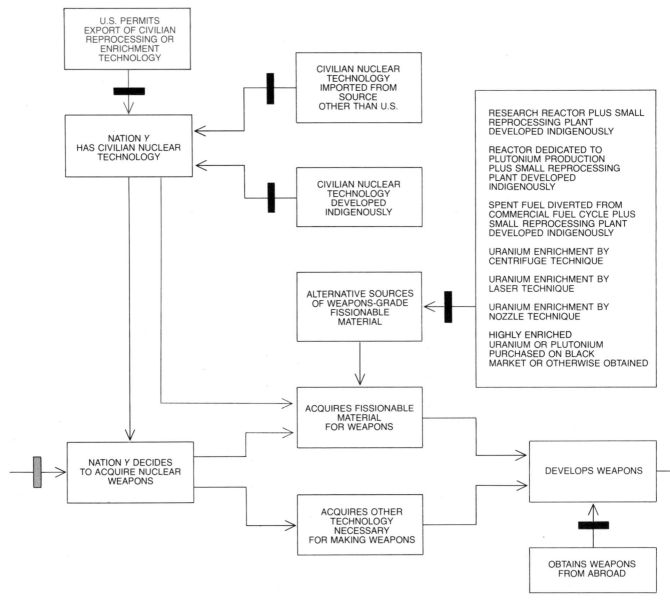

NUCLEAR-PROLIFERATION SCENARIO currently perceived as being worrisome by the U.S. Government is illustrated in the form of this sequential logic diagram. The main horizontal decision path shows the series of steps that could be followed by a non-nuclear-weapons nation, designated *Y*, to acquire nuclear weapons and to use them. The vertical paths show several possible inputs to *Y*'s decision. For example, the U.S. sees the acquisition of a nuclear-fuel-reprocessing capability as a stimulant to weapons proliferation. Accordingly

obligations as a party to the Nonproliferation Treaty by implementing its proposed export criteria. We do observe, however, that the loss of confidence in the effectiveness of international safeguards that has taken place in the U.S. is reflected in many non-nuclear states by a corresponding loss of confidence in the ability of the Nonproliferation Treaty to provide an acceptable legal framework for the international distribution of peaceful and military applications of nuclear energy. In such circumstances the fabric of the global nonproliferation regime is inevitably weakened.

All the considerations we have discussed here show up as destabilizing routes in our second proliferation scenario. Not only does country Y find logical incentives to install domestic nuclear-fuel facilities, but also it perceives a world more fragmented and less secure. Feeling less secure itself, it naturally imagines others feeling the same way and hence it must increase its own security unilaterally. Escalation of uncertainty leads to escalation of international instability; a program originally intended by the U.S. to decrease the dangers of nuclear proliferation inadvertently has the opposite effect. Meanwhile the U.S. isolates itself from the mainstream of world nuclear policy, and its ability to favorably affect that policy diminishes.

More caveats and auxiliary views remain to be displayed. None of this analysis is meant to overstate the role of nuclear power in solving the world's energy problems. That mistake has been made too many times before. Some of the problems currently facing the nuclear-power industry in non-Communist developed countries can probably best be understood in terms of a backlash against earlier technological overoptimism. For the majority of the less developed countries nuclear-generated electricity cannot play a significant part in meeting energy requirements for a long time. Costs have risen alarmingly, and besides, the type and scale of the energy supplied by currently available nuclear-

power stations seem less compatible with the energy-demand structure in many of these countries than the output of other energy-supply systems.

Moreover, in some developing countries where nuclear power is intended to play a major role the overall development targets frequently appear overambitious and unlikely to be realized. Fears have been expressed that nuclear-power technology could critically exacerbate rivalries among the various political, industrial and technocratic elites and increase the gap between such elites and the remainder of the population. Until now suppliers have not discernibly modified any "hard sell" policies because they might ultimately contribute to domestic political instabilities with unpredictable international consequences, and it is unlikely that such self-restraint will be shown in the future. The point here is not, of course, that the industrialized supplier nations should decide what is good for the development of the poorer countries and impose export restrictions accordingly but that nuclear technology may be "sensitive" for many reasons other than the increased access to weapons-grade material it may provide.

We conclude, therefore, that the conventionally defended analyses have been inadequate. The various original motivations virtually disappeared from our discussion, and rightly so; they were more nearly goals than real policies. Both are necessary; to neglect the hard work of developing policy causes much trouble because then the original vision, however high-minded, is washed away by a sea of events, and only consequences remain.

What to do? In answering this question we have had to assume that many other issues related to nuclear power can be resolved. To us the three largest of those issues seem to be reactor safety, the management of nuclear wastes and the prevention of subnational nuclear felonies. Although we have not dealt with these matters, we realize their gravity. Our analysis and recom-

mendations would be irrelevant, however, only if both the U.S. and almost all other countries opted out of nuclear power. Nuclear power might disappear in the U.S., but neither present reactors nor breeders will go away in many other places. If the activities of Western Europe and Japan are unconvincing, one need only consider the U.S.S.R. and its Eastern European allies; they also develop nuclear power, with the most sensitive activities being reserved to the U.S.S.R. The U.S. must stop acting from time to time as though nuclear power was about to go away, or as though its disappearance would have little consequence.

The Administration has begun to discuss some of the necessary changes: for example a gradual shift to a more flexible policy, more emphasis on providing an assured nuclear-fuel supply and a suggestion to set up an international storage facility for spent fuel. Owing to significant reductions in projected overall energy demand and particularly in forecasts of installed nuclear capacity, the next few years seem to us a period of grace, perhaps overperceived in the U.S. and underperceived abroad, during which fundamental repairs can be made to the fabric of nuclear goals and policies. The period may be a decade, but it can hardly be much longer, being limited by the consequences of the inexorable pressure on other energy resources. The time is technically long enough to develop variations in the current fuel cycle, perhaps long enough to devise a brand-new one and even reactors and other facilities to use the fuel, but such activity would require far firmer decisions and much prompter (and more expensive) implementation than we have seen. Both the preparation of fissionable fuel from new or reprocessed material (the "front end" of the fuel cycle) and the reprocessing of spent fuel must be carefully considered, not just the latter. As we have said, we consider the front end of the fuel cycle, the enrichment of natural uranium for example, to be a sensitive proliferation issue. The exis-

DEPLOYS WEAPONS → USES WEAPONS → WIDESPREAD USE OF NUCLEAR WEAPONS

the Administration has sought to restrict the export of commercial reprocessing technology as a way of blocking this access route to the acquisition of nuclear weapons. Other routes exist, however, whereby Y could gain access to weapons-grade material; some of these possible routes are listed in the large box. In order to erect a really effective barrier to weapons proliferation, the U.S. would have to block all the lines marked with a black bar. Decision paths on which the U.S. can have only an indirect influence are marked with gray bars.

tence of about 200 power reactors around the world working on the present fuel cycle must also be considered, and the opportunities for technological innovation that might be applied to them are limited.

All these things make large demands on one decade or a little more. So do institutional accommodations among nations, not only with respect to the currently developed nuclear-fuel cycles but also to many other things. Time is short, whether for technological modification or for international institution-building. Whatever the outcome of the former, the latter is an indispensable part of efforts to deal with the problems of nuclear-weapons proliferation and energy scarcity. Although the Administration's proposals have created many problems, they have succeeded in injecting a new sense of urgency into the

situation. It is essential that this asset not be allowed to evaporate.

Regarding present fuel cycles and other matters directly related to nuclear power in the next decade or two we have five main recommendations to offer:

1. Nuclear power should be kept alive in the U.S. at least as a long-term "insurance" option, and that means not only the continued development of light-water reactors but also progress toward de-

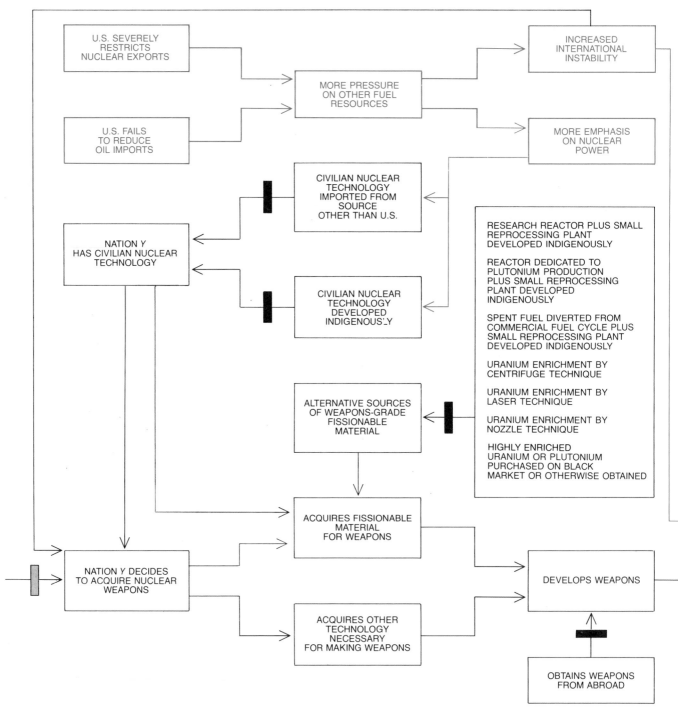

MORE COMPLEX SCENARIO is needed to portray more realistically the likely effects of an overly restrictive U.S. nuclear-export policy on the entire problem of international stability and the proliferation of nuclear weapons. The main decision path followed by **nation *Y* toward the acquisition of a nuclear-weapons capability is the same as it is in the preceding diagram, as are a number of other elements in the diagram. In this case, however, the vertical inputs to *Y*'s decisions start with the new U.S. restrictions on nuclear exports,**

veloping a viable breeder reactor. Central-station solar power and controlled fusion are only long-term possibilities, oil is only a short-term source of energy and we have little faith in coal for the long term.

2. To reduce uncertainty for the U.S. electric-utility industry and others the Federal Government should take several steps. First, it should reaffirm that the reprocessing of spent nuclear fuel is being delayed but not abandoned. Second, the Federal Government should assure the electric utilities of a review of national policy on reprocessing as the debate about it matures, and certainly within five years; that would include an assessment of the projected uranium supply, which would draw on the current national uranium-resource evaluation and other programs. Third, the complexity and prolixity of the licensing process for nuclear-power plants should be eased by making it more difficult for license applications to be recycled again and again to the Nuclear Regulatory Commission. The nuclear licensing bill currently under consideration appears to be facing formidable obstacles at several stages of the legislative process, and in any case it seems to address these issues only partially in its present form. Finally, the Federal Government should take on the entire burden of managing spent fuel, and guarantee to take responsibility for the fuel reasonably soon after its discharge from power reactors. That includes spent-fuel reprocessing, if and when a decision comes to do it. No other sector has an adequately long time perspective to plan and operate the appropriate facilities. In particular, the chemical industry, on which the task might otherwise be expected to fall, traditionally expects a payback on investment in a very few years, and therefore discounts far-future profits too much to match the long-term nature of the tasks, particularly waste handling and storage.

3. The U.S. should offer to explore with other nations the costs and benefits to the international community of completing the reprocessing plant now sitting idle at Barnwell, S.C., and operating it as an international facility. The principal objectives of such a project would be to gain experience with commercial reprocessing technology, to assess the effectiveness of international safeguards and to demonstrate the institutional viability of international cooperation in the provision of fuel-cycle services.

4. Efforts to increase the security of the international supply of uranium and enrichment services should be intensified. Domestically, differences among the various branches of the Government should not be allowed to interfere with the pivotal task of reestablishing the U.S. as a reliable supplier of enriched uranium fuel.

5. In all these activities we note the need for an international agency. We see none better prepared than the International Atomic Energy Agency (IAEA), and we believe it should be greatly strengthened so that it can continuously inspect sensitive facilities. The answer does not, however, lie in the mere strengthening or even proliferation of agencies. For example, restrictions on the export of appropriate nuclear systems may undermine the Nonproliferation Treaty, and the IAEA can do nothing about it.

Beyond all these issues we see others seeming to stand out in the distance. First, it is noteworthy that the diagrams we have drawn differ from the conventionally discussed diagrams of causes and effects. Our discussion has been almost entirely international, as befits the problem. The U.S. approach has been too self-centered, insufficiently sensitive to the problems of other nations and lacking in awareness of its own potentially disruptive character.

Near the beginning of this discussion we mentioned that the Administration has attempted to breathe new life into the larger issues presented by nuclear weapons. If governments and people are so concerned with the risks of future proliferation, how much more should they worry about the huge numbers of nuclear weapons already deployed? One who lives on the edge of an abyss should not squander his effort avoiding small ditches. The real threat of nuclear weapons is seen once again, more clearly than before, in the illuminating perspective provided by the juxtaposition of thousands of existing megatons on the one hand and a few hypothetical kilotons on the other.

This brings us to the more general question of international peace and stability. In the worldwide search for routes to a juster and more sustainable society it has become clear to many observers that a peace in which the world is divided ever more rigorously into haves and have-nots is neither just nor likely to be very sustainable, whether the basis for division is social, economic or (as here) seemingly technological. Such a division not only defeats itself in the long run; even worse, it is wrong.

We propose that the real long-term solution both to the nuclear-power problem and to the larger problems of international instability lies not in fostering division but in its opposite: mutually cooperative international interdependence. Since nations must depend on one another, they lose more by going separately than by staying in partnership. Our analysis shows that this partnership must include the developing countries, since many of them, if they are excluded, are capable of upsetting the international order through the acquisition of nuclear weapons and other acts.

All of this will not be easy, but other approaches have yielded nothing but unstable arms escalation. The partnership should logically involve food, health care and many other sectors where the U.S. can make valuable contributions. Only in that way will we have a chance of answering constructively the question that can no longer be put aside: Why do people want to make nuclear weapons in the first place?

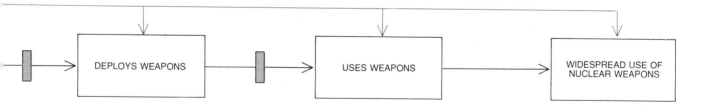

coupled with the failure of the U.S. to reduce significantly its imports of crude oil. Several additional logic paths, shown in color, represent the possible sequence of events that could result in added incentives for nation Y not only to push ahead with its own civilian nuclear-power program, including uranium-enrichment technology, spent-fuel reprocessing plants and breeder reactors, but also perhaps to respond to increased international instability arising out of growing competition for energy resources by joining the nuclear-weapons "club."

10

Superphénix:
A Full-Scale
Breeder Reactor

by Georges A. Vendryes
March 1977

The decision has been made to begin the construction in France of a 1,200-megawatt breeder-reactor power station. The joint European project will be the prototype of future nuclear plants

The need to resort to nuclear fission to help meet the anticipated world demand for energy over the next few decades is widely, if not universally, recognized. What is often not appreciated sufficiently, however, is the fact that if the construction of new nuclear power plants is limited to the same basic types of reactor generally in service today, the respite gained will be only a brief one. Most experts agree that at current prices the world's economically recoverable uranium reserves are inadequate to ensure a lifetime supply of fuel for light-water nuclear reactors built after the year 2000. This means that unless uranium is used in a more efficient way than it is in such reactors, it will turn out to be an energy resource not very different in scale from oil.

Only breeder reactors—nuclear power plants that produce more fuel than they consume—are capable in principle of extracting the maximum amount of fission energy contained in uranium ore, thus offering a practical long-term solution to the uranium-supply problem. Breeder reactors would make it possible to obtain some 50 times more energy from a given amount of natural uranium than can be obtained with present-day light-water reactors. Hence the minimum uranium content of economically recoverable ore could be significantly lowered. For these two reasons (of which the second is by far the more important) the useful supply of natural uranium could be greatly enlarged. Uranium would then constitute a virtually inexhaustible fuel reserve for the world's future energy needs.

Recognizing the importance of these considerations, a number of nations have undertaken intensive research programs aimed at developing an economically competitive breeder reactor before the uranium-supply situation becomes critical. Last fall a consortium of major European electric-utility companies, acting through a joint subsidiary, decided to start the construction of a 1,200-megawatt breeder-reactor power plant at Creys-Malville in France. The new full-scale breeder reactor, named Superphénix, will be described here. First, however, it is necessary to explain just what is meant by the term breeding, which serves to characterize the operation of such plants.

Two types of heavy isotope are present in the active core of every nuclear reactor. One type, called the fissile (or fissionable) isotope, undergoes most of the fission reactions and is the source both of the heat energy released by the reactor and of the neutrons that sustain the chain reaction in the core. The only fissile isotope that exists in nature is uranium 235, which constitutes .7 percent of natural uranium; the nonfissionable isotope uranium 238 accounts for the remaining 99.3 percent. Two other fissile isotopes, plutonium 239 and uranium 233, are expected to play an increasingly important role in the future as substitutes for uranium 235.

The second type of heavy isotope in the core of every reactor is said to be fertile; it undergoes practically no fission reactions, but by capturing a stray neutron a fertile nucleus can be transmuted into a fissile nucleus at the end of a series of radioactive disintegrations. A typical example of a fertile nucleus is uranium 238, which is transmuted by neutron capture into fissile plutonium 239. Similarly, fertile thorium 232, the only form of thorium extracted from the ground, can be transmuted into fissile uranium 233.

In every nuclear reactor, as the fissile nuclei are being consumed new fissile nuclei are being created by the transmutation of fertile nuclei. Most reactors in operation today, however, use either ordinary (light) water or deuterated (heavy) water to moderate, or slow, the neutron flux in the active core. In such a slow-neutron reactor it is impossible to produce as many fissile nuclei by neutron capture as are consumed. As a result the proportion of fissile nuclei in the fuel quickly falls below a certain minimum level, and the depleted fuel must be removed from service with most of the fertile nuclei still not transformed. A set of special conditions must be satisfied to raise the breeding ratio (the ratio of the amount of fissile material produced from fertile material to the amount of fissile material consumed during the same period) to a value greater than 1. The most favorable conditions for breeding are obtained when fissile plutonium 239 and fertile uranium 238 are used together in a fast-neutron reactor, in which the neutrons from the fission reactions are not slowed down by a moderating substance such as water between the time they are emitted by one fission reaction and the time they cause the next reaction. Only under these conditions can the breeding ratio be raised to a value significantly higher than 1.

In a fast-neutron reactor the initial fuel load of plutonium is needed to start the fission chain reactions and the production of power. During this period plutonium is bred from natural uranium (or from uranium depleted in uranium 235) in the reactor core and in the surrounding "breeding blanket." When the fuel subassemblies that make up the core and the blanket have undergone prolonged neutron irradiation, they must be reprocessed chemically in order to separate and remove the fission products. In each reprocessing operation more plutonium is recovered than existed at the start of the irradiation. The excess plutonium is set aside and is replaced in the reactor by natural or depleted uranium. Everything proceeds as though the reactor were consuming only natural or depleted uranium and simultaneously furnishing new plutonium as a by-product of the plant's operation.

The time required for a breeder reactor to produce enough plutonium to fuel a second identical reactor is called the reactor's doubling time. This time factor is inversely proportional to the reactor's breeding ratio. In the future it is expected that breeding ratios on the order of

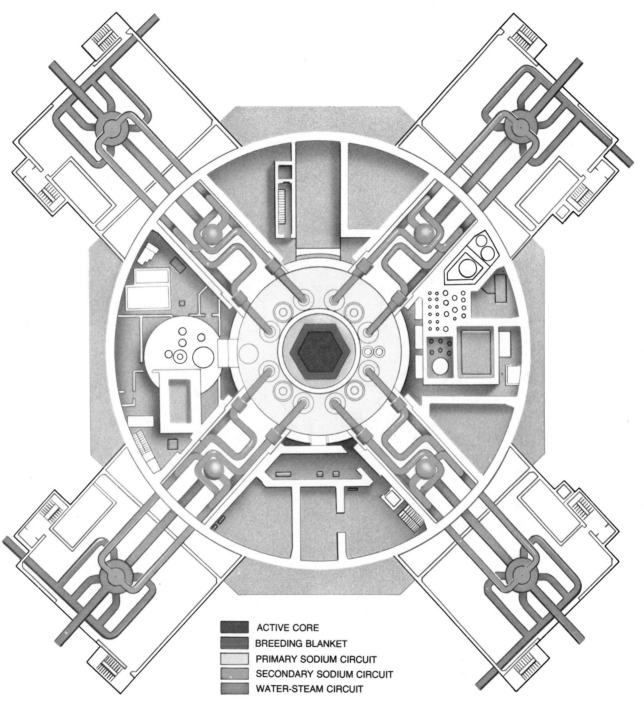

ACTIVE CORE
BREEDING BLANKET
PRIMARY SODIUM CIRCUIT
SECONDARY SODIUM CIRCUIT
WATER-STEAM CIRCUIT

HORIZONTAL SECTION of the proposed Superphénix breeder-reactor power station shows the overall layout of the plant, which will consist essentially of a large circular reactor building with four steam-generator buildings laid out radially around it. The central reactor building, which is designed to house all the plant's nuclear components, will be built of reinforced concrete one meter thick; the building will have an inside diameter of 64 meters and a height of about 80 meters. Each steam-generator building will serve one segment of the secondary sodium circuit. (The associated turbogenerator building is not shown in this view.) The site selected for Superphénix is at Creys-Malville in France. Construction of the plant has received the backing of a consortium of European utilities, representing France (51 percent), Italy (33 percent), West Germany (11.04 percent), the Netherlands (2.36 percent), Belgium (2.36 percent) and the United Kingdom (.24 percent). The color coding adopted for this drawing and the ones on the next three pages is given in the key at the bottom.

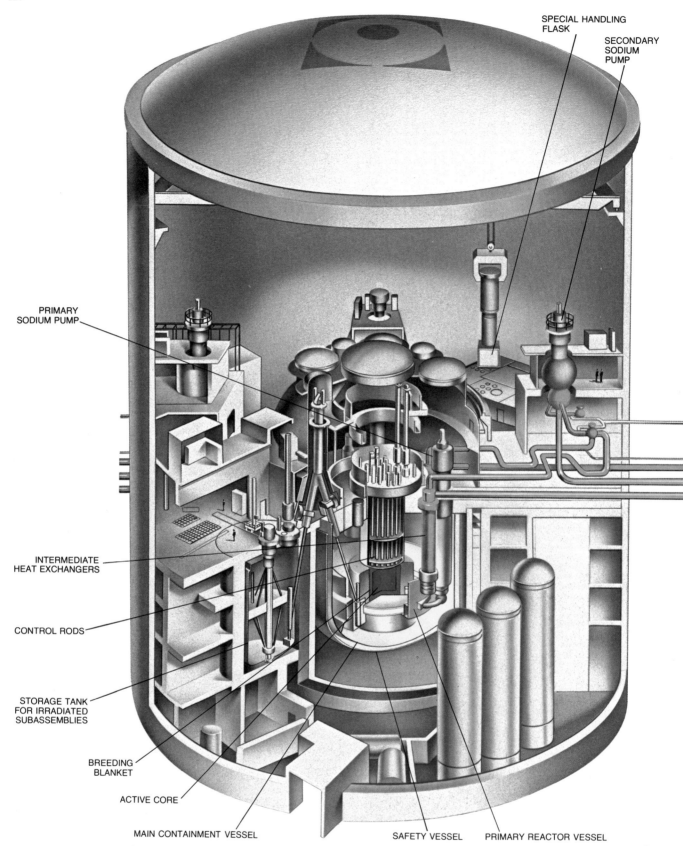

PRIMARY
SODIUM PUMP

SPECIAL HANDLING
FLASK

SECONDARY
SODIUM
PUMP

INTERMEDIATE
HEAT EXCHANGERS

CONTROL RODS

STORAGE TANK
FOR IRRADIATED
SUBASSEMBLIES

BREEDING
BLANKET

ACTIVE CORE

MAIN CONTAINMENT VESSEL

SAFETY VESSEL   PRIMARY REACTOR VESSEL

VERTICAL SECTION of the Superphénix reactor building and one of the four identical steam-generating buildings shows the main operating components of the plant in somewhat greater detail. Superphénix is classified as a pool-type breeder reactor, which means that the active core, the primary sodium pumps and the intermediate heat exchangers are all located within a single large vessel; in this particular design the main steel containment vessel, which is hung from a steel-and-concrete upper slab, is 21 meters across and is filled with 3,300 tons of molten sodium. A cylindrical structure welded to the main vessel supports the control-rod mechanism and the fuel subassemblies, which constitute the active core of the reactor. The four primary pumps convey the sodium upward through the core. The primary reactor vessel separates the "cold" sodium, which enters at the bottom of the subassemblies at a temperature of 395 degrees Celsius, from the "hot" sodium, which leaves at the top at 545 degrees C. The hot sodium then flows downward through the eight intermediate heat

1.4 or so will be achieved, in part by exploiting the concept of the heterogeneous core [*see illustration on next page*]. The corresponding doubling times will then be between 10 and 20 years. Since it is unlikely that the consumption of electricity will double at shorter intervals toward the end of the century, a doubling time in this range will enable fast-neutron reactors to cope with the rising demand for energy unaided, by virtue of their self-fueling feature.

The breeding ratios of the fast-neutron reactors built today are not significant, since for several years the plutonium produced by light-water reactors will constitute the major, if not the exclusive, source of the initial fuel for fast-neutron reactors. Thus a remarkable complementarity exists between these two types of nuclear reactor. Over a fairly long period a two-pronged strategy of nuclear-power generation can be established, with the light-water plants leading the way for the gradual penetration of the market by the fast breeder plants.

Although fast-neutron plants are capable of producing more plutonium than they consume, that potential can be exploited or not. At the discretion of the user plutonium production can be higher or lower than consumption. The amount of plutonium available can be matched exactly to the demand, whether the latter rises, remains stable or even declines; hence a stock of unused plutonium need never be created. In the absence of fast-neutron reactors, on the other hand, it would be impossible to completely burn the plutonium and its transplutonium derivatives produced by the slow-neutron plants. These highly radioactive elements would constitute wastes that would have to be set aside and stored for thousands of years.

The fact that fission reactions are caused by fast neutrons in a breeder reactor makes the dimensions of the core very compact; the core volume of a 1,000-megawatt fast-neutron plant does not have to exceed 10 cubic meters. Fast-neutron reactors by their very nature generate a great deal of heat per unit of volume. To remove this intense heat output from the fuel subassemblies that make up the reactor core it is necessary to use a coolant endowed with outstanding thermal properties. Water is unsuitable because hydrogen is a powerful neutron moderator, and any material of that kind must be avoided.

Of all conceivable fluids molten sodium is the one that combines the most attractive array of properties. A liquid at 98 degrees Celsius, it boils at 882 degrees C. at atmospheric pressure. Since the maximum sodium temperature in the reactor core never exceeds 550 degrees C. in normal operation, it is not necessary to pressurize the vessels and the circuits that contain the sodium. More-

over, the excellent thermal properties of sodium mean that the steam produced in the steam generators has characteristics equivalent to those required to drive the turbines of the most modern fossil-fuel power plants. The overall efficiency of a fast-neutron reactor is equal to or greater than 40 percent, whereas the efficiency of a typical light-water power plant does not exceed 33 percent; the comparatively high efficiency of fast-neutron reactors is a positive feature with respect to thermal discharges into the environment.

Every fast-neutron reactor that has been or is being built in the world today calls for molten sodium as the coolant. The fact that all the countries with active breeder programs (including the U.S., the U.S.S.R., France, Britain, West Germany, the Benelux nations, Italy, Japan and India) have made the same basic technological choice is a very favorable factor. It avoids the spreading of effort mounted along divergent lines and enhances overall efficiency. The approach followed has been much the same in all the countries involved. Reactors built and planned during the still prevailing development phase belong to three categories that follow in a logical succession: experimental reactors, demonstration plants and prototype power stations.

In line with this logical sequence the forerunners of Superphénix were Rapsodie and Phénix. The experimental reactor Rapsodie (the name associates the words *rapide* and sodium) was commissioned in 1967. Its power level is low (40 megawatts of thermal output) and it does not produce any electricity. Nevertheless, its main features are representative of the breeder regime from a technical standpoint with respect to temperature and other factors. Rapsodie has operated in a satisfactory manner for almost 10 years, with an average availability of nearly 90 percent during the operating runs. It is in continuous use as a test facility for investigating the effects of prolonged irradiation on various fuel assemblies.

One year after Rapsodie went into operation the decision was made to build the Phénix demonstration plant, named for the mythological bird that was reborn from its own ashes. The achievement of a high breeding ratio was not of particular concern in the design stage. The principal purpose of Phénix was to confirm the validity and reliability of the entire system by demonstrating the possibility of building a fast-neutron power plant within a reasonable period of time and of running it satisfactorily. Phénix was put into regular operation in July, 1974. The record of the first two years is particularly gratifying. These excellent results do not mean that the demonstration is over. The day-to-day operation of the reactor is being closely watched, and unforeseeable incidents

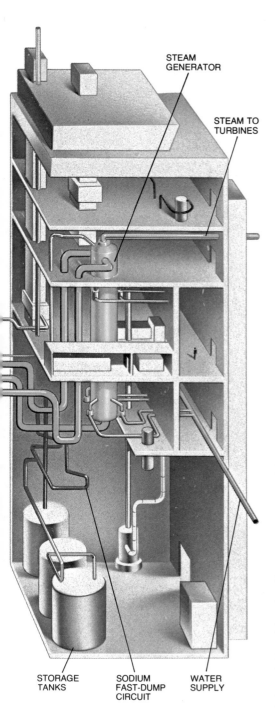

STEAM GENERATOR

STEAM TO TURBINES

STORAGE TANKS

SODIUM FAST-DUMP CIRCUIT

WATER SUPPLY

exchangers, which form part of a secondary circuit of nonradioactive sodium, inserted for reasons of safety between the primary sodium circuit and the water-steam circuit. Each of four secondary loops consists of two intermediate heat exchangers, a secondary pump installed inside spherical expansion tank and a steam generator in the adjacent building.

128

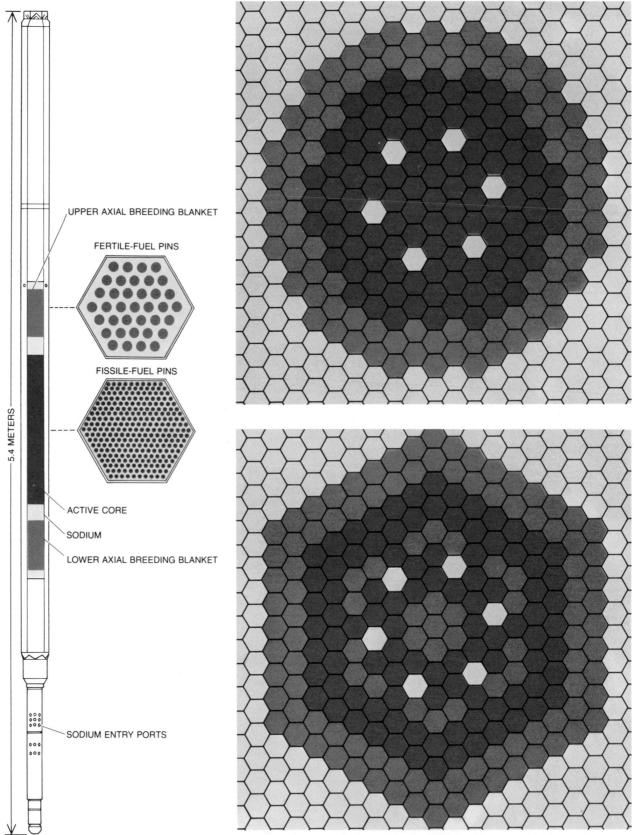

NUCLEAR-FUEL SUBASSEMBLY of the Superphénix reactor is shown in the cutaway vertical diagram at left. In each active-core subassembly the fuel is subdivided into 271 long, thin pins along which the sodium of the primary circuit (*yellow*) flows; the fissile material (*red*) occupies the central portion of the pin, fertile material (*orange*) being placed at both ends. (The fertile subassemblies contain fewer, larger pins.) Two alternative core designs under consideration for the breeder reactors of the future are represented by the sche-matic horizontal sections at the right. The two designs differ in the arrangement of the stainless-steel subassemblies: in the conventional core design (*top*) the central zone of fissile subassemblies is surround-ed by an outer "breeding blanket" of fertile subassemblies; in the new heterogeneous core design (*bottom*) fertile material is inserted into core in the form of clusters of fertile subassemblies. Gray hexagons are control rods. Designs are idealized here; in reality active core and breeding blanket will account for a total of about 600 subassemblies.

could still occur. Small sodium leaks detected during the summer of 1976 in two intermediate heat exchangers have led to the temporary shutdown of the plant for repairs to the observed defects, which are minor and do not call the design into question. The initial results are considered encouraging enough to proceed with confidence.

Superphénix, the next step in the development sequence, will be the prototype for the commercial breeder power plants of the future. In design it is very similar to Phénix. It was thought essential for overall efficiency and success to maintain the continuity of technological choices as far as possible. In spite of this constraint continuous progress in acquired know-how led in some cases to significant changes with respect to Phénix, if only to meet increasingly stringent safety criteria. Creys-Malville, where Superphénix will be built, is in the upper Rhône valley, not far from the electric-power grids of Italy and Germany. The site selected for the plant, on the banks of the Rhône 40 miles east of Lyons, is in a sparsely populated farming region where no other major industrial projects are planned.

From the geological standpoint the Creys-Malville site is in a low-seismicity zone of Degree VI on the international macroseismic scale (which has a range of 12 degrees, with an interval of one degree corresponding to a factor of two in ground acceleration). The Superphénix plant is designed to continue operating after being subjected to a Degree VI earthquake, which corresponds to the maximum intensity already observed in the region. Furthermore, the design guarantees that all essential safety functions of the plant, such as the neutron shutdown of the reactor, the removal of residual power from the core and the integrity of the containment, will be maintained in the event of an earthquake of Degree VII in intensity.

The Superphénix power station will be designed to adapt its operation to variations in demand on the electric-power grid. It will be operated as a base-load plant. The gross power output of the plant has been set at 1,200 megawatts of electricity, which is similar to the power level of light-water nuclear plants scheduled for construction at the same time. In 1985, 1,200 megawatts will represent between 1.5 and 2 percent of the total installed power of the French grid. The choice of this figure for Superphénix results from a compromise. On the one hand there is a trend toward large nuclear power plants on the grounds of economics; on the other hand extrapolation from Phénix to Superphénix must remain within reasonable limits.

A fast breeder plant does not differ greatly in its general layout and operating scheme from any other nuclear pow-

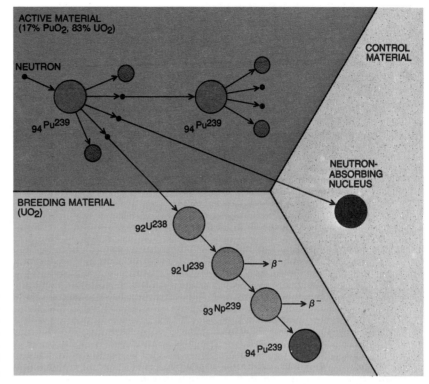

NUCLEAR REACTIONS that take place in the core of a breeder reactor are diagrammed. As in any nuclear reactor, the fissile active-core material (in this case plutonium 239) undergoes a self-perpetuating chain of fission reactions, yielding both the heat energy produced by the reactor and the neutrons that sustain the chain reaction. In the breeder fertile nuclei in the core and in the blanket material (in this case uranium 238) can also be transmuted into fissile nuclei by capturing a stray neutron each, thereby creating new fissile material. The absorption of neutrons by nuclei in the control rods can be adjusted to regulate the rate of the reactions (right). In a "fast" breeder the neutrons are not slowed down by a moderating substance such as water.

er station. Heat produced in the reactor core is conveyed by a fluid (molten sodium in this case) to water, producing steam, which feeds an electric-generating turbine unit. In order to avoid any accidental reaction between the radioactive sodium and the water an intermediate, or secondary, loop of nonradioactive sodium is inserted between the primary circuit conveying the sodium through the core and the water-steam circuit. Instead of one 1,200-megawatt turbogenerator of advanced design, two 600-megawatt units will be used in parallel, incorporating only conventional equipment that has proved its reliability in many oil-fired plants.

The design of the core and the fuel assemblies is a key factor in the realization of Superphénix. The core, as the seat of heat generation, is the most highly stressed of all the parts of a nuclear plant. This is all the more true in a fast-neutron core, where the heat production per unit of volume is exceptionally high (up to 500 kilowatts per liter) and all the structures are subjected to an intense flux of fast neutrons (6×10^{15} neutrons per square centimeter per second). To ensure that the heat is evacuated without giving rise to excessive temperatures the nuclear fuel is subdivided into long, thin pins (less than a centimeter in diameter) along which the sodium flows at a

speed of six meters per second. The fissile material is located in the central portion of the pin, fertile material being placed at both ends. A cluster of 271 pins are fastened together within the hexagonal stainless-steel structure known as a fuel subassembly. All together 364 subassemblies, packed in a regular array, constitute the reactor's active core, which is in turn surrounded by 232 similar subassemblies containing larger pins of fertile material, representing the breeding blanket. The sodium flows upward, entering at the bottom of the subassemblies at 395 degrees C. and leaving at the top at 545 degrees. At the center of the active core 450 watts of heat energy is generated per centimeter of fuel pin.

A fuel mixture with a mean composition of 17 percent plutonium oxide (PuO_2) and 83 percent uranium oxide (UO_2) has been selected as the fissile material; the fertile material consists of uranium oxide alone. Long and satisfactory experience with these materials has been gained in the operation of Rapsodie and Phénix. Of the 25,000 PuO_2-UO_2 fuel pins that have been irradiated so far in Rapsodie, 3,000 survived a burnup of 80,000 megawatt-days per ton and some have reached 150,000 megawatt-days per ton. Less than one pin per 1,000 irradiated failed. So far

PHÉNIX, a 250-megawatt demonstration breeder-reactor power station, is located on the Rhône River near Avignon. The plant began generating electricity at full power in July, 1974.

15,000 fuel pins have been irradiated in the Phénix core. At present subassemblies are taken out of the reactor as soon as they reach a burnup of 50,000 to 65,000 megawatt-days per ton. Not a single pin has failed while in service in Phénix.

Development work is also being devoted to new carbide and nitride fuels, which are likely to exhibit breeding characteristics superior to those of the oxides of plutonium and uranium currently called for. It remains to be seen whether this potential advantage will be offset by increased difficulties in fabrication, irradiation behavior and chemical reprocessing. The use of carbide and nitride fuels in Superphénix is not contemplated at this time.

Another important technical problem concerns the choice of the material for the hexagonal structure of the subassemblies and for the pin tubes, which must meet very stringent requirements. They must maintain good mechanical strength at temperatures approaching 650 degrees C. Furthermore, the internal pressure in the pin tubes may be as high as 30 kilograms per square centimeter, owing to the buildup of gaseous fission products. The pins are also subjected to considerable thermal stresses. Last but not least, they are exposed to a peculiar phenomenon: under prolonged irradiation by fast neutrons, vacancies form in the crystal lattice of the metal and grow into tiny cavities, causing the metal to swell. Some idea of the intensity of neutron bombardment in a high-power fast-neutron reactor can be gained from the fact that every atom of the material "cladding" the fuel pins is struck or at least caused to vibrate once

every 100 hours on the average by the passage of a neutron or another atom recoiling from a neutron collision. Another impressive figure is the cumulative number of fast neutrons crossing any given square centimeter of the cladding material after irradiation in the reactor core: this figure approaches one full gram of neutrons! The swelling of metallic alloys under neutron irradiation must be kept low enough to avoid deformation of the subassembly, which is liable to raise problems in the operation of the reactor, particularly in the fuel-handling maneuvers. A great deal of research-and-development work has already been accomplished but more is required in order to find a complete solution to the problem.

The different types of fast-neutron reactor are distinguished essentially by the organization of the primary sodium circuit. In the pool design the reactor core, the intermediate heat exchangers and the primary sodium pumps are all within a single large vessel. In the loop design only the reactor core is housed within the vessel and the intermediate heat exchangers and pumps are connected to it by loops. It must be stressed that the two systems rely on the same technology, that most development work on components is common to both and that the differences between the two concepts are much less than those between, say, pressurized-water and boiling-water reactors. In most countries loop-type reactors were built first, since the separation of components facilitated construction, operation and maintenance, justifying such a choice at an early stage of development. The first pool-type

breeder reactor in the world was built in the U.S. more than 10 years ago. Following the loop-type construction of Rapsodie, the pool concept was adopted for Phénix and, owing to the excellent record of that plant, it was maintained fundamentally unchanged for Superphénix.

It is clear that both the pool system and the loop system can be built and run, and that both have advantages and drawbacks only long operating experience can distinguish. Among the main reasons for the selection of the pool system for Phénix and Superphénix, following a meticulous comparison with the loop system, was a safety consideration. For a large plant, say 1,000 megawatts or more, it was thought the integrity of the primary sodium circuit could be maintained in all reasonably foreseeable circumstances more readily by enclosing it within a single vessel of simple design than by dispersing it in a highly intricate system of pipes and vessels involving many hundreds of meters of piping up to one meter in diameter. Although the main pool-type vessel is larger than the loop-type reactor vessel (roughly 20 meters in diameter as against 10), the pool-type vessel is much more straightforward in design. As a result construction, inspection and maintenance are far easier. The main problem encountered in the pool design concerned the cover of the main vessel. The solution implemented in Phénix could not be extrapolated to the dimensions of Superphénix. It was decided to hang the steel main vessel directly from the steel-and-concrete upper slab, and to put under the slab a layer of metallic thermal insulation that is in contact with the argon atmosphere above the sodium. The tests performed to date indicate that this arrangement is entirely satisfactory.

Experience with nuclear power plants of every type has shown that the steam generator is a crucial component. In fast-neutron reactors particular care must be taken in design and construction to prevent any violent chemical reaction between the sodium and the water, which would result from a leak in the exchanger tubes. The steam-generator model selected for Phénix, the only one with which extensive experience had been gained at the time, was subdivided into 36 low-power modules (17 megawatts each). The subdivision made it possible to subject three complete full-scale modules to thorough tests in simulated operational conditions. Although this approach was justifiable for an initial project, it could not be maintained for a large power plant because of its prohibitive cost. Research for Superphénix was therefore oriented toward units of different design, with a higher power per unit (several hundred megawatts). The problems presented by the fabrication and the operation of these units did not appear to grow with size, but the large modules do have certain drawbacks, the main ones being the near impossibility of conducting full-scale tests prior to installation in the power plant and the increased electric-power loss in case of the unavailability of a unit.

The tests performed under normal and accidental conditions on two "once through" mock-ups, one with straight ferritic steel tubes and the other with helical Incoloy tubes, provided complete satisfaction and showed good agreement with the design forecasts. The helical-tube model was finally selected for Superphénix, with each secondary loop including a steam generator with a thermal power of 750 megawatts. A steam-reheat stage can be added with either sodium or steam. The sodium system was employed for Phénix, raising the net efficiency of the plant to 42 percent. The steam system was adopted for Superphénix, simplifying the steam generator and the associated circuits, because a cost study showed that the lowered investment cost offset the loss in efficiency.

It is obviously important to prevent the development of even the smallest leak in the tubes separating the water from the sodium and to minimize the effects of any contact of the two fluids that may nevertheless occur. Ultrasensitive hydrogen detectors (capable of detecting a leak of as little as two milligrams of hydrogen per second) will be housed in several locations in each steam generator. Automatic systems designed to limit the consequences of an incipient reaction are also available. Two such systems consist of automatic valves that immediately shut off the sodium circuits and discharge systems designed to remove the products of the reaction and to limit the ensuing pressure surge.

The maneuvers required to convey the fuel subassemblies to their core positions and to withdraw them from the reactor after irradiation will be conducted exclusively during plant shutdown. They will be carried out by a series of devices that manipulate the assemblies in sodium at all times, in order to allow the removal of the residual heat released by the fission products. Two eccentric rotating plugs housed in the upper slab of the reactor will make it possible to position the device that grips the subassembly heads above any point of the core and blanket. This system, which allows direct service above each subassembly, also copes with potential deformations of the subassemblies due to swelling under irradiation. One of the main drawbacks of sodium is its opacity, which makes it impossible to follow

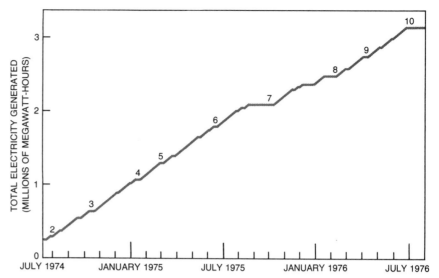

OPERATING RECORD OF PHÉNIX during its first two years was considered encouraging enough to proceed with Superphénix, the next stage in the French breeder-development program. The level parts of this cumulative electric-generating curve correspond to plant shutdowns; the numbers refer to refueling shutdowns. An extended work period scheduled after one year of operation took place during seventh refueling shutdown. In its first two years of operation Phénix generated electricity at full power for the equivalent of 530 days, a better performance than is typical of present-day light-water reactors in their first two years of operation.

the progress of the handling operations visually. Ultrasonic transmitter-receiver units, employing the principle of sonar, have been developed to surmount this obstacle. These devices, operating within the sodium itself, provide a guarantee that the subassemblies manipulated will occupy their correct positions at all times, without the risk of colliding with the handling devices.

The principles underlying the control

of a fast-neutron reactor are identical with those of any other nuclear reactor. The existence of delayed neutrons gives the mechanisms acting on core reactivity the time to act smoothly, whether to raise or lower the power of the plant or to keep it stable. These operations are performed by means of control rods containing a suitable neutron-absorbing material, which move in channels parallel to the fuel subassemblies. Superphé-

	PHÉNIX	SUPERPHÉNIX
GROSS ELECTRICAL RATING (MEGAWATTS)	264	1,240
THERMAL RATING (MEGAWATTS)	590	3,000
GROSS EFFICIENCY (PERCENT)	44.75	41.5
VOLUME OF CORE (LITERS)	1,227	10,820
LENGTH OF FUEL ASSEMBLIES (METERS)	4.3	5.5
NUMBER OF FUEL PINS PER ASSEMBLY	217	271
OUTSIDE DIAMETER OF FUEL PINS (MILLIMETERS)	6.6	8.65
MAXIMUM LINEAR POWER (WATTS PER CENTIMETER)	430	450
RATE OF FUEL BURNUP (MEGAWATT-DAYS PER TON)	50,000	70,000
MAXIMUM TOTAL NEUTRON FLUX (NEUTRONS PER SQUARE CENTIMETER PER SECOND)	7.2×10^{15}	6.2×10^{15}
BREEDING RATIO	1.12	1.24
NOMINAL CLADDING TEMPERATURE (DEGREES CELSIUS)	650	620
INTERVAL BETWEEN REFUELING OPERATIONS (MONTHS)	2	12

PHÉNIX AND SUPERPHÉNIX are compared in this table. The continuity of technological choices was maintained as far as possible in going to the larger plant, although a number of significant changes were incorporated in Superphénix design, partly to meet new safety criteria.

PRIMARY CIRCUIT		
	NUMBER OF PUMPS	4
	TEMPERATURE AT CORE INLET	395 DEGREES CELSIUS
	TEMPERATURE AT CORE OUTLET	545 DEGREES CELSIUS
	OVERALL SODIUM FLOW RATE	16.4 METRIC TONS PER SECOND
	WEIGHT OF SODIUM	3,300 METRIC TONS

SECONDARY CIRCUIT		
	NUMBER OF PUMPS	4
	NUMBER OF INTERMEDIATE HEAT EXCHANGERS	8
	TEMPERATURE AT INLET OF INTERMEDIATE HEAT EXCHANGER	345 DEGREES CELSIUS
	TEMPERATURE AT OUTLET OF INTERMEDIATE HEAT EXCHANGER	525 DEGREES CELSIUS
	OVERALL SODIUM FLOW RATE	13.2 METRIC TONS PER SECOND
	TOTAL WEIGHT OF SODIUM IN THE FOUR LOOPS	1,700 METRIC TONS

TERTIARY CIRCUIT		
	NUMBER OF STEAM GENERATORS	4
	TEMPERATURE OF SUPERHEATED STEAM	490 DEGREES CELSIUS
	PRESSURE OF SUPERHEATED STEAM	180 BARS
	OVERALL STEAM FLOW RATE	1.36 METRIC TONS PER SECOND

SPECIFICATIONS for the steam-generating system of Superphénix are summarized in this table. The use of sodium in the reactor's primary and secondary circuits is expected to give the new electric station a comparatively high overall thermal efficiency of at least 40 percent.

nix will be provided with a highly redundant system of control rods, divided into three independent groups. One of these is specially designed to penetrate the core even in the extreme and improbable case of its undergoing a large deformation. Uninterrupted monitoring of the Superphénix core is provided by a diversified set of detectors, whose output is processed and correlated by computer. The temperature of the sodium leaving each subassembly is measured by three thermocouples; two of them are of the chromel-alumel type and the third uses a sodium-steel couple and responds almost instantaneously. The boiling-sodium detectors, flowmeters and devices for the detection and localization of any cladding failures are improved versions of those employed in Phénix. The neutron detectors and electronic instruments for measuring variations in reactivity have proved their reliability through long experience with them.

The many precautions implemented in the design of Superphénix were subjected to detailed scrutiny by the licensing authorities before their approval was secured. These safety measures reduce the probability of an accident to an extremely low level. The procedure followed went to the extent of considering the case in which a total shutdown of forced sodium circulation through the core at full power is not accompanied by any action of the many control systems designed to shut down the fission chain reaction and energy production imme-

diately. Even in this case the considerable thermal inertia represented by the large mass of sodium present in the primary vessel (3,300 tons) and the interval of several hundred degrees C. between the sodium temperature in normal operation and its boiling point furnish a substantial time interval for manual emergency action. Nevertheless, it is necessary to ensure that even in the highly improbable case of a serious accident involving a core meltdown the consequences will be contained in such a way that no significant quantity of plutonium or fission products can escape into the environment.

The containment system for Superphénix therefore consists of a series of successive enclosures, which can withstand both internal reactor accidents and external aggression such as an airplane crashing into the power plant. Finally, special arrangements have been made to prevent potential sodium fires and to limit their spread should they occur. Sodium fires would not actually jeopardize the safety of the installation, but it is nonetheless necessary to take full precautions to maximize the reliability and the availability of the power station.

In all areas, not just in the priority area of safety, a considerable research-and-development effort has preceded the design and construction of Superphénix. This program, which calls for full-scale tests in sodium of all components where innovations have been made, will continue to back up con-

struction of the reactor in the coming years.

Phénix was built in slightly more than four years. Preliminary site preparation began in the fall of 1968, and the filling of the primary and secondary circuits with 1,400 tons of sodium was started before the end of 1972. For Superphénix a building schedule spread over 68 months has been adopted. Construction deadlines are comparable to those set for other types of nuclear power plant. The fact that breeder reactors are not pressurized and that their components, even the large ones, are made of comparatively thin stainless-steel sheet and pipe makes it possible to perform most of the final assembly on the site. The Phénix experience clearly showed the advantages of this approach and the flexibility that it engendered in adherence to the construction schedule.

The investment cost of the Superphénix power plant will significantly exceed that of a light-water plant of comparable output. This cost differential is unavoidable since Superphénix is the first plant of its kind, whereas light-water plants (of which more than 100 have been built to date in the world) have amply profited from the fruits of technical progress and above all of series production. In spite of the fact that Superphénix is a prototype, it should be emphasized that the cost of a kilowatt-hour of electricity produced by Superphénix will be in the same range as that produced by an oil-fired power station. It is probable that the investment cost of fast breeder plants, which will progressively decline as larger numbers are constructed, will remain for some time higher than that of light-water plants, if only because of the larger amounts of stainless steel employed and the presence of an intermediate sodium circuit added for safety reasons. Economic competitiveness with light-water plants will derive from a cheaper fuel cycle, made possible by fuel breeding, and this factor will become increasingly important with the foreseeable increase in the price of natural uranium.

The expansion program of the French national utility company Électricité de France (EDF) already calls for a series of breeder plants, employing plutonium provided by a large number of pressurized-water reactors built simultaneously. It is reasonable to expect that two pairs of fast-neutron plants will be initiated in France between 1980 and 1985, representing, together with Superphénix, about 8,000 megawatts of electric-generating capacity in service in the early 1990's. Commitments may grow to 2,000 megawatts per year after 1985, so that by the year 2000 fast-neutron plants may account for about a fourth of the installed capacity and a third of the total energy output of all the nuclear plants in France. Simultaneous with the successive launching of these plants will be the construction of plants for the fabrication and reprocessing of fast-breeder fuels, thus closing the fuel cycle. The latter will be high-capacity plants (with an output of about 200 tons of oxides per year) aimed at achieving a low overall fuel-cycle cost.

The importance of Superphénix must be gauged in relation to the coming generation of power plants derived from it. It is in a way the culmination of the technological development phase and the final stage before the commercial series, the technical definition of which will rely directly on the Superphénix experience. If everything proceeds as planned, by the mid-1980's, thanks to Superphé-

nix, one may expect to have at least a preliminary operating record with a large fast-neutron power plant. This experience, which will be shared by several large electric utilities, symbolizes the joining of efforts by the European countries involved in aiming at the earliest possible commercial launching of a type of reactor that is indispensable to their economies.

REACTOR	LOCATION	POWER (MEGAWATTS)	FORM OF OUTPUT	BASIC DESIGN	SCHEDULE
EXPERIMENTAL BREEDER REACTOR 1	U.S.	.2	ELECTRICAL	LOOP	
DOUNREAY FAST REACTOR	U.K.	15	ELECTRICAL	LOOP	
EXPERIMENTAL BREEDER REACTOR 2	U.S.	20	ELECTRICAL	POOL	
B.R. 5	U.S.S.R.	10	THERMAL	LOOP	
ENRICO FERMI FAST BREEDER REACTOR	U.S.	66	ELECTRICAL	LOOP	
RAPSODIE	FRANCE	40	THERMAL	LOOP	
B.N. 350	U.S.S.R.	150	ELECTRICAL	LOOP	
SOUTHWEST EXPERIMENTAL FAST OXIDE REACTOR	U.S.	20	ELECTRICAL	LOOP	
B.O.R. 60	U.S.S.R.	12	THERMAL	LOOP	
PROTOTYPE FAST REACTOR	U.K.	250	ELECTRICAL	POOL	
FAST-FLUX TEST FACILITY	U.S.	400	THERMAL	LOOP	
K.N.K. 2	GERMANY	20	ELECTRICAL	LOOP	
PHÉNIX	FRANCE	250	ELECTRICAL	POOL	
P.E.C.	ITALY	116	THERMAL	LOOP	
S.N.R. 300	GERMANY	300	ELECTRICAL	LOOP	
JOYO	JAPAN	100	THERMAL	LOOP	
B.N. 600	U.S.S.R.	600	ELECTRICAL	POOL	
FAST BREEDER TEST REACTOR	INDIA	15	ELECTRICAL	LOOP	
CLINCH RIVER BREEDER REACTOR	U.S.	380	ELECTRICAL	LOOP	
COMMERCIAL FAST REACTOR	U.K.	1,300	ELECTRICAL	POOL	
SUPERPHÉNIX	FRANCE	1,200	ELECTRICAL	POOL	
MONJU	JAPAN	300	ELECTRICAL	LOOP	
S.N.R. 2	GERMANY	1,300	ELECTRICAL		
B.N. 1500	U.S.S.R.	1,500	ELECTRICAL		
PROTOTYPE LARGE BREEDER REACTOR	U.S.	1,200	ELECTRICAL		

1945 1950 1955 1960 1965 1970 1975 1980

TYPE OF FACILITY	DESIGN PHASE	CONSTRUCTION PHASE	OPERATION PHASE
EXPERIMENTAL REACTOR			
DEMONSTRATION PLANT			
COMMERCIAL PROTOTYPE			

WORLD SURVEY of progress in the development of liquid-metal-cooled fast breeder reactors includes all facilities with a thermal-power output greater than one megawatt. The plants are listed in chronological order according to the beginning of their design stage. Different colors are used to distinguish the three main categories of reactors built or planned so far: experimental reactors, demonstration plants and prototype commercial power stations. Different intensities of color denote design, construction and operation phases. Projections are made only to 1980; bars that stop short of the present represent decommissioned facilities. S.N.R. 300, Superphénix and S.N.R. 2 are multinational European projects. The German K.N.K. reactor has been in operation since 1968 with a slow-neutron core; beginning in 1977 it will be run with a fast-neutron core as K.N.K. 2. The British Prototype Fast Reactor at Dounreay in Scotland and the Russian B.N. 350 at Shevchenko, two demonstration plants comparable in size to the French Phénix, are both completed but have not yet been run at full power, owing to difficulties with their steam-generating equipment. Preliminary site work is about to begin on the closest comparable U.S. plant, the 380-megawatt Clinch River project near Oak Ridge, Tenn., which is expected to be completed in the early 1980's.

11

The Prospects of Fusion Power

by William C. Gough and Bernard J. Eastlund
February 1971

*Recent advances in the performance of several
experimental plasma containers have brought the
fusion-power option very close to the "break even"
level of scientific feasibility*

The achievement of a practical fusion-power reactor would have a profound impact on almost every aspect of human society. In the past few years considerable progress has been made toward that goal. Perhaps the most revealing indication of the significance of this progress is the extent to which the emphasis in recent discussions and meetings involving workers in the field has tended to shift from the question of purely scientific feasibility to a consideration of the technological, economic and social aspects of the power-generation problem. The purpose of this article is to examine the probable effects of the recent advances on the immediate and long-term prospects of the fusion-power program, with particular reference to mankind's future energy needs.

The Role of Energy

The role of energy in determining the economic well-being of a society is often inadequately understood. In terms of *total* energy the main energy source for any society is the sun, which through the cycle of photosynthesis produces the food that is the basic fuel for sustaining the population of that society. The efficiency with which the sun's energy can be put to use, however, is determined by a feedback loop in which auxiliary energy sources form a critical link [*see illustration, page 136*]. The auxiliary energy (derived mainly from fossil fuels, water power and nuclear-fission fuels) "opens the gate" to the efficient use of the sun's energy by helping to produce fertilizers, pesticides, improved seeds, farm machinery and so on. The result is that the food yield (in terms of energy content) produced per unit area of land in a year goes up by orders of magnitude. This auxiliary energy input, when it is transformed into food energy, enables large populations to live in cities and develop new ways to multiply the efficiency of the feedback loop. If a society is to raise its standard of living by increasing the efficiency of its agricultural feedback loop, clearly it must expand its auxiliary energy sources.

The dilemma here is that the economically less developed countries of the world cannot *all* industrialize on the model of the more developed countries, for the simple reason that the latter countries, which contain only a small fraction of the world's population, currently maintain their high standard of living by consuming a disproportionately large share of the world's available supply of auxiliary energy. Just as there is a direct, almost linear, relation between a nation's use of auxiliary energy and its standard of living, so also there is a similar relation between energy consumption and the amount of raw material the nation uses and the amount of waste material it produces. Thus the more developed countries consume a correspondingly oversized share of the world's reserves of material resources and also account for most of the world's environmental pollution.

In order to achieve a more equitable and stable balance between the standards of living in the more developed countries and those in the less developed countries, only two alternatives exist. The more developed countries could reduce their consumption of auxiliary energy (thereby lowering their standard of living as well) or they could contribute to the development of new, vastly greater sources of auxiliary energy in order to help meet the rising demands for a better standard of living on the part of the rapidly growing populations of the less developed countries.

When one projects the world's long-term energy requirements against this background, another important factor must be taken into account. There are finite limits to the world's reserves of material resources and to the ability of the earth's ecological system to absorb pollutants safely. As a consequence future societies will be forced to develop "looped," or "circular," materials economies to replace their present, inherently wasteful "linear" materials economies [*see bottom illustration, page 145*]. In such a "stationary state" system, limits on the materials inventory, and hence on the total wealth of the society, would be set by nature. Within these limits, however, the standard of living of the population would be higher if the rate of flow of materials were lower. This maximizing of the life expectancy of the materials inventory could be accomplished in two ways: increasing the durability of individual commodities and developing the technological means to recycle the limited supply of material resources.

The conclusion appears radical. Future societies must *minimize* their physical flow of production and consumption. Since a society's gross national product for the most part measures the flow of physical things, it too would be reduced.

But all nations now try to *maximize* their gross national product, and hence their rate of flow of materials! The explanation of this paradox is that in the existing linear economies the inputs for increasing production must come from the environment, which leads to depletion, while an almost equal amount of materials in the form of waste must be returned to the environment, which leads to pollution. This primary cause of pollution is augmented by the pollution that is produced by the energy sources used to drive the system.

In order to make the transition to a stationary-state world economy, the wealthier nations will have to develop the technology—and the concomitant auxiliary energy sources—necessary to operate a closed materials economy. This capability could then be transferred to the poorer nations so that they could develop to the level of the wealthier nations without exhausting the world's supply of resources or destroying the environment. Thus some of the causes of international conflict would be removed, thereby reducing the danger of nuclear war.

Of course any effort to bring about a rapid change from linear economies to looped economies will encounter the massive economic, social and political forces that sustain the present system. The question of how to distribute the stock of wealth, including leisure, within a stationary-state economy will remain. In summary, the world's requirements for energy are intimately related to the issues of population expansion, economic development, materials depletion, pollution, war and the organization of human societies.

The Energy Options

What are the available energy options for the future? To begin with there are the known finite and irreplaceable energy sources: the fossil fuels and the better-grade, or easily fissionable, nuclear fuels such as uranium 235. Estimates of the life expectancy of these sources vary, but it is generally agreed that they are being used up at a rapid rate—a rate that will moreover be accelerated by increases in both population and living standards. In addition, environmental considerations could further restrict the use of these energy sources.

Certain other known energy sources, such as water power, tidal power, geothermal power and wind power are "infinite" in the sense of being continuously replenished. The total useful *amount* of energy available from these sources, however, is insufficient to meet the needs of the future.

Direct solar radiation, resulting from the fusion reactions that take place in the core of the sun, is an abundant as well as effectively "infinite" energy source. The immediate practical obstacle to the direct use of the sun's energy as an effective auxiliary energy source is the necessity of finding some way to economically concentrate the available low energy density of solar radiation. Controlled fusion is another potentially "infinite" energy source; its energy output arises from the reduction of the total mass of a nuclear system that accompanies the merger of two light

U.S. TOKAMAK, a toroidal plasma-confinement machine used in fusion research, was recently put into operation at the Plasma Physics Laboratory of Princeton University. Until about a year ago this machine, formerly known as the Model-C stellarator and now called the Model ST tokamak, had been the largest of the stellarator class of experimental plasma containers developed primarily at the Princeton laboratory. The decision to convert it to the closely related tokamak design followed the 1969 announcement by the Russian fusion-research group of some important new results obtained from their Model T-3 machine, the most advanced of the tokamak class of plasma containers developed mainly at the I. V. Kurchatov Institute of Atomic Energy near Moscow. In large part because of the cooperative nature of the world fusion-research program, this conversion was accomplished quickly and the Model ST has already produced results comparable to those obtained by the Russians. Several other tokamak-type machines are being built in this country.

RUSSIAN STELLARATOR is now the largest representative of this class of experimental plasma containers in the world. It is named the Uragan (or "hurricane") stellarator and is located at the Physico-Technical Institute of the Academy of Sciences of the Ukrainian S.S.R. at Kharkov. In both photographs on this page the large circular structures surrounding and almost completely obscuring the toroidal plasma chambers are the primary magnet coils. The main difference between the tokamak design and the stellarator design is that in a tokamak a secondary plasma-stabilizing magnetic field is generated by an electric current flowing axially through the plasma itself, whereas in a stellarator this secondary magnetic field is set up by external helical coils situated just inside the primary coils and hence not visible.

nuclei. The most likely fuel for a fusion-power energy source is deuterium, an abundant heavy isotope of hydrogen easily separated from seawater.

In addition to these two primary "infinite" energy sources, secondary "infinite" energy sources could be made by using neutrons to transmute less useful elements into other elements capable of being used effectively as fuels. Thus for fission systems the vast reserves of uranium 238 could be converted by neutron bombardment into easily fissionable plutonium 239; similarly, thorium 232 could be converted into uranium 233. For fusion systems lithium could be converted into tritium, another heavy isotope of hydrogen with a comparatively low resistance to entering a fusion reaction and a comparatively high energy output once it does.

The prime hope for extending the world's reserves of nuclear-fission fuels is the development of the neutron-rich fast breeder fission reactors [see SCIENTIFIC AMERICAN Offprint 339, "Fast Breeder Reactors," by Glenn T. Seaborg and Justin L. Bloom]. Another potential source of abundant, inexpensive neutrons is a fusion-fission hybrid system, an alternative that will be discussed further below.

Fusion Energy

Nuclear fusion, the basic energy process of the stars, was first reproduced on the earth in 1932 in an experiment involving the collision of artificially accelerated deuterium nuclei. Although it was thereby shown that fusion energy could be released in this way, the use of particle accelerators to provide the nuclei with enough energy to overcome their Coulomb, or mutually repulsive, forces was never considered seriously as a practical method for power generation. The reason is that the large majority of the nuclei that collide in an accelerator scatter without reacting; thus it is impossible to produce more energy than was used to accelerate the nuclei in the first place.

The uncontrolled release of a massive amount of fusion energy was achieved in 1952 with the first thermonuclear test explosion. This test proved that fusion energy could be released on a large scale by raising the temperature of a high-density gas of charged particles (a plas-

ma) to about 50 million degrees Celsius, thereby increasing the probability that fusion reactions will take place within the gas.

Coincident with the development of the hydrogen bomb, the search for a more controlled means of releasing fusion energy was begun independently in the U.S., Britain and the U.S.S.R. Essentially this search involves looking for a practical way to maintain a comparatively low-density plasma at a temperature high enough so that the output of fusion energy derived from the plasma is greater than the input of some other kind of energy supplied to the plasma. Since no solid material can exist at the temperature range required for a useful energy output (on the order of 100 million degrees C.) the principal emphasis from the beginning has been on the use of magnetic fields to confine the plasma.

The variety of magnetic "bottles" designed for this purpose over the years can be arranged in several broad categories in order of increasing plasma density [see illustration, pages 140 and 141]. First there are the basic plasma devices. These are low-density, low-temperature systems used primarily to study the fundamental properties of plasmas. Their configuration can be either linear (open) or toroidal (closed). Linear basic-plasma devices include simple glow-discharge systems (similar in operation to ordinary fluorescent lamps) and the more sophisticated "Q-machines" ("Q" for "quiescent") found in many university plasma-physics laboratories. Toroidal representatives of this class include the "multipole" devices, developed primarily at Gulf Energy & Environmental Systems, Inc. (formerly Gulf General Atomic Inc.) and the University of Wisconsin, and the spherator, developed at the Plasma Physics Laboratory of Princeton University.

Next there are the medium-density plasma containers; these are defined as systems in which the outward pressure of the plasma is much less than the inward pressure of the magnetic field. A typical configuration in this density range is the linear magnetic bottle, which is usually "stoppered" at the ends by magnetic "mirrors": regions of somewhat greater magnetic-field strength that reflect escaping particles back into the bottle. In addition extra current-carrying structures are often used to improve the stability of the plasma. These structures were originally proposed on theoretical grounds in 1955 by Harold Grad of New York University. They were first used successfully in an experimental test in 1962 by the Russian physicist M. S. Ioffe.

ROLE OF AUXILIARY ENERGY in determining the economic well-being of a society is illustrated by these two diagrams of agricultural feedback loops. In an economically less developed country (top) the bulk of the population must be devoted to the agricultural transformation of the sun's energy into food in order to support itself at a subsistence level. In an economically more developed industrial country (bottom) auxiliary energy sources "open the gate" to the more efficient utilization of the sun's energy, making it possible for the entire population to maintain a higher standard of living and freeing many people to live in cities and develop new ways to multiply the efficiency of the feedback loop.

		LIFE EXPECTANCY OF KNOWN RESERVES (YEARS)		LIFE EXPECTANCY OF POTENTIAL RESERVES (YEARS)		LIFE EXPECTANCY OF TOTAL RESERVES (YEARS)	
		AT .17Q	AT 2.8Q	AT .17Q	AT 2.8Q	AT .17Q	AT 2.8Q
FINITE ENERGY SOURCES	FOSSIL FUELS (COAL, OIL, GAS)	132	8	2,700	165	2,832	173
	MORE ACCESSIBLE FISSION FUELS (URANIUM AT $5 TO $30 PER POUND OF U_3O_8 BURNED AT 1.5 PERCENT EFFICIENCY)	66	4	66	4	132	8
	LESS ACCESSIBLE FISSION FUELS (URANIUM AT $30 TO $500 PER POUND OF U_3O_8 BURNED AT 1.5 PERCENT EFFICIENCY	43,000	2,600	129,000	7,800	172,000	10,400
"INFINITE" NATURAL ENERGY SOURCES	WATER POWER, TIDAL POWER, GEOTHERMAL POWER, WIND POWER	INSUFFICIENT		INSUFFICIENT		INSUFFICIENT	
	SOLAR RADIATION	10 BILLION	10 BILLION			10 BILLION	10 BILLION
	FUSION FUELS (DEUTERIUM FROM OCEAN)	45 BILLION	2.7 BILLION			45 BILLION	2.7 BILLION
"INFINITE" ARTIFICIAL ENERGY SOURCES (ELEMENTS TRANSMUTED FROM OTHER ELEMENTS BY NEUTRON BOMBARDMENT)	FISSION FUELS (PLUTONIUM 239 FROM URANIUM 238; URANIUM 233 FROM THORIUM 232)	8.8 MILLION	536,000	21 MILLION	1.3 MILLION	30 MILLION	1.8 MILLION
	FUSION FUELS (TRITIUM FROM LITHIUM) a) ON LAND b) IN OCEAN	48,000 120 MILLION	2,900 7.3 MILLION	UNKNOWN	UNKNOWN	48,000+ 120 MILLION	2,900+ 7.3 MILLION

WORLD ENERGY RESERVES are listed in this table in terms of their life expectancy estimated on the basis of two extreme assumptions, which were chosen so as to bracket a reasonable range of values. First, the assumption was made that the world population would remain constant at its 1968 level of 3.5 billion persons and that the energy-consumption rate of this population would remain constant at the estimated 1968 rate of .17 Q (Q is a unit of heat measurement equal to 10^{18} BTU, or British Thermal Units). Second, the assumption was made that the world population would eventually reach seven billion and that this population would consume energy at a per capita rate of 400 million BTU per year (about 20 percent higher than the present U.S. rate), giving a total world energy-consumption rate of 2.8 Q per year. (A commonly projected world energy-consumption rate for the year 2000 is one Q.) Current fission-converter reactors use only between 1 and 2 percent of the uranium's potential energy content, since the com-

ponent of the ore that is burned as fuel is primarily high-grade, or easily fissionable, uranium 235. The world fission-fuel reserves were derived by multiplying the U.S. reserves times the ratio of world land area to the U.S. land area (approximately 16.2 to one). For fusion-converter reactors lithium-utilization studies show that natural lithium, a mixture of lithium 6 and lithium 7, would be superior to pure lithium 6 in a tritium-breeding reactor "blanket" and would yield an energy output of about 86.4 million BTU per gram. The figure for known world lithium reserves is based on a study carried out last year by James J. Norton of the U.S. Geological Survey. The potential reserves of lithium are unknown, since there has been no exploration program comparable to that undertaken for, say, uranium. Lithium, however, is between five and 15 times more abundant in the earth's crust than uranium. Finally, the life expectancy of the earth—and hence that of potentially useful solar radiation—is predicted to be at most 10 billion years.

The straight rods used by Ioffe in his experiment have come to be called Ioffe bars, but such stabilizing structures can assume various other shapes. For example, in one series of medium-density linear devices they resemble the seam of a baseball; accordingly these devices, developed at the Lawrence Radiation Laboratory of the University of California at Livermore, are named Baseball I and Baseball II.

Medium-density plasma containers with a toroidal geometry include the stellarators, originally developed at the Princeton Plasma Physics Laboratory, and the tokamaks, originally developed at the I. V. Kurchatov Institute of Atomic Energy near Moscow. The only essential difference between these two machines is that in a stellarator a secondary, plasma-stabilizing magnetic field is set up by external helical coils, whereas in a tokamak this field is generated by an electric current flowing through the plasma itself. The close similarity between these two designs was emphasized recently by the fact that the Princeton

Model-C stellarator was rather quickly converted to a tokamak system following the recent announcement by the Russians of some important new results from their Tokamak-3 machine.

The astron concept, also developed at the Lawrence Radiation Laboratory at Livermore, is another example of a medium-density plasma container. In overall geometry it shares some characteristics of both the linear and the toroidal designs.

Higher-density plasma containers, defined as those in which the plasma pressure is comparable to the magnetic-field pressure, have also been built in both the linear and the toroidal forms. In one such class of devices, called the "theta pinch" machines, the electric current is in the theta, or azimuthal, direction (around the axis) and the resulting magnetic field is in the zeta, or axial, direction (along the axis). The Scylla and Scyllac machines at the Los Alamos Scientific Laboratory are respectively examples of a linear theta-pinch design and a toroidal theta-pinch design.

As the density of the plasma is increased further, one reaches a technological limit imposed by the inability of the materials used in the magnet coils to withstand the pressure of the magnetic field. Consequently very-high-density plasma systems are often fast-pulsed and obtain their principal confining forces from "self-generated" magnetic fields (fields set up by electric currents in the plasma itself), from electrostatic fields or from inertial pressures. In this very-high-density category are the "zeta pinch" machines, devices in which the electric current is in the zeta direction and the resulting magnetic field is in the theta direction. An example of this type of configuration is the Columba device at Los Alamos.

Other very-high-density, fast-pulsed systems include the "strong focus" designs, in which a stream of plasma in a cylindrical, coaxial pipe is heated rapidly by shock waves as it is brought to a sharp focus by self-generated magnetic forces, and laser designs, in which a pellet of fuel is ionized instantaneously by a pulse from a high-power laser, producing an "inertially confined" plasma. Still another confinement scheme that has been investigated in this general density range includes an electrostatic device in which the plasma is confined by inertial forces generated by concentric spherical electrodes.

The Fusion-Power Balance

What are the fundamental requirements for a meaningful release of fusion energy in a reactor? First, the plasma must be hot enough for the production of fusion energy to exceed the energy loss due to bremsstrahlung radiation (radiation resulting from near-collisions between electrons and nuclei in the plasma). The temperature at which this transition occurs is called the ignition temperature. For a fuel cycle based on fusion reactions between deuterium and tritium nuclei the ignition temperature is about 40 million degrees C. Second, the plasma must be confined long enough to release a significant net output of energy. Third, the energy must be recovered in a useful form.

In the first years of the controlled-fusion research program one of the major goals was to achieve the ignition temperature in a fairly dense laboratory plasma. Steady progress was made toward this goal, culminating in 1963, when the ignition temperature (for a deuterium-tritium fuel mixture) was reached in one of the Scylla devices at Los Alamos. This test, which was performed in a pure deuterium plasma to avoid the generation of excessive neutron flux, resulted in the release of fusion energy: about a thousandth of a joule per pulse, or 370 watts of fusion power during the three-microsecond duration of the pulse. If the test had been performed using a deuterium-tritium mixture, it would have released approximately a half-joule of fusion energy per pulse, or 180,000 watts of fusion power.

Today a large number of different devices have either achieved the deuterium-tritium ignition temperature or are very close to it [*see bottom illustration on opposite page*]. The main difficulties encountered in reaching this goal were comparatively straightforward energy-loss processes involving impurity atoms that entered the plasma from the walls of the container. A large research effort in the areas of vacuum and surface technology was a major factor in surmounting the ignition-temperature barrier.

The problem of confining a plasma long enough to release a significant net amount of energy has proved to be even more difficult than the problem of achieving the ignition temperature. Extremely rapid energy-loss processes—known collectively as "anomalous diffusion" processes—appeared to prevent the attainment of adequate confinement times. Plasma instabilities were the primary cause of this rapid plasma leakage [see "The Leakage Problem in Fusion Reactors," by Francis F. Chen; SCIENTIFIC AMERICAN, July, 1967]. Within the past few years, however, several large containment devices have reduced these instabilities to such a low amplitude that other more subtle effects, such as convective plasma losses and magnetic-field imperfections, can be studied. As a result it has been shown that there is no basic law of physics (such as an instability-initiated anomalous plasma loss) that prevents plasma confinement for times long enough to release significant net fusion energy. In fact, "classical," or ideal, plasma confinement has been achieved in several machines; this is the best confinement possible and yields a plasma-loss rate much lower than that required for a fusion reactor.

The twin achievements of ignition temperature and adequate confinement time, it should be noted, have taken place in quite different machines, each

Reaction	Energy
DEUTERIUM + DEUTERIUM → HELIUM 3 + NEUTRON	3.2 MEV
DEUTERIUM + DEUTERIUM → TRITIUM + PROTON	4.0 MEV
DEUTERIUM + TRITIUM → HELIUM 4 + NEUTRON	17.6 MEV
DEUTERIUM + HELIUM 3 → HELIUM 4 + PROTON	18.3 MEV
LITHIUM 6 + PROTON → HELIUM 3 + HELIUM 4	4.0 MEV
LITHIUM 6 + HELIUM 3 → HELIUM 4 + HELIUM 4 + PROTON	16.9 MEV
LITHIUM 6 + DEUTERIUM → LITHIUM 7 + PROTON	5.0 MEV
LITHIUM 6 + DEUTERIUM → PROTON + HELIUM 3 + HELIUM 4	2.6 MEV
LITHIUM 6 + DEUTERIUM → HELIUM 4 + HELIUM 4	22.4 MEV
LITHIUM 7 + PROTON → HELIUM 4 + HELIUM 4	17.5 MEV

FUSION REACTIONS regarded as potentially useful in full-scale fusion reactors are represented in this partial list. The two possible deuterium-deuterium reactions occur with equal probability. The deuterium-tritium fuel cycle has been considered particularly attractive because this mixture has the lowest ignition temperature known (about 40 million degrees Celsius). Other fuel cycles, including many not shown in this list, have been attracting increased attention lately, since certain plasma-confinement schemes actually operate better at higher temperatures and offer the advantage of direct conversion to electricity. The energy released by each reaction is given at right in millions of electron volts (*MeV*).

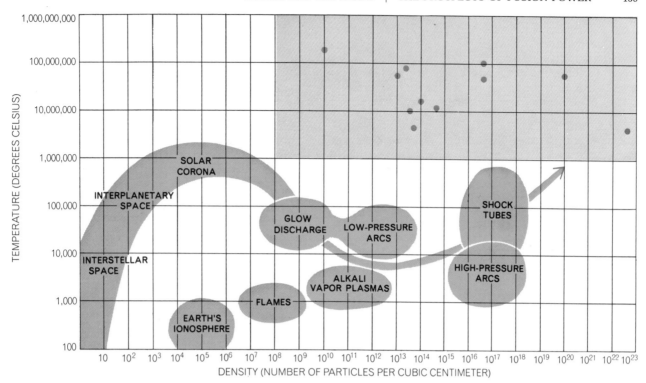

INDUSTRIALLY UNEXPLORED RANGE of plasma temperatures and densities has already been made available by the fusion-power research program. These experimental plasmas (*colored dots*), which range in temperature from 500,000 to a billion degrees C. and in density from 10^9 to 10^{22} ions per cubic centimeter, are compared here with various other industrial and natural plasmas.

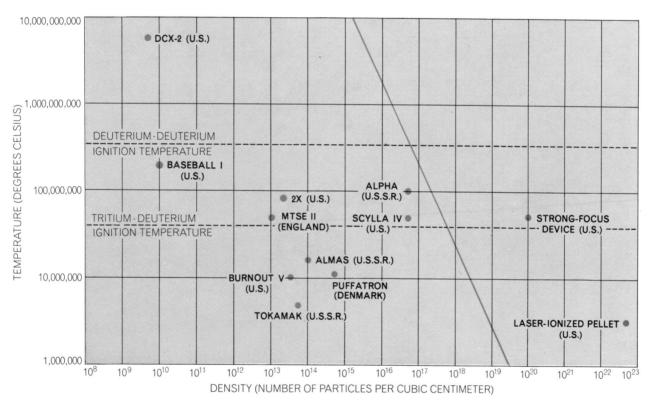

PLASMA EXPERIMENTS that have achieved temperatures near or above the fusion ignition temperatures of a deuterium-tritium fuel (*bottom horizontal line*) and a deuterium-deuterium fuel (*top horizontal line*) are identified by the name of the experimental device and the country in which the experiment took place in this enlargement of the upper right-hand section of the illustration at top. The diagonal colored line represents the limit beyond which the materials used to construct the magnet coils can no longer withstand the magnetic-field pressure required to confine the plasma (assumed to be 300,000 gauss in this case). Beyond this limit only fast-pulsed systems (in which the magnetic fields are generated by intense currents inside the plasma itself) or systems operating on entirely different principles (such as laser-produced, inertially confined plasmas) are possible. The record of six billion degrees C. was achieved with the aid of a high-energy ion-injection system associated with DCX-2 device at the Oak Ridge National Laboratory.

140

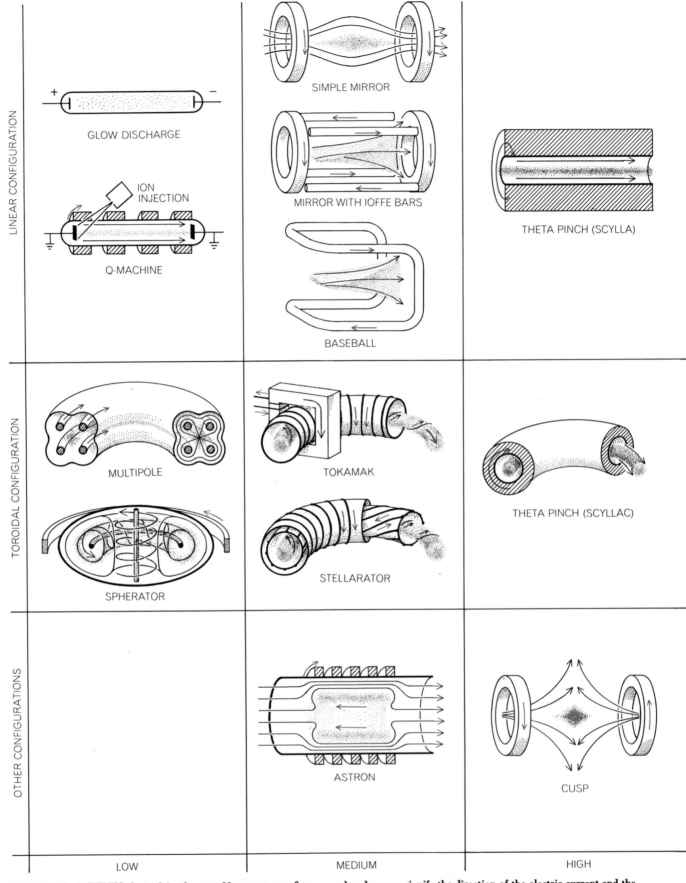

LINEAR CONFIGURATION

GLOW DISCHARGE

ION INJECTION

Q-MACHINE

SIMPLE MIRROR

MIRROR WITH IOFFE BARS

BASEBALL

THETA PINCH (SCYLLA)

TOROIDAL CONFIGURATION

MULTIPOLE

SPHERATOR

TOKAMAK

STELLARATOR

THETA PINCH (SCYLLAC)

OTHER CONFIGURATIONS

ASTRON

CUSP

LOW

MEDIUM

HIGH

PRINCIPAL SCHEMES devised in the past 18 years to confine plasmas for fusion research are arranged in the illustration on these two pages in order of increasing plasma density (*left to right*) and overall geometry (*top to bottom*). Only a few examples are depicted in each category. In every case the plasma is in color, the colored arrows signify the direction of the electric current and the black arrows denote the direction of the resultant magnetic field. Various structural details have been omitted for clarity. For each example shown there are a large number of variations either already in existence or in the conceptual stage. Furthermore, the

ZETA PINCH

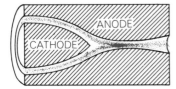

ANODE

CATHODE

PLASMA FOCUS

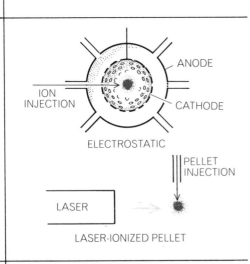

ZETA PINCH

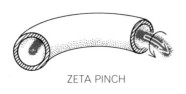

ANODE

ION
INJECTION

CATHODE

ELECTROSTATIC

PELLET
INJECTION

LASER

LASER-IONIZED PELLET

VERY HIGH

fact that an example is given in one category does not necessarily mean that that configuration is not applicable to some other category; there are, for instance, toroidal Q-machines and medium-density cusp designs.

specially designed to maximize the conditions for reaching one goal or the other. How does one compare the performances of these machines in order to gauge how near one is to the combined conditions needed to operate a practical fusion-power reactor? The basic criterion for determining the length of time a plasma must be confined at a given density and temperature to produce a "break even" point in the power balance was laid down in 1957 by the British physicist J. D. Lawson. Combining data on the physics of fusion reactions with some estimates of the efficiency of energy recovery from a hypothetical fusion reactor, Lawson derived a factor, which he called R, that denoted the ratio of energy output to energy input needed to compensate for all possible plasma losses. Lawson's criterion is still in general use as a convenient yardstick for measuring the extent to which losses must be controlled in order to make possible the construction of a fusion reactor. Although more recent calculations consider many other physical constraints in order to arrive at the break-even power balance, these criteria still give values very close to those derived by Lawson.

For a deuterium-tritium fuel mixture Lawson found that at temperatures higher than the ignition temperature the product of density and confinement time must be equal to 10^{14} seconds per cubic centimeter in order to achieve the break-even condition. This criterion defines a surface in three-dimensional space, the coordinates being the logarithmic values of density, temperature and confinement time [see illustration on page 143]. The goal of a break-even release of energy will have been achieved when the set of conditions for a given machine reaches this surface. It should be emphasized that the exact location and shape of the surface is a function of both the fuel cycle used and the recovery efficiency of the hypothetical reactor system. Fuels other than the deuterium-tritium mixture would increase the temperature needed to achieve a break-even power balance.

The extraordinary progress made recently by various groups in learning how to raise the combination of density, temperature and confinement time to a set of values approaching this break-even surface can be appreciated by referring to the illustration of the Lawson-criterion surface. The several plasma systems shown range in density from about 10^9 ions per cubic centimeter to 5×10^{22} ions per cubic centimeter. (Below a density of about 10^{11} ions per cubic centimeter the power density would be so low that it would require an impractically

large reactor.) The particular density range chosen for investigation in each case is a function of the scientific preferences of the investigators concerning the best route to fusion power and the available technology (magnets, power supplies, lasers and so forth). Thus there are various trajectories to the break-even surface being followed through the three-dimensional "parameter space" of the illustration. Closing the gap between where each trajectory is now and the break-even surface depends in some cases (for example the tokamak devices) on obtaining a better understanding of the physical principles required to develop reliable scaling rules, whereas in other cases (such as the linear theta-pinch devices) all that may be required is an economic solution to the engineering problem of building a large enough system.

Fusion-Reactor Designs

How would a full-scale fusion reactor operate? In the first place fusion reactors, like fission reactors, could be run on a variety of fuels. The nature of the fuel used in the core of a fusion reactor would, however, have a decisive effect on the method used to recover the fusion energy and the uses to which the recovered energy might be put. Most research on reactor technology has centered on the use of a deuterium-tritium mixture as a fuel. The reason is that the mixture has the lowest ignition temperature, and hence the lowest rate of energy loss by radiation, of any possible fusion fuel. Nonetheless, other combinations of light nuclei have been considered for many years as potential fusion fuels. Prominent among these are reactions involving a deuterium nucleus and a helium-3 nucleus and reactions involving a single proton (a hydrogen nucleus) and a lithium-6 nucleus. Because containment based on the magnetic-mirror concept actually operates better at higher temperatures, a number of other fuels have been attracting increased attention [see illustration on page 138].

Depending on the fuel used, a fusion reactor could release its energy in several ways. For example, neutrons, which are produced at various rates by different fusion reactions, can cross magnetic fields and penetrate matter quite easily. A reactor based on, say, a deuterium-tritium fuel cycle would release approximately 80 percent of its energy in the form of highly energetic neutrons. Such a reactor could be made to produce electricity by absorbing the neutron energy in a liquid-lithium shield, circulating the

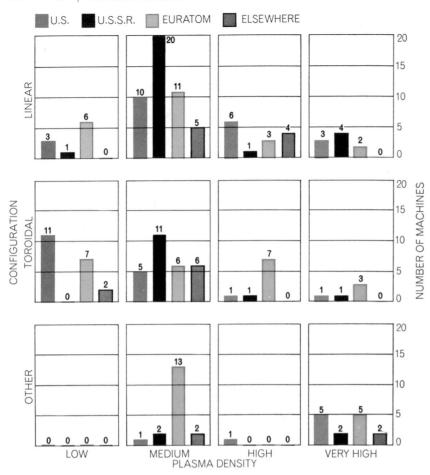

INVENTORY of the number of machines now operating throughout the world in each of the broad categories represented in the illustration on the preceding two pages is given in this table. The total number in each category is broken down into subtotals for the U.S., the U.S.S.R., the European Atomic Energy Community, or Euratom, countries (Belgium, France, Germany, Italy, Luxembourg and the Netherlands) and the rest of the world (principally Japan, Sweden and Australia). Britain, although not officially a member of Euratom, is included in the Euratom subtotal. The figures are drawn mainly from a recent survey compiled by Amasa S. Bishop and published by the International Atomic Energy Commission. The U.S. fusion-power program currently represents about a fifth of the world total.

liquid lithium to a heat exchanger and there heating water to produce steam and so drive a conventional steam-generator electric power plant [see top illustration on page 144].

This general approach could also lead to an attractive new technique for converting the world's reserves of uranium 238 and thorium 232 to suitable fuels for fission reactors—the fusion-fission hybrid system mentioned above. By employing the abundance of inexpensive, energetic neutrons produced by the deuterium-tritium fuel cycle to synthesize fissionable heavy nuclei, a fusion reactor could act as a new type of breeder reactor. This could have the effect of lowering the break-even surface defined by Lawson's criterion, bringing the fusion-breeding scheme actually closer to feasibility than the generation of electricity solely by fusion reactions. Cheap fuel might thus be made for existing fission

reactors in systems that could be inherently safe. A "neutron-rich" economy created by fusion reactors would have other potential benefits. For example, it has been suggested that large quantities of neutrons could be useful for "burning" various fission products, thereby alleviating the problem of disposing of radioactive wastes.

Fuel cycles that release most of their energy in the form of charged particles offer still other avenues for the recovery of fusion energy. For example, Richard F. Post of the Lawrence Radiation Laboratory at Livermore has proposed a direct energy-conversion scheme in which the energetic charged particles produced in a fusion-reactor core are slowed directly by an electrostatic field set up by an array of large electrically charged plates [see bottom illustration on page 144]. By a judicious arrangement of the voltages applied to the

plates such a system could theoretically be made to operate at a conversion efficiency of 90 percent.

J. Rand McNally, Jr., of the Oak Ridge National Laboratory has suggested that a long sequence of fusion reactions similar to those that power the stars could be reproduced in a fusion reactor. The data necessary to evaluate fuel cycles operating in this manner, however, do not exist at present.

The characteristics of a full-scale fusion reactor would depend not only on the fuel cycle but also on the particular plasma-confinement configuration and density range chosen. Thus it is probable that there eventually will exist a number of different forms of fusion reactor. For example, medium-density magnetic-mirror reactors and very-high-density laser-ignited reactors could be expected to operate at power levels as low as between five and 50 megawatts, which could make them potentially useful for fusion-propulsion schemes.

For central-station power generation the medium-density reactors would most likely operate on a deuterium-tritium fuel cycle in order to take advantage of the mixture's low ignition temperature. Because of the high neutron output associated with this fuel, a heat-cycle conversion system would be appropriate.

A reactor of this type would operate most efficiently with a power output in the billion-watt range. Before such a reactor can be built, it will be necessary to prove that the plasma will remain stable as present devices are scaled to reactor sizes and temperatures. Problems likely to be encountered in this effort involve the long-term equilibrium of the plasma, the interaction of the plasma with the walls of the container and the necessity of pumping large quantities of liquid lithium across the magnetic field.

Medium-density linear reactors would be better suited for fuel cycles that yield a major part of their energy output in the form of charged particles, since this approach would allow the direct recovery of the kinetic energy of these reaction products through schemes such as Post's. Such fuel cycles could be based on a deuterium-deuterium reaction, a deuterium-helium reaction or a proton-lithium reaction. A system operating on this principle could be made to produce direct-current electricity at a potential of about 400 kilovolts, which would be ideal for long-distance cryogenic (supercooled) power transmission.

Although the break-even conditions would be lowered in this case (because of the high energy-conversion efficiency), it still remains to be shown that existing experiments can be scaled to large sizes and higher temperatures. Some major technological obstacles that need to be overcome include the construction of large atomic-beam injectors and extremely strong magnetic mirrors.

For reactors operating on the basis of any of the higher-density schemes, such as the theta-pinch machines or the fast-pulsed systems, major technological hurdles include the development of efficient energy-storage and energy-transfer techniques and problems related to heating techniques such as lasers.

In addition to generating electric power and possibly serving in a propulsion system, fusion reactors are potentially useful for other applications. For example, fusion research has already made available plasmas that range in temperature from 500,000 to a billion degrees C. and in density from 10^9 to 10^{22} ions per cubic centimeter. Almost all industrial processes that use plasmas fall outside this range [see top illustration on page 139]. In order to suggest how this industrially unexplored range might be exploited, we recently put forward the concept of the "fusion torch." The gen-

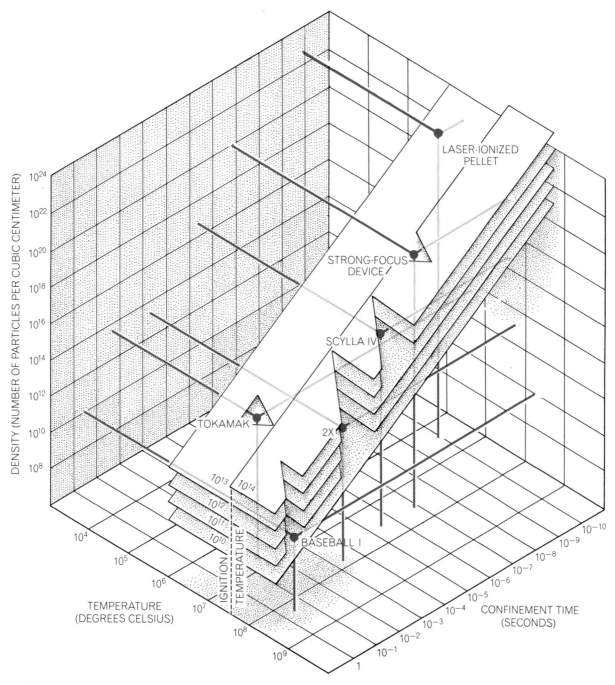

BASIC CRITERION for determining the length of time a plasma must be confined at a given density and temperature to achieve a "break even" point in the fusion-power balance is represented in this three-dimensional graph. The graph is based on a method of analysis devised in 1957 by the British physicist J. D. Lawson. For a deuterium-tritium fuel mixture in the temperature range from 40 million degrees C. to 500 million degrees C., Lawson found that the product of density and confinement time must be close to 10^{14} seconds per cubic centimeter to achieve the break-even condition (based on an assumed energy-conversion efficiency of 33 percent). This criterion corresponds to the top layer in the stack of planes in the illustration. The lower planes, which correspond to successively smaller values of density times confinement time, are included in order to give some idea of the positions of the best confirmed results from several experimental devices with respect to the combination of parameters needed to operate a full-scale fusion reactor.

144

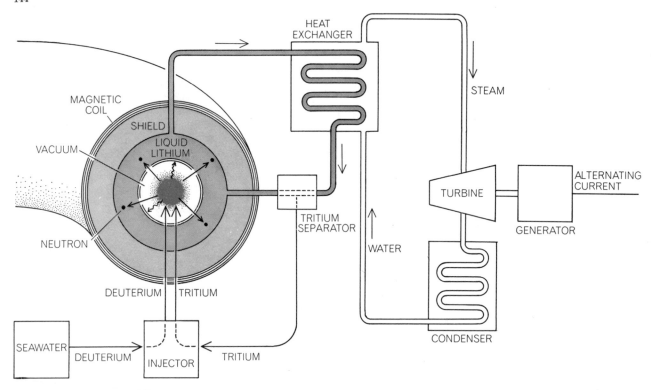

THERMAL ENERGY CONVERSION would be most effective in a fusion reactor based on a deuterium-tritium fuel cycle, since such a fuel would release approximately 80 percent of its energy in the form of highly energetic neutrons. The reactor could produce elec-

tricity by absorbing the neutron energy in a liquid-lithium shield, circulating the liquid lithium to a heat exchanger and there heating water to produce steam and thus drive a conventional steam-generator plant. The reactor core could be either linear or toroidal.

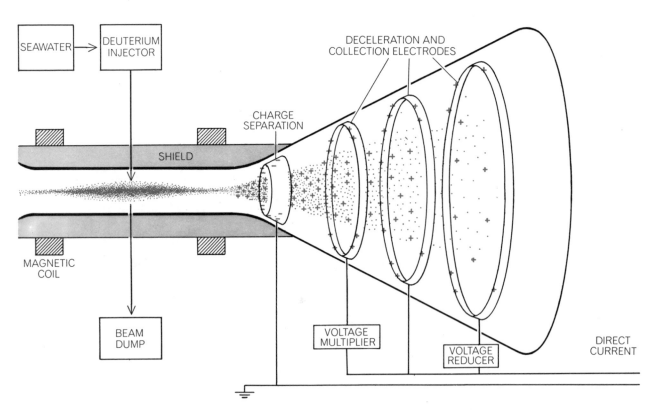

DIRECT ENERGY CONVERSION would be more suitable for fusion fuel cycles that release most of their energy in the form of charged particles. In this novel direct energy-conversion scheme, first proposed by Richard F. Post of the Lawrence Radiation Laboratory of the University of California at Livermore, the energetic charged particles (primarily electrons, protons and alpha particles) produced in the core of a linear fusion reactor would be released

through diverging magnetic fields at the ends of the magnetic bottle, lowering the density of the plasma by a factor of as much as a million. A large electrically grounded collector plate would then be used to remove only the electrons. The positive reaction products (at energies in the vicinity of 400 kilovolts) would finally be collected on a series of high-voltage electrodes, resulting in a direct transfer of the kinetic energy of the particles to an external circuit.

eral idea here is to use these ultrahigh-density plasmas, possibly directly from the exhaust of a fusion reactor, to vaporize, dissociate and ionize any solid or liquid material [*see top illustration at right*]. The potential uses of such a fusion-torch capability are intriguing. For one thing, an operational fusion torch in its ultimate form could be used to reduce all kinds of wastes to their constituent atoms for separation, thereby closing the materials loop and making technologically possible a stationary-state economy. On a shorter term the fusion torch offers the possibility of processing mineral ores or producing portable liquid fuels by means of a high-temperature plasma system.

The fusion-torch concept could also be useful in transforming the kinetic energy of a plasma into ultraviolet radiation or X rays by the injection of trace amounts of heavy atoms into the plasma. The large quantity of radiative energy generated in this way could then be used for various purposes, including bulk heating, the desalting of seawater, the production of hydrogen or new chemical-processing techniques. Because such new industrial processes would make use of energy in the form of plasmas rather than in the form of, say, chemical solvents, they would be far less likely to pollute the environment. Although the various fusion-torch possibilities are largely untested and many aspects may turn out to be impractical, the concept is intended to stimulate new ideas for the industrial use of the ultrahigh-temperature plasmas that have already been developed in the fusion program as well as those plasmas that would be produced in large quantities by future fusion reactors.

Environmental Considerations

The environmental advantages of fusion power can be broken down into two categories: those advantages that are inherent in all fusion systems and those that are dependent on particular fuel cycles and reactor designs. Among the inherent advantages, one of the most important is the fact that the use of fusion fuel requires no burning of the world's oxygen or hydrocarbon resources and hence releases no carbon dioxide or other combustion products to the atmosphere. This advantage is shared with nuclear-fission plants.

Another advantage of fusion power is that no radioactive wastes are produced as the result of the fuel cycles contemplated. The principal reaction products would be neutrons, nonradioactive heli-

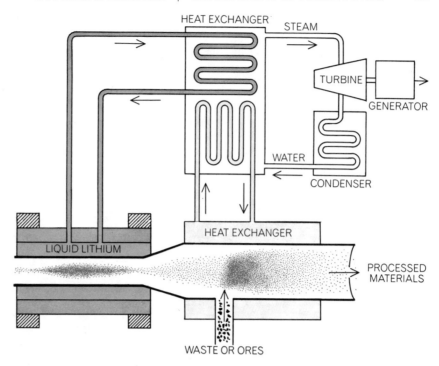

POTENTIAL NONPOWER USE of fusion energy is represented by the concept of the "fusion torch," which was put forward recently by the authors as a suggestion intended to stimulate new ideas for the industrial exploitation of the ultrahigh-temperature plasmas already made available by the fusion-research program as well as those that would be produced by fusion reactors. The general idea is to use some of the energy from these plasmas to vaporize, dissociate and ionize any solid or liquid material. In its ultimate form the fusion torch could be used to reduce any kind of waste to its constituent atoms for separation.

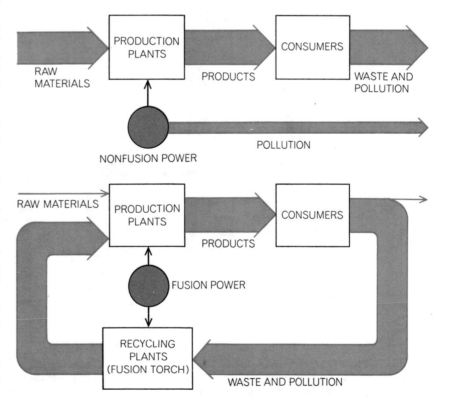

CLOSED MATERIALS ECONOMY could be achieved with the aid of the fusion-torch concept illustrated at the top of this page. In contrast to present systems, which are based on inherently wasteful linear materials economies (*top*), such a stationary-state system would be able to recycle the limited supply of material resources (*bottom*), thus alleviating most of the environmental pollution associated with present methods of energy utilization.

um and hydrogen nuclei, and radioactive tritium nuclei. It is true that tritium emits low-energy ionizing radiation in the form of beta particles (electrons), but since tritium is also a fusion fuel, it could be returned to the system to be burned. This situation is strongly contrasted with that in nuclear fission, which by its very nature must produce a multitude of highly radioactive waste elements.

Fusion reactors are also inherently incapable of a "runaway" accident. There is no "critical mass" required for fusion. In fact, the fusioning plasma is so tenuous (even in the "high density" machines) that there is never enough fuel present at any one time to support a nuclear excursion. This situation is also in contrast to nuclear-fission reactors, which must contain a critical mass of fissionable material and hence an extremely large amount of potential nuclear energy.

Among the system-dependent environmental advantages of fusion power must be counted the fact that the only radioactive fusion fuel considered so far is tritium. The amount of tritium present in a fusion reactor can range from near zero for a proton-lithium fuel cycle to a maximum for a deuterium-tritium cycle, where a "blanket" for the production of tritium must be included. Tritium, however, is one of the least toxic of the radioactive isotopes, whereas the fission fuel plutonium is one of the most toxic radioactive materials known.

The most serious radiological prob-

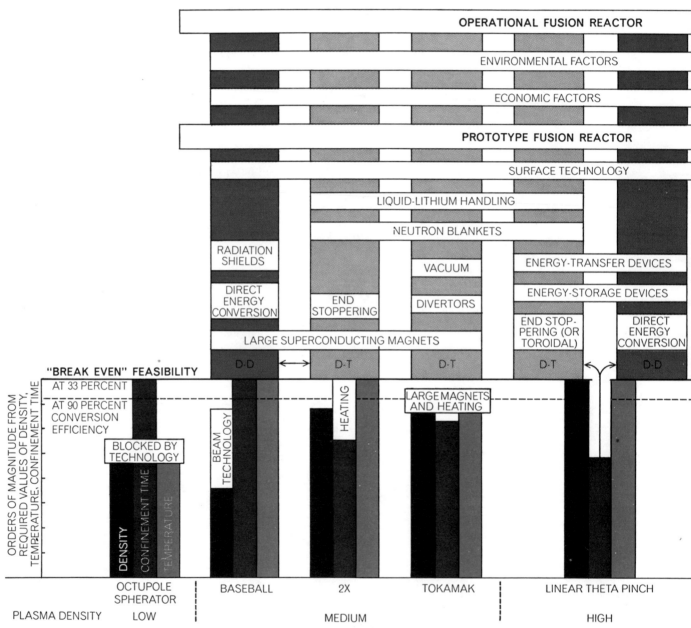

REMAINING PROBLEMS that must be solved before the goal of useful, economic fusion power can be achieved are depicted schematically in this illustration. The major experimental routes to the goal are ordered according to plasma density. Various experimental devices are represented by bars indicating the best combination of plasma density, confinement time and temperature achieved by each device; the logarithmic scale at lower left gauges how far each of these essential parameters is from the values needed to attain break-even feasibility. Technological problems that must be solved in each case are labeled. The achievement of a prototype reactor will be a function not only of plasma technology but also of the fuel cycle and the method of energy conversion chosen. Thus medium-density magnetic-mirror devices could be built to operate with either a deuterium-tritium (D-T) fuel mixture or a deuterium-deuterium (D-D) fuel mixture; the arrows signify these alternatives. If a D-D cycle is chosen, then direct energy conversion is possible, and once the converters are developed very few obstacles would remain to delay the construction of a prototype reactor. If, on the other hand, a D-T cycle is chosen, then conventional thermal energy conversion would be needed, and the listed technological

lems for fusion would exist in a reactor burning and producing tritium. A representative rate of tritium consumption for a 2,000-megawatt deuterium-tritium thermal plant would be about 260 grams per day. Tritium "holdup" in the blanket and other elements of the tritium loop would dictate the tritium inventory. Holdup is estimated to be about 1,000 grams in a 2,000-megawatt plant. If necessary, the doubling time of breeding tritium could be less than two months in order to meet the needs of an expanding economy. The amount of tritium produced by the plant is controllable, however, and need not exceed the fuel requirements of the plant.

Careful design to prevent the leakage of tritium fuel from a deuterium-tritium reactor is mandatory. Engineering studies that take into account economic considerations indicate that the leakage rate can be reduced to .0001 percent per day. The conclusion is that even for an all-deuterium-tritium fusion economy the genetic dose rate from worldwide tritium distribution would be negligible.

In fact, for a given total power output the tritium inventory for an all-deuterium-tritium fusion economy (including both the inventory within the plant and that dispersed in the biosphere) would be between one and 100 times what it would be for an all-fission economy. It is true that tritium would be produced in a deuterium-tritium fusion reactor at a rate of from 1,000 to 100,000 times faster than in various types of fission reactor. Since tritium is burned as a fuel, however, it has an effective half-life of only about three days rather than the normal 12 years.

A technology-dependent but possibly serious limitation on deuterium-tritium fusion plants could be the release of tritium to the local environment. The level would be quite low but the long-term consequences from tritium emission to the environment in the vicinity of a deuterium-tritium reactor needs to be explored. In general the biological-hazard potential of the tritium fuel inventory in a deuterium-tritium reactor is lower by a factor of about a million than that of the volatile isotope iodine 131 contained in a fission reactor. Of course there is no expectation in either case that such a release would occur.

The radioactivity induced in the surrounding structures by a fusion reactor is dependent on both the fuel cycle and the engineering design of the plant. This radioactivity could range from zero for a fuel cycle that produces no neutrons up to very high values for a deuterium-tritium cycle if the engineering design is such that the type and amount of structural materials could become highly activated under neutron bombardment. Cooling for "after heat" will be required for systems that have intense induced radioactivity. Even if the cooling system should fail, however, there could be no nuclear excursion that would disperse the radioactivity outside the plant.

Other system-dependent environmental advantages of fusion power include safety in the event of sabotage or natural disaster, reduced potential for the diversion of weapons-grade materials and low waste heat. In fact, the potential exists for fusion systems to essentially eliminate the problem of thermal pollution by going to charged-particle fuel cycles that result in direct energy conversion. Finally, there is the advantage of the materials-recycling potential of the fusion-torch concept.

The Timetable to Fusion Power

The construction and operation of a power-producing controlled-fusion reactor will be the end product of a chain of events that is already to a certain extent discernible. For controlled fusion, however, there can never be an instant equivalent to the one that demonstrated the "feasibility" of a fission-power reactor (the Stagg Field experiment of Enrico Fermi in 1942). To reach the plasma conditions required for a net release of fusion power it is necessary to first develop many new technologies. In this context the term "scientific feasibility" cannot be precisely defined. To some investigators it means simply the achievement of the basic plasma conditions necessary to reach the break-even surface in the illustration on page 313. To others it represents reaching the same surface—but with a system that can be enlarged to a full-scale, economic power plant. To a few it represents the attainment of a full understanding of all the phenomena involved.

Although these differing interpretations of what is needed to give confidence in our ability to construct a fusion reactor may be somewhat confusing, each interpretation nevertheless contains a modicum of truth. To depict the complexity of this drive toward the goal of fusion power we have prepared the highly schematic illustration on the opposite page. The goal is to achieve useful, economic fusion power. The major routes to the goal are ordered in the illustration according to plasma density. Various individual experiments have climbed past various obstacles to reach positions close to the break-even level. In fact, in some instances two of the three essential parameters (density, temperature and confinement time) have already been achieved. The ignition temperature has been achieved in a number of cases. The rest of the climb to the break-even level in some cases involves a better understanding of the physics of the plasma-confinement system, but in others it may involve only engineering problems. Indeed, the location of the break-even level is a function of the technology used. Direct energy conversion, for example, would lower this level.

The next portion of the climb, the

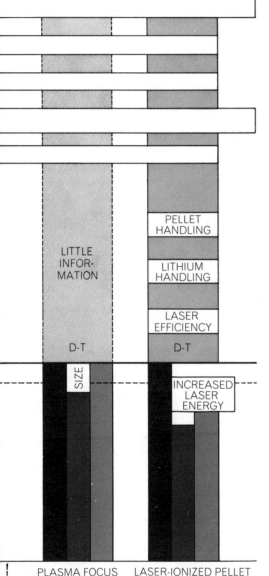

PLASMA FOCUS   LASER-IONIZED PELLET

VERY HIGH

hurdles would have to be overcome. Other systems operating on the D-T cycle would have to climb past similar hurdles. The high-density linear theta-pinch device could take either the thermal-conversion path or the direct-conversion path (arrows). The final step, from a prototype to an operational reactor, would proceed through a region in which economic and environmental considerations can be expected to be paramount.

construction of a prototype reactor, will be a function of the route taken to scientific feasibility. For instance, if a deuterium-tritium mixture is the fuel, this would require the development of components such as lithium blankets, large superconducting magnets, radiation-resistant vacuum liners, fueling techniques and heat-transfer technology. If fuels that release most of their energy in the form of charged particles are considered, however, then in the case of mirror reactors direct-conversion equipment may be part of a device used to demonstrate break-even feasibility. The step from that device to a prototype reactor could then be very short because the conversion equipment would be already developed. Other devices would face similar problems of differing magnitude in prototype construction. The final step from a prototype to an operational reactor would proceed through a much more nebulous region in which economic and environmental considerations would influence the comparative desirability of different power plants.

At present the main factor limiting the rate of progress toward fusion power is financial. The annual operating and equipment expenditures for the U.S.

fusion program, when one uses the consumer price indexes to adjust these dollars for inflation, has remained fairly constant for the past eight years [*see illustration below*]. The total amount spent on the program since its inception is the cost equivalent of a single Apollo moon shot. The annual funding rate of about $30 million per year is the equivalent of 15 cents per person per year in the U.S.

The road to fusion power is a cumulative one in that successive advances can be built on earlier advances. At present the U.S. has a fairly broad program of investigations approaching the break-even surface for net energy release. It is essential that larger (and thus more expensive) devices be built if the goal of the break-even surface is to be reached. The surface should be broken through in a number of places so that the relative advantages of the possible routes beyond that surface to an eventual fusion-power reactor can be assessed.

Clearly the timetable to fusion power is difficult to predict. If the level of effort on fusion research remains constant or decreases slightly, the requirement for larger devices and advanced engineering will automatically cause a premature

narrowing of the density range under investigation. This increases the risk of reaching the goal in a given time scale. To put it another way, it extends one's estimate of the probable time scale. If the level of fusion research expands sufficiently to maintain a fairly broad program across the entire density range, the probability of success increases and the probable time scale decreases. If fusion power is pursued as a "national objective," expanded programs could be carried out across the entire density range accompanied by parallel strong programs of research on the remaining engineering and materials problems to determine as quickly as possible the best routes to practical fusion-power systems. Therefore, depending on one's underlying assumptions on the level of effort and the difficulties ahead, the time it would take to produce a large prototype reactor could range from as much as 50 years to as little as 10 years.

There is at least one case in which the fusion break-even surface could be reached without making any new scientific advances and without developing any new technologies. This "brute force" approach, which might not be the optimum route to an eventual power reactor, would involve simply extending the length of the existing theta-pinch linear devices. It has been estimated that to reach the break-even surface by this method such a system would have to be about 2,000 feet long—less than a fifth of the length of the Stanford Linear Accelerator. This one fusion device, however, would cost an order of magnitude more than any experimental fusion device built to date. Even though a simple scaling of this type would introduce no new problems in plasma physics, one could not exclude the possibility of unexpected difficulties arising solely from the extended length of the system.

The length of such a device could be shortened by as much as 90 percent by installing magnetic mirrors at the ends, by increasing the diameter of the plasma or by making the system toroidal, but these steps would introduce new physical conditions. The system could also be shortened by the use of a direct energy-conversion approach, but this would introduce an unproved technology. At present a significant portion of the fusion-power program is concentrating on developing the new physics and technology that would reduce the cost of such break-even experiments. This continuing effort is sustained by the growing conviction that the eventual attainment of a practical fusion-power reactor is not blocked by the laws of nature.

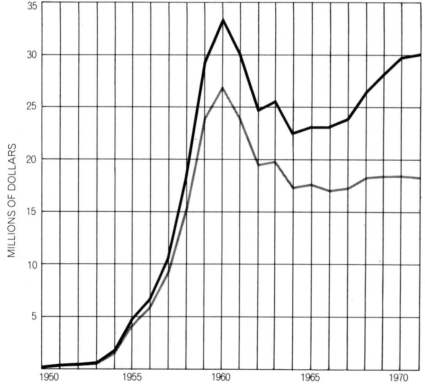

FINANCIAL SUPPORT is currently the main factor limiting the rate of progress toward the goal of fusion power. The solid curve shows the annual operating and equipment expenditures for the U.S. fusion program. The gray curve shows these expenditures adjusted for inflation. The adjustment shows that fusion research has been funded by the Atomic Energy Commission at an essentially constant rate for the past eight years. Smaller research programs have been funded by both private industry and other Government agencies.

CONSERVATION, POLLUTION PROBLEMS, SOLAR ENERGY, AND BEYOND

IV

IV

CONSERVATION, POLLUTION PROBLEMS, SOLAR ENERGY, AND BEYOND

INTRODUCTION

The cheapest new energy source is conservation, especially when reckoned on the basis of marginal (that is, incremental) cost. This point is so important that it deserves a fuller exposition. Up until about 1973 energy costs had been declining. Cheaper Middle East oil was gradually displacing more costly domestic oil in the United States; the lower fuel costs and higher efficiencies were bringing down the price of electric power. But then the world price of oil increased dramatically, while at the same time pollution-control requirements, inflation, and construction delays increased the cost of new electric plants. As a result, the incremental cost of new energy rose sharply; but since its cost was "rolled into" a much larger supply of older and cheaper energy, the average cost was not affected.*

Because of this high marginal cost of new energy to the nation, the government has every reason to encourage, subsidize, and sometimes even legislate conservation. Such is the case for automobiles, which account for 15 percent of U.S. energy use (and about one-third of all of the oil). Congress has legislated improvements in fuel economy. John Pierce describes in some detail how such economies can be achieved. But another important way to conserve involves a change in lifestyle: either by driving less or by carpooling.

The problem of environmental protection can be viewed in a similar manner. Here various kinds of social costs are imposed on other members of society by those who pollute. Again, this situation provides a rationale for government action, first by setting ambient air- and water-quality standards (based on considerations of human health and welfare), and next by setting emission standards for automobiles, power plants, various industries, and so on, which are designed to achieve these ambient standards. The article by S. Fred Singer describes the pollution effects of energy production in particular. Coal-fired power plants have the greatest negative impact on both air and land, gas-fired plants the least. Nuclear plants are capable of very low pollution impacts, with uranium mining being a major item. Breeder reactors would reduce the amount of land disturbance considerably.

All forms of energy generation produce large quantities of waste heat, which has local and eventually global impacts. Solar energy is the major exception, but it often requires the construction of dams and windmills, or the covering of large land areas with solar collectors. Solar energy also has the advantage of not adding to the carbon dioxide in the atmosphere. (The same is true for nuclear

---

*For example, if the new energy sources cost 10 times as much (that is, if marginal costs increase by 1000 percent) but constitute only 0.5 percent of the total energy supply, then the average cost increases only by $(100\% - 0.5\%) + (0.5\% \times 10)$ or 4.5 percent. If new energy sources constitute 5 percent of the total, then the increase in average cost is 45 percent; if 50 percent, the increase jumps to 500 percent.

energy.) Fossil fuel burning, on the other hand, is increasing atmospheric carbon dioxide (CO_2) concentrations. The consequences are not certain, nor can we fully explain where all the CO_2 comes from and goes to in the environment. In the past, the oceans and the biosphere (especially trees) were believed to have absorbed about 50 percent of the CO_2 generated by fossil fuel burning. Some ecologists now believe that, because of extensive deforestation, the biosphere actually adds to atmospheric CO_2.

Solar energy comes in many forms: hydropower, windpower, biomass (including fossil fuels), and direct radiation. Radiation can be degraded into heat—and, by focusing, into high-temperature heat. But the radiation can also be converted directly into electricity through solar "batteries" using the photovoltaic effect. Currently, this field of research and development is most active, the aim being to reduce the cost of solar electric power. This effort is described in the article by Bruce Chalmers.

It seems likely that solar electric power, will first become competitive in decentralized applications where central-station power is either unavailable or very expensive. It is useful to keep in mind that electricity is cheap in the United States; the *average* cost is 3 to 10 cents per kilowatt-hour (kwh). In less developed countries the price of urban electricity can be up to ten times as great. Incremental power costs usually are much greater. Diesel-generated electric power can run up to $1 to $2 per kwh, and primary battery power on the order of $10 per kwh. Solar energy would therefore be even more competitive if we were to compare its cost with the incremental rather than the average cost of conventionally produced electricity.

The final article in the series is rapidly becoming a classic. Freeman J. Dyson explains in nontechnical language why we have energy in the universe at all. Why hasn't the universe run down its initial endowment of energy? And what of the future?

12   The Fuel Consumption
of Automobiles

by John R. Pierce
January 1975

*The biggest target for energy conservation is the poor
fuel economy of American cars. Here is how their
efficiency can be increased at least 40 percent by 1980*

A fourth of all the energy used in the U.S. is devoted to transportation, and of that fraction close to 60 percent is supplied in the form of gasoline to roughly 100 million automobiles and small personal trucks. Americans use more energy to fuel their cars than they do for any other single purpose. At the current price of some 55 cents per gallon, the average family is obliged to spend more than $600 a year just on gasoline. The fuel used by American cars and personal trucks would approximately fill all the energy needs of Japan, a nation of 108 million and the world's largest consumer of energy after the U.S. and the U.S.S.R. In the urgent effort to reduce U.S. consumption of an increasingly costly fuel whose chief reserves lie overseas, the American automobile and current habits of its utilization are a prime target.

One does not have to be a partisan of the automobile to recognize that virtually every aspect of American life—industrial, commercial, cultural and recreational—is now organized around the existence of motor vehicles. Whether or not they provide the most rational means of transportation in an advanced technological society is, of course, a matter of debate. In order to illuminate that debate a colleague and I organized a series of six two-day seminars on the subject "Energy Consumption in Private Transportation." The series, supported by the U.S. Department of Transportation, was held at the California Institute of Technology between December, 1973, and April, 1974, a period that coincided with the Arab oil embargo, with President Nixon's call for "Project Independence 1980" ("To ensure that by the end of this decade Americans will not have to rely on any source of energy beyond our own") and with the quadrupling of oil prices after the embargo was lifted.

The seminar participants addressed themselves to the following questions (among others): In any rational energy program what role will the private automobile play? Can it be replaced by more economical forms of transportation? To what extent can communication replace transportation? Will our pattern of life change in such a manner that we simply do not travel as much?

Although many fascinating and even plausible alternatives to the gasoline-powered automobile were discussed, it became evident to the participants that no dramatic change in transportation methods or habits can be expected or effectuated in the short run, say before 1990. In this article, therefore, I shall deal only with existing or readily foreseeable technologies for improving the fuel economy of automobiles as we know them.

In President Ford's address before a joint session of Congress last fall he announced his determination to obtain "either by agreement or by law a firm program aimed at achieving a 40 percent increase in [automobile] gasoline mileage within a four-year development deadline." Subsequently, in late October, the Department of Transportation (DOT) and the Environmental Protection Agency (EPA) submitted a report to Congress ("Potential for Motor Vehicle Fuel Economy Improvement") that provided a careful review of feasible engineering changes that should make it possible for 1980-model cars to go 40 percent farther on a gallon of gasoline than the average 1974 model did.

According to the DOT-EPA report, the average 1974 model, adjusted for the sales of different brands and models, achieved 14 miles per gallon on a "composite" fuel-economy cycle based on EPA dynamometer tests that simulate city driving conditions and highway driving conditions in a 55 : 45 ratio (a ratio chosen as being typical of American car use). The new 1975 models, on a projected sales-weighted basis, achieve 15.9 m.p.g., an improvement of 13.5 percent [*see top illustration on page 154*].

The improvement is chiefly attributable to engineering changes that regained much of the efficiency previously lost in adjusting engines to meet Federal exhaust-emission standards. Many of the 1975 models have catalytic converters to clean up exhaust emissions, making it possible for the engine to be retuned for higher efficiency. Partly because of emission controls and partly because of vehicle weight and other factors, the fuel economy of American cars dropped about 12 percent between 1967 and 1974, climaxing a long, steady decline that began as early as 1951 [*see top illustration on page 155*].

If the performance of 1974 models is taken as the base line, as the DOT-EPA report recommends, the industry has already moved a third of the way to the 40 percent improvement asked by President Ford. A 40 percent improvement would mean that the average car built in the 1980-model year would have to achieve a minimum fuel economy of 19.6 m.p.g. (Bills that are now being drafted for presentation to Congress will undoubtedly pick a round number for minimum 1980 performance, probably 20 or 21 m.p.g.)

According to the DOT-EPA report, the 40 percent improvement by 1980 should be attainable with the present Otto-cycle (four-stroke) gasoline engine, in combination with improved transmissions, reduced weight and aerodynamic drag and improved accessories. If the composition of sales can also be altered to include a much higher proportion of compact and subcompact models than

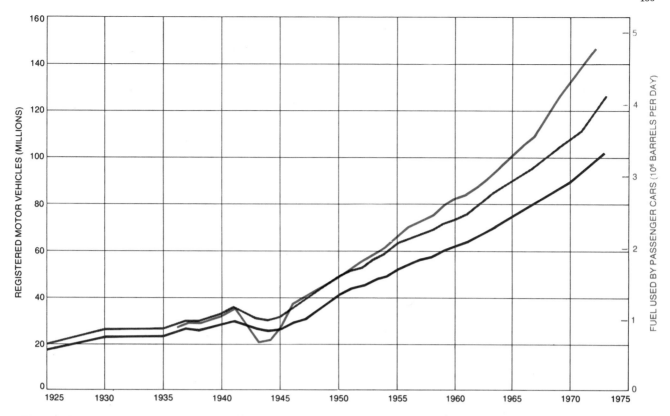

NUMBER OF MOTOR VEHICLES IN THE U.S. (*gray curve*) has been doubling approximately every 20 years, corresponding to an increase of 3.5 percent per year. Private cars (*black curve*) make up about 82 percent of the total motor-vehicle population; the balance consists chiefly of trucks. Gasoline consumption by private cars (*curve in color*) has been increasing more rapidly than the car population, owing primarily to a steady downward drift in average fuel economy (*see top illustration on page 155*). Because of duplicate registrations when cars change hands or owners move the registration figures may overstate the car population by about 10 percent.

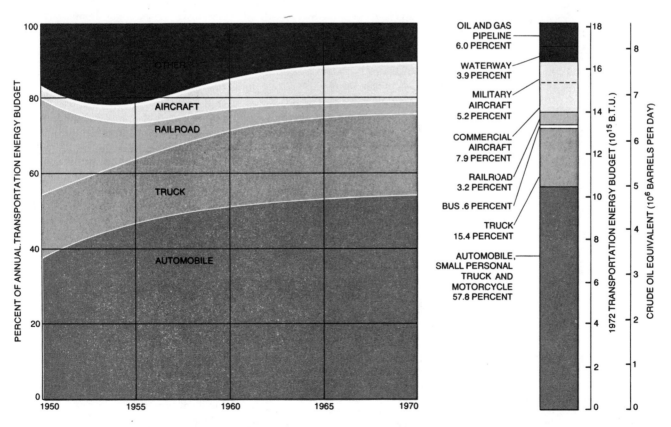

TRANSPORTATION ENERGY BUDGET depicts the decline of railroads and the ascendancy of trucks, automobiles and airplanes over the 20-year period 1950–1970 (*left*). "Other" includes mass-transit systems, buses, waterways and oil pipelines. The detailed breakdown for 1972 (*right*) is based chiefly on a study by the Rand Corporation. In 1972 the nation's total energy consumption was 72.1×10^{15} British thermal units, of which transportation took 25.2 percent. Oil supplied 45.5 percent of all energy used in 1972.

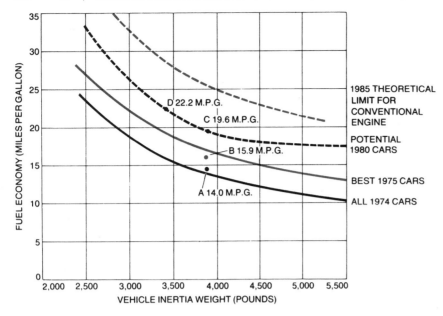

FUEL-ECONOMY RESULTS, computed by the Environmental Protection Agency (EPA) from dynamometer tests, show a substantial increase in the performance of 1975-model cars over 1974 models. The sales-weighted average for the 1974 "fleet" was 14 miles per gallon (A); for the 1975 fleet, assuming the same model mix, it is 15.9 m.p.g. (B), or an average improvement of 13.5 percent. The change in performance between the two model years, however, varied considerably among different car lines (see illustration below). The curves shown here are from a report recently presented to Congress by the EPA and the U.S. Department of Transportation (DOT). The fuel-economy figures represent a new "composite" cycle, consisting of a 55 : 45 combination of two test cycles conducted by the EPA: a suburban-urban cycle with several stops and starts at an average speed of 20 miles per hour and an uninterrupted highway cycle at an average speed of 49 m.p.h. (The EPA publishes the two figures separately for 1975 cars; these are the figures often seen in car advertisements.) The DOT-EPA report to Congress suggests that it should be possible for 1980 cars to achieve an average of 19.6 m.p.g. (C), an improvement of 40 percent over 1974 models. With a shift to smaller, lighter cars, average of 1980 automobile fleet might even reach 22.2 m.p.g. (D).

it does at present, it may even be possible for the 1980 "fleet" of new cars to exceed 22 m.p.g., an improvement of nearly 60 percent over 1974.

To achieve still greater advances in fuel economy for 1985 and beyond, it will probably be necessary to introduce new types of engines. The DOT-EPA report suggests the use of diesel engines in medium-size and large cars and, for smaller cars, gasoline engines designed to operate on a "stratified charge," engines in which the air-fuel mixture is made intentionally nonhomogeneous to provide an average lean mixture, with a consequent improvement in efficiency.

The present Otto-cycle automobile engine typically achieves a thermal efficiency of between 22 and 27 percent. Under the normal range of driving conditions, however, the net efficiency of power delivered to the wheels is only about 10 percent. Gasoline-fueled aircraft engines attain efficiencies of about 30 percent. The efficiencies of diesel engines range from 35 percent to as high as 38 percent. (In EPA tests the 1975 Mercedes-Benz 300D, which has a five-cylinder, 77-horsepower diesel engine, gets 24 m.p.g. in the simulated city-driving cycle and 31 m.p.g. in the highway-driving cycle, yielding a composite fuel economy of 27.2 m.p.g. These values are 50 percent higher than those of the comparable gasoline-engine model of the Mercedes: the four-cylinder, 93-horsepower Model 230. It should be noted, however, that diesel fuel contains about 10 percent more energy than ordinary gasoline.)

Those who are as old as I am can remember a time when mass transit dominated urban life in the U.S. In St. Paul toward the end of World War I my parents did not own an automobile. The roads between towns were unmarked and often badly rutted, making travel unattractive for those who did drive. Everything was within easy walking distance of the streetcar or interurban line, even the cottage at White Bear Lake where we stayed during the summer. Small stores and shops were within easy walking distance of our home, and in St. Paul itself the stores and offices in the central business district were accessible by streetcar.

Between 1920 and 1930 the number of passenger cars registered in the U.S. nearly tripled, from eight million to 23 million. In spite of the Great Depression the public desire for private transportation continued to grow (even though most streetcar lines were still running), until by 1940 there were more than 27

MANUFACTURER	FUEL ECONOMY (MILES PER GALLON)		CHANGE (PERCENT)
	1974	1975	
GENERAL MOTORS	10.60	13.30	+25.5
FORD	12.16	11.92	-2.0
CHRYSLER	11.56	12.45	+7.7
AMERICAN MOTORS	14.64	17.05	+16.5
VOLKSWAGEN	22.11	21.95	-.7
TOYOTA	16.89	16.37	-3.1
NISSAN	20.63	22.04	+6.8
VOLVO	16.55	15.61	-5.7
AUDI	19.14	20.17	+5.3
PEUGEOT	17.33	19.09	+10.2
SAAB	17.28	21.41	+23.9
DAIMLER-BENZ	10.93	11.80	+8.0
BMW	17.44	15.47	-11.3
FLEET	11.95	13.33	+11.5

FUEL-ECONOMY RESULTS BY MANUFACTURER for 1974 and 1975 automobiles, assuming the 1974 sales-weighting for various models for both years, appear in a paper by Thomas C. Austin and Karl H. Hellman of the EPA. The m.p.g. values are for the suburban-urban cycle only and reflect the changes in the two model years that can be attributed solely to engineering "system" changes (for example changes due to engine emission-control-system calibrations and changes in transmission and axle ratios). The table thus omits the effect of new engine sizes and new engine-vehicle combinations that contributed to a total improvement of 13.8 percent in the performance of the 1975 fleet over the 1974 fleet in the suburban-urban test cycle. (The gain is 13.5 percent when the highway cycle is included.)

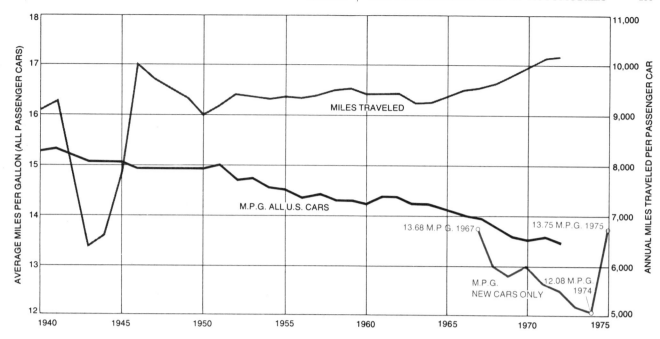

DETERIORATION IN FUEL ECONOMY of American automobiles has been virtually uninterrupted since 1940 (*black curve*). The figures are compiled by the Federal Highway Administration on the basis of gasoline sales and state-by-state surveys of annual miles traveled per vehicle (*gray curve; scale at right*). The sharp drop in miles traveled between 1941 and 1946 represents the effects of strict World War II gasoline rationing. If the average car on the road still obtained 15 m.p.g., as it did in 1951, the U.S. consumption of gasoline would be 450,000 barrels per day (or nearly 10 percent) less than it actually is. Between 1967 and 1974 the fuel economy of

new cars fell sharply owing to a number of factors: increased body weight, the growing popularity of air conditioning and particularly engine resettings to meet new Federal standards on exhaust emissions. The curve in color, based on the paper by Austin and Hellman, shows that for the suburban-urban test cycle a 13.8 percent improvement between 1974 and 1975 models has returned the fuel economy of new cars to approximately the level that prevailed in 1967. Austin and Hellman point out that the suburban-urban test cycle yields fuel-economy figures that agree closely with the national figures based on gasoline sales and total miles traveled.

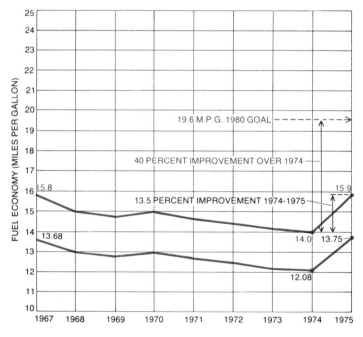

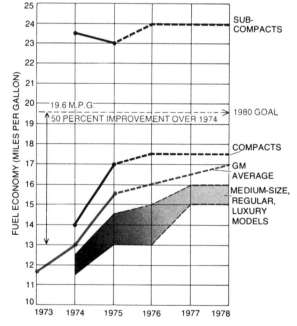

1980 FUEL-ECONOMY GOAL may be embodied in a law requiring a 40 percent improvement over the performance of 1974 models, which averaged 12.08 m.p.g. as measured by the EPA suburban-urban test cycle (*lower curve at left*) or 14 m.p.g. as measured by the composite cycle of city and highway driving (*upper curve*). By the second measurement the 1980 goal would be 19.6 m.p.g., which will probably be rounded upward to 20 or even 21 m.p.g. It will obviously be easier for makers of light, small cars to meet that goal than for a manufacturer such as General Motors. The solid curve in color at right shows the fuel economy for the GM fleet in 1973,

1974 and 1975 as measured by GM's own "city-urban" cycle conducted on the road. The GM method gives results that compare fairly well with the EPA composite cycle. The 1974 GM fleet averaged about 12.2 m.p.g. by the EPA method and 13 m.p.g. by the GM method. (For the 1975 fleet the two methods yielded 15.7 and 15.5 m.p.g. respectively.) Thus by 1980 GM must improve the performance of its new car fleet by 6.6 m.p.g. (50 percent) or 7.4 m.p.g. (60 percent), depending on the test method selected. GM has already announced that its 1978 fleet will achieve 17 m.p.g. (*broken curve*) if no more changes are made in safety or emission standards.

million automobiles registered in the U.S., or one car for every 1.3 families. When the production of automobiles resumed after World War II, automobile registrations climbed swiftly to 40 million in 1950, to 61.7 million in 1960 (approaching 1.2 per family), to 89 million in 1970 (1.4 per family) and to an estimated 105 million in 1974 (1.5 per family). Today more than eight families in 10 own automobiles, and one family in three has two or more vehicles, if small personal trucks are included.

Nationwide studies show that automobiles are used primarily for short trips: about half of all trips are five miles or less and three-fourths are less than 10 miles. These short trips account for nearly a third of all vehicle miles traveled and for a substantially larger fraction of the total gasoline consumption. Moreover, about 40 percent of all automobile travel is work-related, chiefly commuting trips (with an average occupancy of 1.2 per car) at hours of high traffic density and resulting low efficiency of operation. Except for the special case of Manhattan Island, where 79 percent of all workers reach their jobs by public transit, the automobile provides the principal means for getting to work. According to the 1970 census, 60 million Americans commute by private automobile (51 million travel alone and nine million in car pools); 4.2 million use a bus or streetcar; 1.8 million use a subway or an elevated-railway line; 500,000 use railroads; 300,000 use taxis and 5.7 million live close enough to their jobs to walk to them. Clearly an enormous national effort, extending over several decades and costing many billions of dollars, would be required to provide public-transit facilities attractive, convenient and extensive enough to persuade a large number of Americans to leave their cars at home. In the absence of such a commitment any substantial reduction of gasoline in automobile usage can come only through changing the efficiency of the use of private automobiles or changing the efficiency of the automobiles themselves. National statistics show that if automobiles could achieve 25 to 30 miles per gallon, they would be about as efficient in moving people, at least thermodynamically, as present-day bus transit systems are.

The efficiency of the use of automobiles of a given construction and state of repair is chiefly affected by freeways and traffic-control systems. Our suburban pattern of living preceded the construction of freeways and it exists where freeways are few or absent. Once freeways have been built, however, they frequently lead to the creation of new locations of work and residence. Thus it is not clear whether in the long run freeways increase distances to and from work or decrease them.

Apart from (usually) reducing the travel time between two points, freeways have other important consequences. They have lower accident rates than country roads or city streets. The number of fatal accidents per million vehicle miles are: city streets, 2.47; country roads, 1.64; urban streets, 1.63, and urban freeways, .44. In addition the stop-and-go character of city-street traffic causes high gasoline consumption. For an equivalent trip it is usually more economical of gasoline to travel by freeway. It is doubtful, however, that construction of freeways can be justified as a way of reducing the energy demand. Such construction should be decided on other grounds.

Like freeways, traffic control can affect energy consumption as well as safety and convenience. Repeated starting and stopping wastes fuel. Although traffic lights are old, computer-controlled signal systems designed to facilitate the flow of traffic are fairly new. Computerized traffic control can reduce the number of starts and stops. Clogged free-

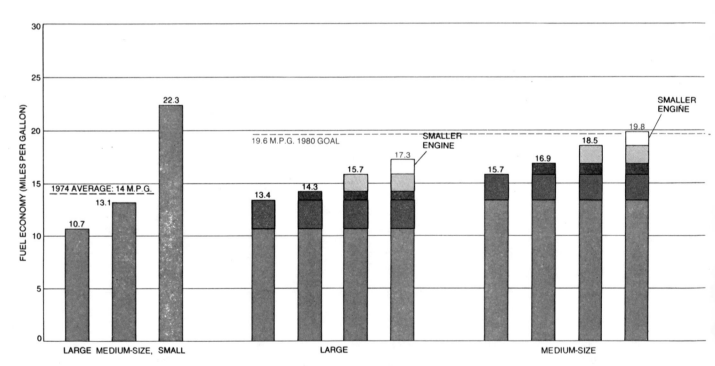

REASONABLE TECHNICAL IMPROVEMENTS that should enable car makers to reach the fuel-economy goal of 19.6 m.p.g. by the 1980-model year are described in the DOT-EPA report to Congress. The first set of three bars shows the average fuel economy of 1974 models (city-highway cycle) in three size categories. The large size is represented by present standard-size and luxury cars, which carry six people in comfort. The medium size corresponds to present compact and intermediate size, which carry four or five people. The small size corresponds to subcompacts and small imported cars designed for four passengers. The market share of each size category in 1974 was respectively 27, 45 and 28 percent. The largest single increase in fuel economy (from 15 to 25 percent) will come from engine modifications, some of which have already been incorporated in the 1975 models of some manufacturers. Further gains, amounting to about 9 percent in each size class, can be expected from more efficient four-speed automatic transmissions with

ways are wasteful of energy. Ramp control can reduce the congestion and increase the traffic flow on freeways. Indeed, it is estimated that fuel consumption can be reduced by about 10 percent in the controlled area.

The cost of computerized traffic control is considerable. Its principal benefits are reductions in travel time, increased traffic flow without the construction of additional roads and reduction in accidents. Fuel savings are an attractive added benefit. The fact remains that computer control of traffic must be justified by its overall benefits and not solely as a means of saving fuel.

Apart from a reduction in automobile usage (and a strict enforcement of the 55-mile-per-hour speed limit), more efficient automobiles seem the only sure way to achieve substantial savings in petroleum consumption in any future we can now foresee and make plans for. Fortunately there are many opportunities for making the private car more efficient.

As the Federal figures show, there are already large differences in gas mileage among current automobiles, ranging in economy tests of 1975 models from 27 m.p.g. in simulated city driving for the subcompact Datsun B-210 in the 2,250-

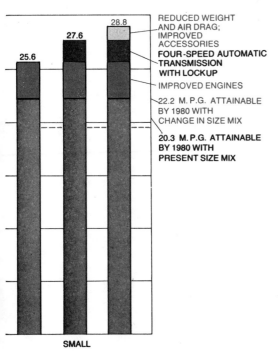

a lockup in high to prevent slippage. The third category of improvements (of 5 to 12 percent) will result from reductions in curb weight, aerodynamic drag, rolling resistance and power required by accessories. For large and medium-size cars a still further gain of 10 to 15 percent in fuel economy can be achieved by reducing the sizes of engines.

pound inertia-weight class to 12 m.p.g. for typical American cars in the 4,500-pound inertia-weight class down to 10 m.p.g. or less for the largest models in the 5,500-pound inertia-weight class. ("Inertia weight" is curb weight plus an allowance of 300 pounds for occupants.) In simulated highway driving the performance in the three weight categories rises to between 33 and 39 m.p.g., 15 and 18 m.p.g. and 14 and 16 m.p.g.

If one compares the average fuel economy of the lightest cars with that of the heaviest, one finds that each additional 100 pounds of car weight requires an extra 15 to 17 gallons of gasoline per year of average driving (10,000 miles). This does not mean, however, that simply by removing 100 pounds of weight from a heavy car one can achieve a comparable fuel saving. What it does mean is that the owner of one of the smaller, more economical cars will need to buy 500 to 550 fewer gallons of gasoline in the course of a year than the driver of a big car. Looked at another way, if all the automobiles now on the road averaged 23.5 m.p.g. instead of the estimated 13.5, U.S. gasoline consumption would drop more than 40 percent, or some two million barrels of gasoline per day below the current demand of about five million barrels. With crude oil at $11 per barrel this would translate into an annual saving of more than $8 billion.

The most obvious way, therefore, to save energy is to cut the weight of the average car sold. At low speeds, where aerodynamic drag is not a major factor, rolling resistance and the energy needed to overcome it are proportional to weight. On the EPA composite cycle of city and highway driving, rolling resistance and aerodynamic drag each absorb 24.7 percent of the useful power delivered by the engine of the typical American car [see top illustration on page 160]. Existing cars, American as well as foreign, show that lighter cars can be as quiet, easy-riding, roomy and comfortable as heavier cars, but to attain these qualities in a lighter car requires good engineering.

Although lighter cars tend to have less rolling resistance per unit of weight than heavier cars, the rolling resistance depends on the nature of the tire and its pressure. The energy loss in rubber tires is caused by the flexing of rubber; because of hysteresis the tire does not give back all the energy that went into deforming it. At the same pressure radial tires offer significantly less rolling resistance than conventional bias-ply tires. In mixed city and highway driving existing cars equipped with steel-belted

radials will go about 2.5 percent farther on a gallon of gasoline than a car with bias-ply tires. In a car with less air resistance than current cars, the percentage gain in going to radial tires would be even larger. With either kind of tire the rolling resistance can be cut about a fourth by raising the tire pressure from 20 pounds per square inch to 40 pounds. Beyond that a worn tire toward the end of its life has only a little more than half the rolling resistance of the same tire when it was new. Perhaps we have incorporated too much rubber in tires in seeking long tire life rather than good gasoline mileage.

Shock absorbers absorb energy only in going over bumps. Tires absorb energy in rolling on a smooth surface. Lighter wheels combined with suspension systems carrying less unsprung weight would make it possible to mount harder, lower-loss tires. Thus it appears that substantial energy savings could be attained through better suspensions and smaller, harder tires, particularly if a rubber could be developed with lower hysteresis at ambient temperatures than present rubbers have.

Next to building (and persuading Americans to buy) smaller and lighter cars, important gains in fuel economy can be made in cars of every size category by improving the performance of present engines, by reducing transmission losses, by reducing weight (without sacrificing safety or passenger comfort) and, not least important, by reducing aerodynamic drag. With 1974 car performance as a base line, the DOT-EPA report to Congress estimates that engine improvements should yield economy gains of between 15 and 25 percent, depending on car size. The adoption of four-speed transmissions that would eliminate slippage losses at cruising speed by "locking up" at high gear ratios should yield gains of about 9 percent in each size class. Reductions in curb weight, air resistance, rolling resistance and the power required by accessories would provide another fuel-economy gain of at least 12 percent for large and medium-size cars and 5 percent for small vehicles. In addition the fuel economy of medium-size and large cars could be raised another 10 to 15 percent by reducing engine size so that their power-to-weight ratio is brought into line with the ratio of current small cars [see illustration on these two pages].

It is somewhat surprising that the single factor of aerodynamic drag has received so little attention from American automobile makers. As we have seen, overcoming air resistance absorbs about 25 percent of the engine's output in present-day cars in city-highway driving. Between the late 1920's, when most cars were still shaped like boxes, and the late 1940's the drag coefficient of American cars was reduced about 25 percent, from .70 to .52. (A drag coefficient of 1 corresponds roughly to the air resistance of a rectangular block.) Twenty-five years later the drag coefficient of the typical American car has declined only another 10 percent, to .47. The Citroën, perhaps the most highly streamlined car in large-scale production, has a drag coefficient of about .33.

Since air drag increases as the square of vehicle speed, it has substantial importance at speeds above 45 m.p.h. In the speed range between 45 and 65 m.p.h. each additional 10 m.p.h. above 45 m.p.h. subtracts between 1.5 and two miles per gallon from fuel economy [*see illustration on page 161*]. We can look at a reduction in air drag either as saving energy or as enabling us to go faster with the same expenditure of energy.

It should be possible to reduce the air resistance of present-day large cars by 40 to 50 percent and of compacts by about a third. Such reductions can be achieved by designing cars to have a sloping front, smooth contours, a fairly flat back and a "dam" extending below the front bumper. For cars of roughly the current weight, size and construction but with an engine appropriately reduced in size to hold performance constant, a one-third reduction in aerodynamic drag should yield an improvement of about 10 percent in fuel economy under typical city-highway driving conditions. The percent improvement could be larger in a car with reduced rolling resistance.

There are various power drains in addition to the energy required to propel the car. Air conditioning, now installed in about 75 percent of all new cars, takes about six horsepower in a car traveling at 55 m.p.h. when the air temperature is 100 degrees F. Other accessories such as the engine fan, water pump, air pump and power steering will collectively absorb another five to 15 horsepower, depending on engine speed. Such accessories have to be designed to operate satisfactorily when the automobile is operating at low speeds, and commonly no provision is made to avoid un-necessary power consumption when it is operating at high speeds. A few cars now have electrically driven fans that operate only when the coolant temperature is high. Avoidance of unnecessary power loss at high speeds would cost something in design and complexity but would save energy.

One matter brought up during our Cal Tech seminars is the surprising inefficiency of cars for short trips without warm-up. Starting from an ambient temperature of 70 degrees F., a car gets an average of only 50 percent of its warmed-up gas mileage in a one-mile trip and only about 60 percent in a two-mile trip. In very cold weather the efficiency is much worse [*see bottom illustration on page 160*]. For full gasoline mileage the tires, the grease in the differential-gear box and the oil in the transmission system as well as in the engine must all be warmed up.

Thus substantial savings in fuel consumption could be achieved by reducing vehicle weight and air resistance, by using better and harder tires together with better suspensions, by cutting the waste of energy by accessories at high speeds, by designing better transmissions and, if possible, by achieving efficient operation with a shorter warm-up period. Beyond these possibilities we must consider engine efficiency.

In the past engines have been chosen on the basis of cost and performance rather than efficiency. The diesel engine, as we have seen, is at least 40 percent more efficient than comparable gasoline engines and gives proportionately better fuel economy. Diesel engines, however, are heavy, costly and tend to be somewhat noisy. Moreover, the acceleration of diesel cars is below that of gasoline cars in the same price range. Thus for all their advantages diesel cars have not been notably popular.

The diesel engine is efficient partly because of its high compression ratio (21 : 1 in the Mercedes-Benz diesel) and partly because it operates with a lean fuel mixture (that is, with an excess of air). Power is controlled not by throttling, as it is in most gasoline engines, but by varying the amount of fuel injected into the combustion chambers. When a diesel engine is idling, it consumes only about 15 percent as much fuel as an idling gasoline engine. Operation with a lean mixture has the added advantage of reducing the emissions of hydrocarbons and carbon monoxide. Indeed, if the mixture is lean enough, the oxides of nitrogen are reduced as well. Ideally one could meet emission stan-

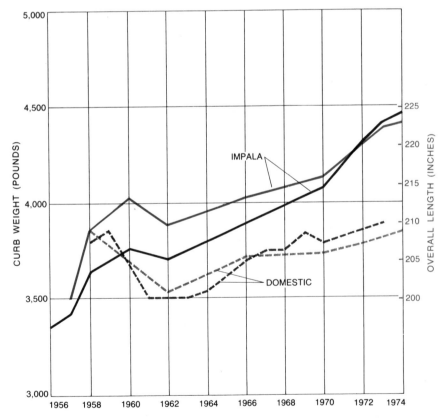

SIZE OF STANDARD AMERICAN CAR has grown sharply in curb weight and overall length since 1956. The best-selling car in the U.S., the Chevrolet Impala, has increased more than half a ton in weight (1,100 pounds) and nearly two feet in overall length. Changes in the standard Ford, the Galaxie 500, are comparable. Because of the introduction of intermediate, compact and subcompact model lines, however, the sales-weighted average of all domestic cars has shown only small changes in curb weight and overall length (**broken curves**).

dards without costly emission-control devices and with high engine efficiency.

The advantages of lean burning have been sought in a modified form of the present gasoline engine that is receiving much attention. This is the stratified-charge engine in which the air-fuel mixture is made intentionally nonhomogeneous. Near the spark plug it is initially rich enough for ignition, but on the average the mixture is lean, with a resulting improvement in efficiency and reduction in emissions. The charge can be stratified in a single combustion chamber by injection of the fuel, as in a diesel. In the Honda engine it is stratified by using an auxiliary combustion chamber in which the mixture is rich. Ignition by means of a special spark plug allows operation at a lower compression ratio than in a diesel engine and gives good starting in cold weather.

One form of stratified-charge engine, developed by Texaco, employs what is called the Texaco controlled-combustion system. A converted 1950 Plymouth using the Texaco system showed a 37 percent improvement in miles per gallon, as compared with the original engine, at speeds between 40 and 60 m.p.h. More recently a converted four-cylinder engine for a military jeep has shown improvements in fuel economy ranging between 40 and 70 percent in road tests. Texaco is now trying to see how much of this gain can be retained while meeting the 1977 emission standards. The Texaco engines operate equally well on gasoline, diesel fuel or jet fuel. Potential disadvantages of the Texaco system include the need for specially shaped pistons and a fuel injector and the tendency of the engine to produce particulate emissions under some conditions. Another form of stratified-charge engine, the PROCO engine (for programmed combustion process), has been under development by the Ford Motor Company. It is stated that the PROCO engine would improve fuel economy about 25 percent in medium-size and large cars and 15 percent in small cars.

Although very lean nonstratified mixtures of gasoline and air cannot be ignited, mixtures of hydrogen and air can be ignited even at an air-hydrogen ratio of 40 : 1, which is about twice as lean as the leanest air-gasoline ratio. Such ratios suggest the addition of hydrogen to the fuel mixture as a way of achieving lean burning both to raise efficiency and reduce emissions. This stratagem has been demonstrated in a Chevrolet V-8 engine in work at the Jet Propulsion Laboratory of Cal Tech. In dynamometer tests, using a fuel consisting of gasoline and bot-

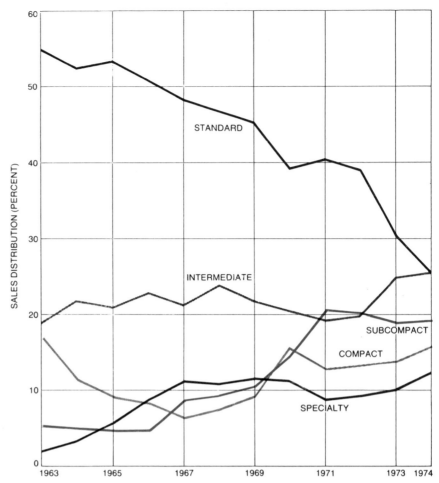

DISTRIBUTION OF CAR SALES by market category has been changing sharply. Sales of subcompact and standard-size models are now about equal. After a drop, sales of compact cars have started to climb again. The curves are based on compilations by *Automotive News*.

tled hydrogen, the equivalent miles per gallon increased from 9.4 for the unmodified engine to 12 for the modified engine, or more than 27 percent. A generator to produce hydrogen from gasoline has been built and operated, but not in an automobile.

All three approaches—diesel, stratified charge, hydrogen admixture—achieve higher efficiency through use of a leaner air-fuel mixture. As we have seen, they also reduce emissions of hydrocarbons and carbon monoxide in the exhaust, perhaps to the point of meeting present emission standards. Proposed standards on emission of oxides of nitrogen, however, are difficult to meet. The production of oxides of nitrogen can be reduced only if the mixture is made lean enough to lower the temperature of combustion substantially. It may be that the proposed standards on emission of oxides of nitrogen are unrealistically stringent.

To sum up, reductions in fuel consumption ranging from 20 to 40 percent have been claimed for lean-mixture

engines under the most efficient operating conditions. Much of the improvement in the past has been lost, however, in adjustments for meeting emission standards. One can hope that further development will lead to a lean-mixture engine with both high efficiency and low exhaust emissions.

Federal emission standards have indirectly had the effect of blocking efficiency improvements that could be achieved with conventional automobile engines simply by raising their compression ratio. Before 1970, when the first Federal standards went into effect, the engines in many American cars had compression ratios as high as 10 : 1 and in a few cases even 10.5 : 1. For efficient antiknock performance an engine with a 10 : 1 compression ratio needs a gasoline with a research octane number of about 100. Oil companies have traditionally added lead alkyl compounds (such as tetraethyl lead) to raise by some five to seven points the octane number of gasoline as it is produced at the refinery; thus a gasoline of 93 to 95 re-

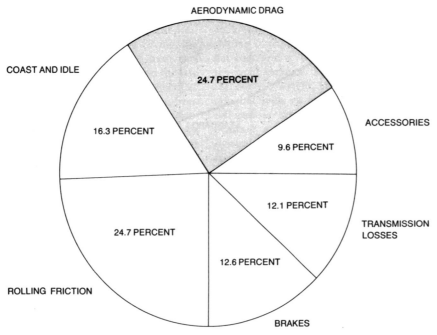

DIVISION OF ENERGY REQUIREMENTS is shown for a 3,500-pound automobile when operated on the EPA composite city-highway test cycle. The energy lost in braking corresponds roughly to the amount of energy previously used to accelerate the car's inertial mass. In steady high-speed cruising most of the engine power is required to overcome aerodynamic drag. In low-speed cruising most of the power is needed to overcome the rolling resistance. In general reducing the power-to-weight ratio will increase the fuel economy.

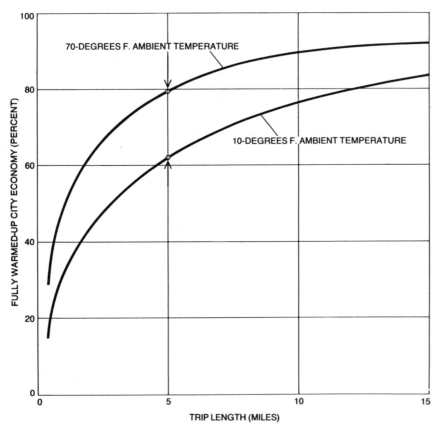

COLD STARTS are a more important cause of poor fuel economy than most drivers realize. These General Motors curves show, for example, that a car that gets, say, 10 m.p.g. when it is fully warmed up in city driving would average only 8 m.p.g. for a five-mile trip when the initial car temperature was 70 degrees Fahrenheit and would average about 6.3 m.p.g. for the same trip on a cold winter morning. About half of all car trips are five miles or less.

search octane can be raised to 100 octane by adding "lead."

With the adoption several years ago of emission standards that would probably require the use of catalytic converters on 1975-model cars it was recognized that gasoline containing lead would poison the catalyst. As a result car makers began to lower compression ratios to between 8:1 and 8.5:1 in order to make it possible for 1975 models to run on unleaded fuel of 91 research octane.

Exxon engineers have calculated that for a typical engine of 350 cubic inches' displacement in a 4,000-pound automobile traveling at 40 m.p.h., raising the compression ratio from 8:1 to 10:1 would yield a 10 percent improvement in fuel economy; at a ratio of 12:1 the improvement would be about 18 percent. The Exxon study shows, however, that since the cost of producing unleaded gasoline rises steeply with octane number (about three cents per gallon to go from 95 octane to 100 at the time the study was made in 1971), the lowest transportation cost to the consumer is achieved with a research octane number of 97, which corresponds to an engine compression ratio of about 9.75:1. (The Exxon study assumes that three grades of gasoline, with an average octane number of 97, would be made available at the pumps.)

Although cars with power plants other than the internal-combustion engine have often been suggested, most of the power plants proposed (steam, gas-turbine, Stirling-cycle) would still need fuel from petroleum. In principle an electric car could get its energy from central power stations running on coal or on nuclear fuels. Electric vans powered by conventional lead-acid batteries have been operated in the U.S. and in other countries for many years. As a private car, however, an electric vehicle with lead-acid batteries seems only marginally promising. A Datsun converted by a Los Angeles engineer, Wally Rippel, gives some idea of the attainable performance.

The car has a range of 70 miles and a top speed of 61 m.p.h. The original transmission is retained, and acceleration is reasonable at the lower gear ratios. By using regenerative braking to recharge the batteries when the vehicle is slowing down or going downhill, electric consumption is reduced 15 percent for a mixture of city-street and freeway driving and as much as 25 percent when the route is a hilly one. The car will travel 3.5 miles per kilowatt-hour of charging power. When this performance is converted to equivalent miles per gallon, it is seen to be quite remarkable: about 52 m.p.g., assuming an efficiency of conversion of fuel to electricity of 40 percent at the power plant. (The energy in a gallon of gasoline is about 125,000 British thermal units, or 37 times the energy in a kilowatt hour; 37 times 3.5 times .40 is 52.) Even though energy in the form of electricity is considerably more expensive than energy in the form of gasoline, the Rippel car still gets about 38.5 miles for the price of a gallon of gasoline, assuming five cents per kilowatt-hour and 55 cents a gallon for gasoline. (It is true, of course, that gasoline carries Federal and state taxes ranging from nine cents to 14 cents per gallon, which should be taken into account in such a calculation, either by subtracting the tax from the price of gasoline or by adding an equivalent tax to the price of electricity.) Fundamentally the efficien-

cy of an electric automobile will depend on the efficiency of electric generating plants (about 40 percent for the best present fossil-fuel plants) minus losses involved in power distribution and the charging of batteries. The batteries in the Rippel car weigh about 1,400 pounds and have a retail value of about $1,200.

An improvement of 2:1 in power-to-weight ratio of batteries would make electric cars more attractive. The nickel-zinc battery or some other kind of battery may provide such an improvement. A more revolutionary regime that might do even better is to store energy in a composite flywheel made of lightweight materials, as has been proposed by Richard F. Post and Stephen F. Post [see "Flywheels," by Richard F. Post and Stephen F. Post; SCIENTIFIC AMERICAN, December, 1973].

There is a formidable obstacle to the production of a satisfactory electric car.

Any gasoline-driven vehicle can be given acceptable performance by putting in a big enough engine. Fuel economy can then be improved gradually by reducing rolling resistance through low weight, hard tires and low unsprung weight, low-friction bearings and good streamlining. If an electric car is to perform satisfactorily, the engineering must be first-class right from the start; a low-loss control system, perhaps with regenerative braking and a high-efficiency motor, must be used.

Such sophisticated engineering design is contrary to the tradition of American automobile manufacturing. Commercial success is not assured even if the engineering is good. Who will take the chance? If we do have electric cars, they may come first as government-purchased vehicles, as high-cost novelties (such as sports cars) or as low-performance vehicles for special uses.

If we continue to use internal-com-

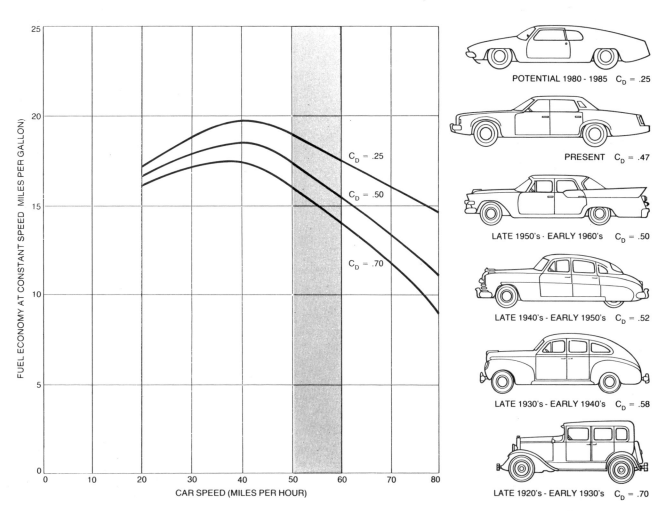

AERODYNAMIC DRAG begins to increase fuel consumption at car speeds of between 40 and 45 miles per hour. The drag coefficients for typical automobile designs of the past 50 years are given at right. As much drag reduction was achieved in the decade between the late 1920's and the late 1930's as in the next three and a half decades. The top car in the group shows a hypothetical design by William H. Bettes of the California Institute of Technology. The car would be a "fastback" with a sloping front, tapered fenders, hard edges and a "dam" below the front bumper. The three curves are computed for a car with the wheelbase and frontal area of present-day cars. The smaller Bettes car, with less frontal area, should considerably exceed the performance represented by the top curve.

bustion engines rather than electric power, we are faced with an inherent but remediable inefficiency. Cars are overpowered for driving on the level in order to provide satisfactory acceleration. Thus the engine operates far below its most efficient power level most of the time. The remedy for this inefficiency is to provide some way of storing energy for use during acceleration. If this were done, a very modest engine could provide lively performance, at least on a level road. Early steam cars attained such performance with steam stored in a boiler.

In the 1930's Robert C. Burt installed a pneumatic transmission in a Plymouth. The gasoline engine pumped air into a tank; the compressed air drove the wheels by means of a converted steam engine. Energy stored in the air tank provided acceleration; the gasoline engine provided steady power. Such a drive system allows regenerative braking (through use of the air engine to compress air in deceleration) and the utilization of waste (exhaust) heat in heating the compressed air.

Recently tests have been made on vans where a small gasoline engine serves to charge batteries. The batteries provide extra power for acceleration; the gasoline engine provides sufficient power to propel the van on the level. A group at the Technical University at Aachen in Germany has constructed a power system in which energy for acceleration is stored in a flywheel; a small gasoline engine drives the flywheel and propels the van during constant-speed driving. The system has been installed in a Volkswagen Microbus. Acceleration better than that of a standard Microbus and a 60 percent fuel saving have been reported.

Vehicles combining a low-power internal-combustion engine and some means for storing energy are generally called hybrid vehicles. Perhaps they are the wave of the future. Perhaps they are too complex for private cars. Perhaps effective means for storing energy, either improved batteries or flywheels, will take us all the way to vehicles driven by electric power. It is sometimes overlooked, however, that all-electric private transportation would impose an enormous new load on the electric-utility industry.

Today the nation's 100-odd-million private cars and small trucks consume nearly 60 percent as many energy units as all the nation's electric-power plants. To increase the capacity of the electric-power system between now and, say, the year 2000, to provide power for a national fleet of vehicles swollen to perhaps 160 million—in addition to expanding generating capacity 5 or 6 percent per year for all other purposes—would be an immensely costly undertaking.

Our view of the future is full of perhapses. What we do know with certainty is that in the near future private cars will continue to consume a great deal of gasoline and that there are many ways in which they could be made more efficient. Plausible projections by the Department of Transportation indicate that automobile fuel consumption could be stopped from growing before 1980 and thereafter even reduced [*see illustration below*]. Whether or not cars become more efficient will depend not only on technical ingenuity and enterprise but also on the economic pressure of the price of gasoline, and on any other pressure that may come into being.

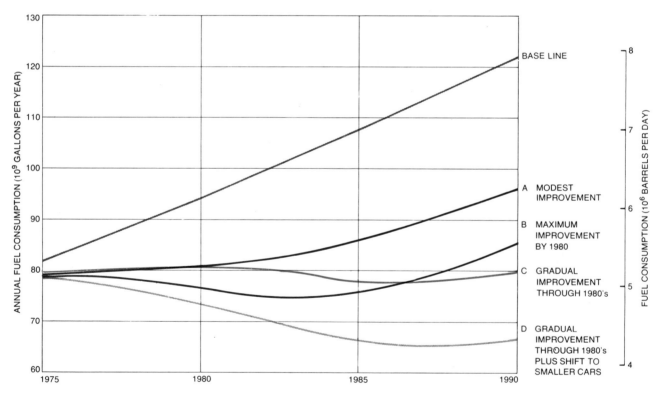

POTENTIAL FUEL SAVINGS between now and 1990, corresponding to four different levels of private-car improvement, were presented to Congress in the DOT-EPA report. The base-line curve represents a steady growth in vehicle miles of 2.6 percent per year. Curve *A* is based on announced industry goals (some engine changes, use of radial tires, slight weight and air-drag reduction) with no significant improvements beyond 1978. Curve *B* is based on maximum improvements through 1980 with little change thereafter. Changes would include rapid weight and air-drag reduction, improved transmissions and optimization of conventional engines. Curve *C* visualizes somewhat slower changes before 1980 but substantial improvements thereafter, including the phasing in of diesel engines for larger cars between 1981 and 1989 and adoption of stratified-charge engines for smaller cars. Curve *D* includes all the changes projected in curve *C* combined with a sales mix after 1980 consisting of 10 percent large cars, 25 percent intermediates, 25 percent compacts and 40 percent subcompacts. Curve *D* would yield a total saving of 3.6 million barrels of fuel per day by 1990.

13

Human Energy Production as a Process in the Biosphere

by S. Fred Singer
September 1970

*In releasing the energy stored in fossil and nuclear
fuels man accelerates slow cycles of nature. The
waste products of power generation then interact with
the fast cycles of the biosphere*

The earth in general and the biosphere in particular have grand-scale pathways of energy metobolism. For example, solar energy falls on the earth, green plants utilize a tiny fraction of it to manufacture energy-rich compounds and some of these compounds are stored in the earth's crust as what we have come to call fossil fuels. The primary fission fuel uranium and the potential fusion fuel deuterium were originally "cooked" in the interior of stars. In releasing the energy of these chemical and nuclear fuels man is in effect racing the slow cycles of nature, with inevitable effects on the cycles themselves.

Before 1800 the power available to human societies was limited to solar energy that had only recently been radiated to the earth. The most direct form of such power was human or animal power; the energy came from the metabolism of food, which is to say from the biological oxidation of compounds storing solar energy. The burning of wood or oils of animal or vegetable origin to provide light and heat also represented the conversion of recently stored solar energy. By the same token the use of moving air or falling water to drive mills or pumps constituted the use of recently arrived solar energy. Among the other limitations of such power sources was the fact that they could not be readily transported and that their energy could not be transmitted any considerable distance.

This picture has of course changed completely since 1800, and it has assumed significant new dimensions in the past two decades with the advent of nuclear power. The most striking measure of these changes is the increased per capita consumption of energy in the developed countries. Indeed, the correlation between a nation's per capita use of energy and its level of economic development is almost linear [*see illustration on page 165*]. The minimum per capita consumption of energy is what is required in food for a man to stay alive, namely about 2,000 kilocalories per day or 100 watts (thermal). Today the per capita use of energy in the U.S. is 10,000 watts, and the figure is rising by some 2.5 percent per year.

Hand in hand with the advance in the rate of energy consumption has gone the introduction of the new sources of energy: fossil and nuclear fuels. In contrast to the sources used before 1800, fossil and nuclear fuels represent energy that reached the earth millions and even some billions of years ago. Except occasionally for political reasons, it matters little where the new fuels are found; they can be transported readily, and the energy produced from them can be transmitted over great distances.

On first consideration it might seem that fossil and nuclear fuels are fundamentally different, in that the energy of one is released by oxidation, or burning, and the energy of the other is released by fission or fusion. In a deeper sense, however, the two kinds of fuel are closely related. Fossil fuels store the radiant energy originally produced by nuclear reactions in the interior of the sun. Nuclear fuels store energy produced by another set of nuclear reactions in the interior of certain stars. When such stars exploded, they showered into space the elements that had been synthesized within them. These elements then went into the formation of younger stars such as the sun, together with its family of planets.

The production of fossil fuels is based on the carbon cycle. In the process of photosynthesis plants use radiant energy from the sun to convert carbon dioxide and water into carbohydrates, at the same time releasing oxygen into the atmosphere. When the plant materials decompose or are eaten by animals, the process is reversed: oxygen is used to convert carbohydrates into energy plus carbon dioxide and water.

The amount of carbon dioxide involved in photosynthesis annually is about 110 billion tons, or roughly 5 percent of the carbon dioxide in the atmosphere. The consumption of carbon dioxide through photosynthesis is matched to one part in 10,000 by the annual release of carbon dioxide to the atmosphere through oxidation. Under normal conditions the amounts of carbon dioxide and oxygen in the atmosphere remain approximately in equilibrium from year to year.

There are, however, small long-term imbalances in the carbon cycle, and it is owing to them that the fossil fuels being exploited today all derive from plants and animals that lived long ago. Over a span of geologic history extending back into the Cambrian period of some 500 million years ago, a small fraction of these organisms have been buried in sediments or mud under conditions that prevented complete oxidation. Various chemical changes have transformed them into fossil fuels: coal, oil, natural gas, lignite, tar and asphalt. Although the same geological processes are still operative, they function over vast periods of time, and so the amount of new fossil fuel that is likely to be produced during the next few thousand years is inconsequential. Therefore one can assume that the existing fossil fuels constitute a nonrenewable resource.

Coal has been burned for some eight centuries, but it was consumed in negligible amounts until early in the 19th century. Since the middle of that century the rise in the consumption of coal has been spectacular: in 1870 the world production rate of coal was about 250 million metric tons per year, whereas this year it will be about 2.8 billion tons. The rate of increase, however, is lower now than it was at the beginning of the period, having declined from an average of 4.4 percent per year to 3.6 percent,

largely because of the rapid increase in the fraction of total industrial energy contributed by oil and gas. In the U.S. that fraction rose from 7.9 percent in 1900 to 67.9 percent in 1965, whereas the contribution of coal declined from 89 percent to 27.9 percent.

World production of crude oil was negligible as recently as 1890; now it is close to 12 billion barrels per year. The rise in the rate of production has been nearly 7 percent per year, so that the amount of oil extracted has doubled every 10 years. As yet there is no sign of a deceleration in this rate.

Nonetheless, the finiteness of the earth's fossil fuel supplies gives rise to the question of how long they will last. M. King Hubbert of the U.S. Geological

Survey has estimated that the earth's coal supply can serve as a major source of industrial energy for another two or three centuries. His estimate for petroleum is 70 to 80 years. However much these periods may be stretched by unforeseen discoveries and improved technology, the end of the fossil fuel era will inevitably come. From the perspective of that time—perhaps the 23rd century—the period of exploitation of fossil fuels will be seen as only a brief episode in the span of human history.

This year the U.S. will consume some 68,500 million million B.T.U. of energy, most of it derived from fossil fuels. (One short ton of coal has a thermal value of 25.8 million B.T.U. The thermal

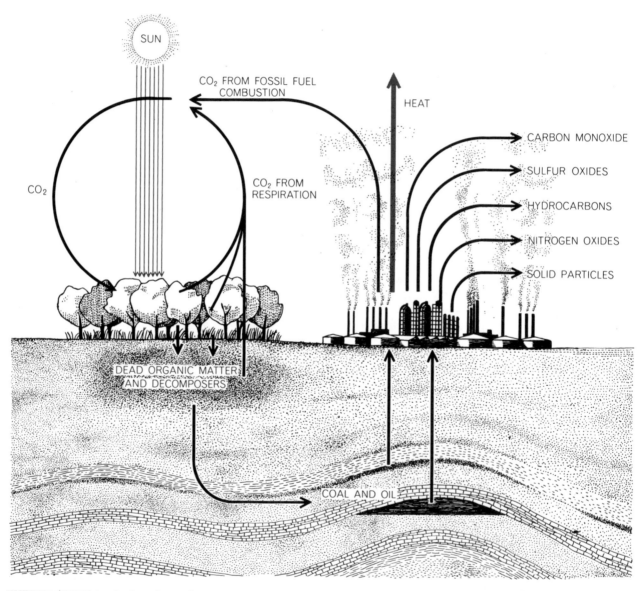

ENERGY CYCLE involved in the combustion of fossil fuels begins with solar energy employed in photosynthesis millions of years ago. A small fraction of the plants is buried under conditions that prevent complete oxidation. The material undergoes chemical changes that transform it into coal, oil and other fuels. When they are burned to release their stored energy, only part of the energy goes into useful work. Much of the energy is returned to the atmosphere as heat, together with such by-products of combustion as carbon dioxide and water vapor. Other emissions in fossil fuel combustion are listed at right in the relative order of their volume.

value of one barrel of oil is 5.8 million B.T.U.) Industry takes more than 35 percent of the total energy consumption. About a third of industry's share is in the form of electricity, which, as of 1960, was generated roughly 50 percent from coal, 20 percent from water power, 20 percent from natural gas and 10 percent from oil.

The nation's homes use almost as much energy as industry does. A major consumer is space heating, which for the average home requires as much energy as the average family car: about 70 million B.T.U. per year, or the equivalent of 900 gallons of oil. The other domestic uses are for cooking, heating water, lighting and air conditioning.

Transportation accounts for 20 percent of the annual energy consumption, mainly in the form of gasoline for automobiles. Another 10 percent goes for commercial consumption in stores, offices, hotels, apartment houses and the like. Agriculture probably consumes no more than 1 percent of all the energy, chiefly for the operation of tractors and for running irrigation and drainage equipment.

Looking at the use of fossil fuels from another viewpoint, one finds that most of the coal goes into the generation of electricity. Oil and natural gas tend to be used directly, either for heating purposes or to provide motive power for vehicles. Fossil fuels are also used as the raw materials for the petrochemical industry. Notwithstanding that industry's rapid growth, however, it still accounts for less than 2 percent of the annual consumption of fossil fuels.

Clearly the production of energy from fossil fuels on the scale typical of a modern industrial nation represents an enormous amount of combustion, with attendant effects on the biosphere. By far the greatest effect is the emission of carbon dioxide. Combustion also injects a number of pollutants into the air. In the U.S. the five most common air pollutants, listed in the order of their annual tonnage, are carbon monoxide, sulfur oxides, hydrocarbons, nitrogen oxides and solid particles. The major sources are automobiles, industry, electric power plants, space heating and refuse disposal. The burning of fossil fuels also produces effects on water: chemical effects when the air pollutants are washed down by water and thermal effects arising from the dispersal of waste heat from thermal power plants.

Carbon dioxide is the only combustion product whose increase has been documented on a worldwide basis. The

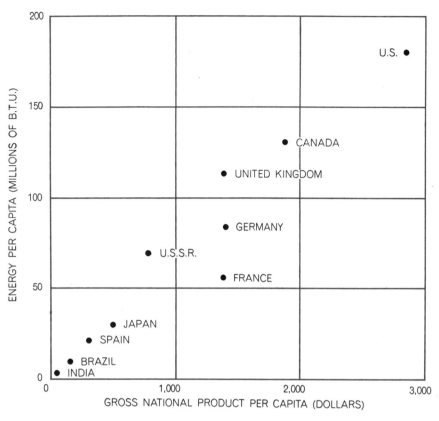

CLOSE RELATION between a nation's consumption of energy and its gross national product is depicted on the basis of a study made by the Office of Science and Technology in 1961. Most of the nations covered beyond the 10 shown would be in the lower left-hand rectangle.

injection of large quantities of carbon dioxide into the atmosphere in the past few decades has been extremely sudden in relation to important natural time scales. For example, although the surface of the sea can adjust to changes in the level of carbon dioxide in the atmosphere in about five years, the deeper layers require some hundreds or thousands of years to adjust. If the oceans were perfectly mixed at all times, carbon dioxide added to the atmosphere would distribute itself about five-sixths in the water and about one-sixth in the air. In actuality the distribution is about equal.

It appears that between 1860 and the present the concentration of carbon dioxide in the atmosphere has increased from about 290 parts per million to about 320 parts per million, an increase of more than 10 percent. Precise measurements by Charles D. Keeling of the Scripps Institution of Oceanography have established that the carbon dioxide content increased by six parts per million between 1958 and 1968. Reasonable projections indicate an increase of 30 percent (over 1860) to about 375 parts per million by the turn of the century and to between 450 and 500 parts per million by 2020.

The most widely discussed matter related to these increases is the possibility that they will lead to a worldwide rise in temperature. The molecule of carbon dioxide has strong absorption bands, particularly in the infrared region of the spectrum at wavelengths of from 12 to 18 microns. This is the spectral region where most of the thermal energy radiating from the earth into space is concentrated. By increasing the absorption of this radiation and by reradiating it at a lower temperature corresponding to the temperature of the upper atmosphere the carbon dioxide reduces the amount of heat energy lost by the earth to outer space. The phenomenon has been called the "greenhouse effect," although the analogy is inexact because a real greenhouse achieves its results less from the fact that the glass blocks reradiation in the infrared than from the fact that it cuts down the convective transfer of heat.

The possibility that additional carbon dioxide from the burning of fossil fuels could produce a worldwide increase in temperature seems to have been raised initially by the American geologist P. C. Chamberlain in 1899. In 1956 Gilbert N. Plass calculated that a doubling of

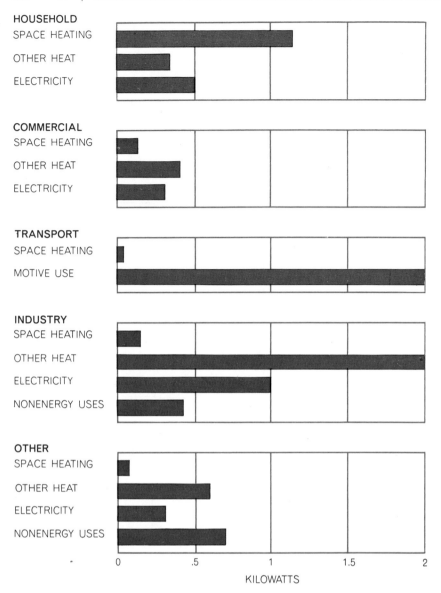

HOUSEHOLD
SPACE HEATING
OTHER HEAT
ELECTRICITY

COMMERCIAL
SPACE HEATING
OTHER HEAT
ELECTRICITY

TRANSPORT
SPACE HEATING
MOTIVE USE

INDUSTRY
SPACE HEATING
OTHER HEAT
ELECTRICITY
NONENERGY USES

OTHER
SPACE HEATING
OTHER HEAT
ELECTRICITY
NONENERGY USES

0   .5   1   1.5   2
KILOWATTS

USE OF ENERGY in the U.S. is expressed in terms of thermal kilowatts per capita in 1967 in five major categories. All together the consumption averages 10,000 watts per person, which is 100 times the food-intake level of 100 watts that is barely exceeded in many nations.

the carbon dioxide content of the atmosphere would result in a rise of 6.5 degrees Fahrenheit at the earth's surface. In 1963 Fritz Möller calculated that a 25 percent increase in atmospheric carbon dioxide would increase the average temperature by one to seven degrees F., depending on the effects of water vapor in the atmosphere. The most extensive calculations have been made by Syukuro Manabe and R. T. Wetherald, who estimate that a rise in atmospheric carbon dioxide from 300 to 600 parts per million would increase the average surface temperature by 4.25 degrees, assuming average cloudiness, and by 5.25 degrees, assuming no clouds.

Unfortunately the problem is more complicated than these calculations imply. An increase of temperature at the surface of the earth and in the lower levels of the atmosphere not only increases evaporation but also changes cloudiness. Changes of cloudiness alter the albedo, or average reflecting power, of the earth. The normal average albedo is about 30 percent, meaning that 30 percent of the sunlight reaching the earth is immediately reflected back into space. Changes in cloudiness, therefore, can have a pronounced effect on the atmospheric temperature and on climate.

The situation is further complicated by atmospheric turbidity. J. Murray Mitchell, Jr., of the Environmental Science Services Administration has determined that atmospheric temperatures

rose generally between 1860 and 1940. Between 1940 and 1960, although warming occurred in northern Europe and North America, there was a slight lowering of temperature for the world as a whole. Mitchell finds that a cooling trend has set in; he believes it is owing partly to the dust of volcanic eruptions and partly to such human activities as agricultural burning in the Tropics. (In the future the condensation trails left by jet airplanes may contribute to this problem.) (See graph on p. 170.)

In sum, the fact that the carbon dioxide content of the atmosphere has increased is firmly established by reliable measurements. The effect of the increase on climate is uncertain, partly because no good worldwide measurements of radiation are available and partly because of the counteractive effects of changes in cloudiness and in the turbidity of the atmosphere. An exciting technological possibility is the use of a weather satellite to keep track of the energy radiated back into space by the earth. The data would provide a basis for the first reliable and standardized measurement of the "global radiation climate."

In any event, the higher levels of carbon dioxide may not persist for long. For one thing, the oceans, which contain 60 times as much carbon dioxide as the atmosphere does, will begin to absorb the excess as the mixing of the intermediate and deeper levels of water proceeds. For another, the increased atmospheric content of carbon dioxide will stimulate a more rapid growth of plants—a phenomenon that has been utilized in greenhouses. It is true that the carbon dioxide thus removed from the atmosphere will be returned when the plants decay. Forests, however, account for about two-thirds of the photosynthesis taking place on land (and therefore for nearly half of the world total), and since forests are long-lived, they tend to spread over a long period of time the return of carbon dioxide to the atmosphere.

The five major air pollutants resulting from the combustion of fossil fuels also interact with the biosphere in various ways, not all of them clearly understood. One tends to think of pollutants as harmful, but the situation is not that simple, as becomes apparent in a consideration of the pollutants and their known effects.

Carbon monoxide appears to be almost entirely a man-made pollutant. The only significant source known is the imperfect combustion of fossil fuels, resulting in incomplete oxidation of the

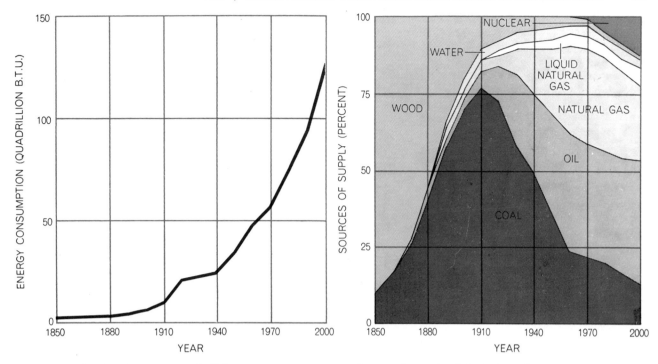

CHANGING SOURCES of energy in the U.S. since 1850 are compared (*right*) with total consumption (*left*) over the same period. At right one can see that in 1850 fuel wood was the source of 90 percent of the energy and coal accounted for 10 percent. By 2000 it is foreseen that coal will be back to almost 10 percent and that other sources will be oil, natural gas, liquid natural gas, hydroelectric power, fuel wood and nuclear energy. The estimates were made by Hans H. Landsberg of Resources for the Future, Inc.

carbon. Although carbon monoxide is emitted in large amounts, it does not seem to accumulate in the atmosphere. The mechanism of removal is not known, but it is probably a biological sink, such as soil bacteria.

Sulfur, which occurs as an impurity in fossil fuels, is among the most troublesome of the air pollutants. Although there are natural sources of sulfur dioxides, such as volcanic gases, more than 80 percent is estimated to come from the combustion of fuels that contain sulfur. The sulfur dioxide may form sulfuric acid, which often becomes associated with atmospheric aerosols, or it may react further to form ammonium sulfate. A typical lifetime in the atmosphere is about a week.

When the sulfur products are removed from the atmosphere by precipitation, they increase the acidity of the rainfall. Values of pH of about 4 have been found in the Netherlands and Sweden, probably because of the extensive industrial activity in western Europe. As a result small lakes and rivers have begun to show increased acidity that endangers the stability of their ecosystems. Certain aquatic animals, such as salmon, cannot survive if the pH falls below 5.5.

Nothing is known about the global effects of sulfur emission, but they are believed to be small. In any case most of the sulfur ends up in the oceans. It is possible, however, that sulfur compounds are accumulating in a layer of sulfate particles in the stratosphere. The layer's mechanism of formation, its effects and its relation to man-made emissions are not clear. The fine particles of the layer could have an effect on radiation from the upper atmosphere, thereby affecting mean global temperatures.

Hydrocarbons are emitted naturally into the atmosphere from forests and

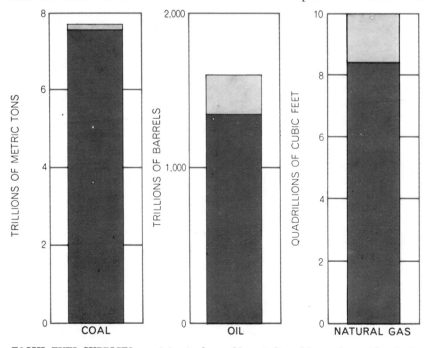

FOSSIL FUEL SUPPLIES remaining in the world are indicated by a scheme wherein the entire gray bar represents original resources, light gray portion shows how much has been extracted and dark gray area shows what remains. Figures reflect estimates by M. King Hubbert of the U.S. Geological Survey and could be changed by unforeseen discoveries.

vegetation and in the form of methane from the bacterial decomposition of organic matter. Human activities account for only about 15 percent of the emissions, but these contributions are concentrated in urban areas. The main contributor is the processing and combustion of petroleum, particularly gasoline for the internal-combustion engine.

The reactions of hydrocarbons with nitrogen oxides in the presence of ultraviolet radiation produce the photochemical smog that appears so often over Los Angeles and other cities. The biological effects of several of the products of the reactions, including ozone and complex organic molecules, can be quite severe. Some of the products are thought to be carcinogenic. Ozone has highly detrimental effects on vegetation, but fortunately they are localized. As yet no regional or worldwide effects have been discovered.

Hydrocarbon pollutants in the form of oil spills are well known to have drastic ecological effects. The spill in the Santa Barbara Channel last year, which involved some 10,000 tons, and the *Torrey Canyon* spill in 1967, involving about 100,000 tons, produced intense local concentrations of oil, which is toxic to many marine organisms. Besides these well-publicized events there is a yearly worldwide spillage from various oil operations that adds up to about a million tons, even though most of the individual spills are small. There are also natural oil seeps of unknown magnitude. Added to all of these is the dumping of waste motor oil; in the U.S. alone about a million tons of such oil is discarded annually. Up to the present time no worldwide effects of these various oil spills are detectable. It can therefore be assumed that bacteria degrade the oil rapidly.

Nitrogen oxides occur naturally in the atmosphere as nitrous oxide (N_2O), nitric oxide (NO) and nitrogen dioxide (NO_2). Nitrous oxide is the most plentiful at .25 part per million and is relatively inert. Nitrogen dioxide is a strong absorber of ultraviolet radiation and triggers photochemical reactions that produce smog. In combination with water it can form nitric acid.

The production of nitrogen oxides in combustion is highly sensitive to temperature. It is particularly likely to result from the explosive combustion taking place in the internal-combustion engine. If this engine is ever replaced by an external-combustion engine that operates at a steady and relatively low temperature rather than at high peaks, the emission of nitrogen oxides will be greatly reduced.

Solid particles are injected into the lower atmosphere from a number of sources, with the combustion of fossil fuels making a major contribution. The technology of pollution control is adequate for limiting such emissions. If it is applied, solid particles will become insignificant pollutants.

Although the fossil fuels still predominate as sources of power, the introduction of nuclear fuels into the generation of power is changing both the scale of energy conversion and the effects of that conversion on the biosphere. Nuclear energy can be considered as a heat source differing from coal or oil, but once the energy has been released in the form of heat it is used in the same way as heat from other sources. Therefore the problem of waste heat is the same. The pollution characteristics of nuclear energy, however, differ from those of the fossil fuels, being radioactive rather than chemical.

Two processes are of concern: the fission of heavy nuclei such as uranium and the fusion of light nuclei such as deuterium. The fission reaction has to start with uranium 235, because that is the only naturally occurring isotope that is fissioned by the capture of slow neutrons. On fissioning the uranium 235 supplies the neutrons needed to carry out other reactions.

Each fission event of uranium 235 releases some 200 million electron volts of energy. One gram of uranium 235 therefore corresponds to 81,900 million joules, an energy equivalent of 2.7 metric tons of coal or 13.7 barrels of crude oil. A nuclear power plant producing

SOURCES OF WASTE HEAT are evident in a thermal infrared image, made at an altitude of 2,000 feet, of an industrial concen- tration along the Detroit River in Detroit. The whiter an object is, the hotter it was when the image was made. The complex at left

1,000 electrical megawatts with a thermal efficiency of 33 percent would consume about three kilograms of uranium 235 per day.

A nuclear "burner" uses up large amounts of uranium 235, which is in short supply since it has an abundance of only .7 percent of the uranium in natural ore. If reactor development proceeds as foreseen by the Atomic Energy Commission, inexpensive reserves of uranium (costing less than $10 per pound) would be used up within about 15 years and medium-priced fuel (up to $30 per pound) would be used up by the year 2000. Hence there has been concern that present reactors will deplete these supplies of uranium before converter and breeder reactors are developed to make fissionable plutonium 239 and uranium 233. Either of these isotopes can be used as a catalyst to burn uranium 238 or thorium 232, which are relatively abundant. Thorium and uranium together have an abundance of about 15 parts per million in the earth's crust, representing therefore a source of energy millions of times larger than all known reserves of fossil fuel.

The possibility of generating energy by nuclear fusion is more remote. Of the two processes being considered—the deuterium-deuterium reaction and the deuterium-tritium reaction—the latter is somewhat easier because it proceeds at a lower temperature. In it lithium 6 is the basic fuel, because it is needed to make tritium by nuclear bombardment. The amount of energy available in this way is limited by the abundance of lithium 6 in the earth's crust, namely about two parts per million. The deuterium-deuterium reaction, on the other hand, would represent a practically inexhaustible source of energy, since one part in 5,000 of the hydrogen in the oceans is deuterium.

One must hope, then, that breeder reactors and perhaps fusion reactors will be developed commercially before the supplies of fossil fuel and uranium 235 are exhausted. With inexhaustible (but not cheap) supplies of nuclear energy, automobiles may run on artificially produced ammonia or methane; coal and oil shale will be used as the basis for chemicals, and electricity generated in large breeder or fusion reactors will be used for such purposes as the manufacture of ammonia and methane, the reduction of ores and the production of fertilizers.

It is difficult at this stage to predict the effects of large-scale use of nuclear energy on the biosphere. One must make certain assumptions about the disposal of radioactive wastes. A reasonable assumption is that they will be rendered harmless by techniques whereby long-lived radioactive isotopes are made into solids and buried. (They are potentially dangerous now because of the technique of storing them as liquids in underground tanks.) Short-lived radioactive wastes can presumably be stored safely until they decay.

For both nuclear energy and for processes involving fossil fuels the major problem and the major impact of human energy production is the dissipation of waste heat. The heat has direct effects on the biosphere and could have indirect effects on climate. It is useful to distinguish between local problems of thermal pollution, meaning the problems that arise in the immediate vicinity of a power plant, and the global problem of thermal balance created by the transformation of steadily rising amounts of energy.

The efficiency of a power plant is determined by the laws of thermodynamics. No matter what the fuel is, one tries to create high-temperature steam for driving the turbines and to condense the steam at the lowest possible temperature. Water is the only practical medium for carrying the heat away. Hence more than 80 percent of the cooling water used by U.S. industry is accounted for by electric power plants. For every kilowatt-hour of energy produced about 5,000 B.T.U. in heat must be dissipated from a fossil fuel plant and about 7,000 B.T.U. from a contemporary nuclear plant.

In the U.S., where the consumption of power has been doubling every eight to 10 years, the increase in the number and size of electric power plants is putting a severe strain on the supply of cooling water. By 1980 about half of the normal runoff of fresh water will be needed for this purpose. Even though some 95 percent of the water thus used is returned to the stream, it is not the same: its increased temperature has a number of harmful effects. Higher temperatures decrease the amount of dissolved oxygen and therefore the capacity of the stream to assimilate organic wastes. Bacterial decomposition is accelerated, further depressing the oxygen level. The reduction of oxygen decreases the viability of aquatic organisms while at the same time the higher temperature raises their metabolic rate and therefore their need for oxygen.

In the face of stringent requirements being laid down by the states and the Department of the Interior, power companies are installing devices that cool water before it is returned to the stream. The devices include cooling ponds, spray ponds and cooling towers. They function by evaporating some of the cooling

center, identifiable by a distinctly warm effluent entering the river, is a power plant. Group of hot buildings at right is a steel mill. Cool land area at bottom is part of Grosse Ile.

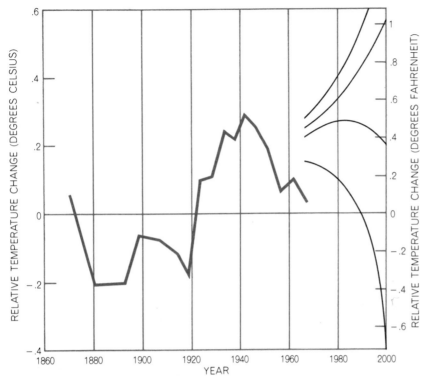

TEMPERATURE TREND in Northern Hemisphere is portrayed as observed (*color*) and as predicted under various conditions (*black*). The top black curve assumes an effect from carbon dioxide only; the other black curves also take account of dust. Second and third curves assume doubling of atmospheric dust in 20 and 10 years respectively; bottom curve, doubling in 10 years with twice the thermal effect thought most probable. Chart is based on work of J. Murray Mitchell, Jr., of the Environmental Science Services Administration.

surrounding countryside affects the ecology and biospheric activity in metropolitan areas in numerous ways. For example, the release of heat in a relatively small local area causes a change in the convective pattern of the atmosphere. The addition of large amounts of particulate matter from industry, space heating and refuse disposal provides nuclei for the condensation of clouds. A study in the state of Washington showed an increase of approximately 30 percent in average precipitation over long periods of time as a result of air pollution from pulp and paper mills.

The worldwide consumption of energy can be estimated from the fact that the U.S. accounts for about a third of this consumption. The U.S. consumption of 68,500 million million B.T.U. per year is equivalent to 2.2 million megawatts. World consumption is therefore some 6.6 million megawatts. Put another way, the present situation is that the per capita consumption of energy in the U.S. of 10,000 watts compares with somewhat more than 100 watts (barely above the food-intake level) in most of the rest of the world.

Projections for the future depend on the assumptions made. If one assumes that in 50 years the rest of the world will reach the present U.S. level of energy consumption and that the population will be 10 billion, the total man-made energy rate would be 110 million megawatts. The energy would of course be distributed in a patchy manner reflecting the location of population centers and the distributing effects of the atmosphere and the oceans.

That figure is numerically small compared with the amount of solar energy the earth radiates back into space. Over the entire earth the average heat loss is about 120,000 million megawats, or more than 1,000 times the energy that

water, so that the excess heat is dissipated into the atmosphere rather than into the stream.

This strategy of spreading waste heat has to be reexamined as the scale of the problem increases. It is already apparent that the "heat islands" characteristic of metropolitan areas have definite meteorological effects—not necessarily all bad. The fact that a city is warmer than the

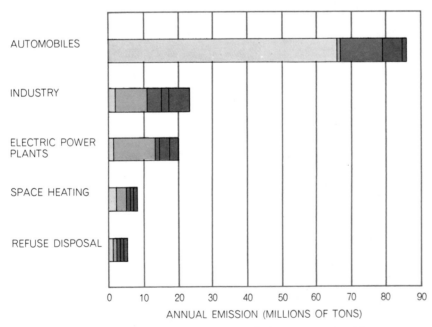

SOURCES OF EMISSIONS from combustion are ranked. Five parts of each bar represent (*from left*) carbon monoxide, sulfur oxides, hydrocarbons, nitrogen oxides and particles.

WASTE HEAT that is an inevitable accompaniment of the human use of energy is evident in the thermal infrared image of New York on the opposite page. The thermogram was made with a Barnes thermograph that depicts emissions of energy on a color scale ranging from black for the coolest objects through green, yellow and red to red-purple for the hottest ones. Some of the emissions represent solar energy stored in the walls of buildings, but a large fraction is waste heat from the human consumption of energy. The rectangular elements of the image result from digitized output of thermograph. Empire State Building is at center.

would be dissipated by human activity if the level of energy consumption projected for 2020 were reached. It would be incautious to assume, however, that the heat put into the biosphere as a result of human energy consumption can be neglected because it is so much smaller than the solar input. The atmospheric engine is subtle in its operation and delicate in its adjustments. Extra inputs of energy in particular places can have significant and far-reaching consequences.

14

The Photovoltaic Generation of Electricity

by Bruce Chalmers
October 1976

*One way to harness the energy of the sun is
to convert it into electricity with solar cells of the kind
carried by spacecraft. Such cells are expensive,
but new kinds of fabrication may make them
economically competitive*

It is paradoxical that Americans should be concerned about their supply of energy even as the U.S. is receiving energy from the sun at 500 times the rate at which they use it for all purposes, but perhaps the paradox can be resolved. Men have only just begun to turn their scientific and technological ingenuity to the task of converting their most abundant and reliable energy source, sunlight, into electricity, the most convenient and adaptable form of energy.

When the sun is high in a clear sky, the energy it radiates reaches the surface of the earth at the rate of a little more than a kilowatt per square meter. That is a very substantial amount of energy; while the sun is shining the energy equivalent of a gallon of gasoline (36

kilowatt-hours) falls on an area the size of a tennis court about every 10 minutes, and an area of only 80 square meters in the least sunny parts of the 48 contiguous states of the U.S. receives in the course of a year enough energy to supply the needs of an average American family. The sun is also a reliable energy source, in the sense that a given place will receive about the same amount of energy in the form of sunshine every year. It is an unreliable source, however, in the sense that sunshine is intermittent: it is not available at night or in unfavorable weather.

Before solar energy can provide electricity on a large scale there are two technical problems to be solved. The first is the problem of converting the energy of the sun's radiation into electrici-

ty. The second has to do with the fact that the electricity must be available whenever and wherever it is needed, regardless of the weather, the season or the time of day. It must also be kept in mind that in the real world both objectives must be achieved at prices that are competitive with those of other sources of energy.

It has been known for well over 100 years that light can generate electricity; the utilization of this effect to measure the strength of light in the photographic exposure meter is familiar to almost everyone. The efficiency of such devices, however, is very low. They could never be considered as a practical means of generating useful amounts of electricity. Then the discovery in the 1950's of the extraordinary electrical properties of

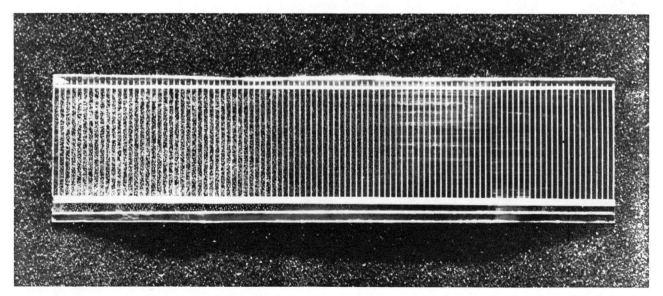

EXPERIMENTAL SOLAR CELL made of silicon grown by a new process known as edge-defined film-fed growth (EFG) was made at the Mobil Tyco Solar Energy Corporation in Waltham, Mass. The EFG process is one method that shows promise of lowering the cost of building photovoltaic power stations to a level comparable to the cost of building fuel-burning stations. This particular solar cell is 10 centimeters long and 2.5 centimeters wide. In full sunlight it can generate between a quarter and a third of a watt of electricity. The white lines at regular intervals are the wires of the current collector on the cell's front surface. Fine speckles are dust on surface of cell.

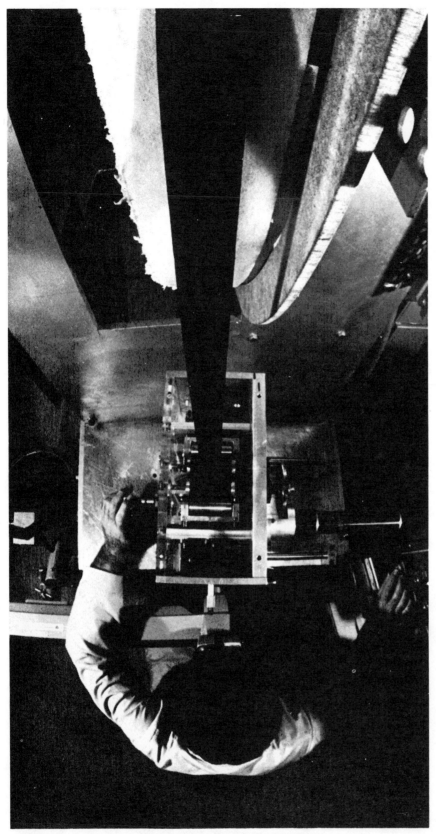

RIBBON OF SILICON CRYSTAL is grown from a pool of molten silicon at Mobil Tyco. The ribbon is 2.5 centimeters wide and .15 millimeter thick, and it emerges through a slot in a graphite die above the melt at the rate of 2.5 centimeters per minute. For the manufacture of solar cells it is cut into pieces the same length as the cell shown in the photograph on the opposite page. The silicon crystal produced by the EFG process is not as chemically pure and crystallographically perfect as that grown by Czochralski process, used to make crystals for microelectronic devices. It is, however, adequate for making solar cells of good efficiency and is cheaper.

the group of materials that came to be called semiconductors led the way to the development of devices in which the efficiency of conversion of light into electricity is much higher. As has happened in a number of other branches of technology, the space program provided the incentive for the development and production of the first practical photovoltaic devices, which have provided electric power in many space vehicles, manned and unmanned, with extremely high reliability and for an apparently unlimited time.

The solar cells that have been employed so far in the space program are made of silicon. This material is not the only one that is suitable for photovoltaic devices, and in the future some alternative material may be found to be superior to silicon. Since the principles that underlie all photovoltaic devices are the same, however, it will be convenient to describe their construction and operation in terms of the silicon solar cell.

An electric current is synonymous with a flow of electrons in a circuit. We are therefore concerned with the behavior of electrons in a crystal of silicon. Silicon has a chemical valence of four, that is, of the 14 electrons that are normally found in each silicon atom four are available to interact with other atoms. They can combine with the atoms of other elements to form chemical compounds or they can interact with other silicon atoms to stabilize the structure of a crystal.

When silicon atoms are free to take up their lowest energy configuration, as when molten silicon is solidifying into a crystal, each atom surrounds itself with four equidistant neighbors. It shares two electrons with each neighbor, one of its own and one of the neighbor's, and therefore it lies at the center of a tetrahedron, each corner of which is the site of another silicon atom. The pattern can be extended ad infinitum, so that each atom, not just the central one, is surrounded by four equidistant neighbors. The result of this geometrical arrangement is the cubic crystal structure characteristic of diamond.

If the structure of the silicon crystal were perfect, each electron would be strongly held in position by the electrostatic forces between it and the two atoms it helps to bind together. Such a crystal would be a perfect insulator because no electrons would be free to move if a voltage were applied to it. In metals, which are good conductors of electricity, the atoms are also bound together by electrons, one electron or more being supplied by each atom. In a metal, however, the electrons are not localized in such a way that they bind specific pairs of atoms together; they are free to move within the crystal as an electric current whenever a voltage is applied.

In a crystal of silicon a substantial amount of energy is needed to break the bond between an electron and the two atoms it binds together so that the electron can become available for conducting electricity. There are several ways in which this energy, which amounts to about 1.1 electron volts, can be supplied. In one of them it is supplied spontaneously whenever an electron receives a sufficiently high concentration of thermal energy. As a result at any temperature except absolute zero there are always some electrons in silicon available for conducting electricity, and the number of electrons increases as the temperature rises. Thus a crystal of silicon is not a perfect insulator but has an intrinsic conductivity. The conductivity is low compared with that of a metal, but it is an important property of the crystal.

When an electron is promoted into the energy level known as the conduction band, where it can play a role in the conduction of electricity, it leaves behind a "hole": a location that lacks an electron. A bound electron from a neighboring site can move into a hole and so exchange places with it. Both electrons and holes can therefore move within the crystal, and their joint motion

constitutes an electric current. When a voltage is applied, the electrons, which are the basic unit of negative charge, move toward the positive potential and the holes move toward the negative potential, as if they were positively charged.

A second way to introduce conduction electrons or holes into a crystal is to build into it atoms that supply either more or fewer electrons than the four that are required for the ideal structure. Here the energy that is needed to create holes or conduction electrons comes from the electrical misfit between the foreign atom and the silicon crystal. For example, phosphorus is an element that has a valence of five, that is, five of its electrons can interact with other atoms. A silicon crystal in which a small proportion of the atomic sites are occupied by phosphorus atoms instead of silicon atoms still has the basic structure of silicon. Nevertheless, it will have extra electrons that are in the conduction band because there are no bonding sites available to them. A phosphorus-doped silicon crystal is called an n-type semiconductor because it has an excess of the negatively charged electrons. Conversely, a silicon crystal can be doped with boron, which has a valence of

three. Then the crystal is called a p-type semiconductor because it has an excess of the positively charged holes.

Doping a crystal not only increases its electrical conductivity but also makes the crystal preferentially receptive to electrons or holes. An n-type crystal, which already has more electrons than it needs for bonding atoms (but not more than it needs for electrical neutrality) can readily absorb extra electrons. Conversely, the p-type crystal has an affinity for holes.

Light falling on the crystal is another source of the energy that can move electrons into the conduction state, creating electrons and holes in equal numbers. A photon of light striking any material is absorbed when its energy is transferred to an electron. If the electron absorbs the photon's energy near the surface of a metal, the electron may be emitted from the surface; that is the well-known photoelectric effect. If the photon penetrates within a semiconductor, and if its energy is equal to or greater than the amount of energy needed to move the electron into the conduction band, it gives rise to a conduction electron and a hole.

Thus if light of sufficient energy impinges on a perfect crystal of silicon, electrons and holes are produced and are free to move within the crystal. If the

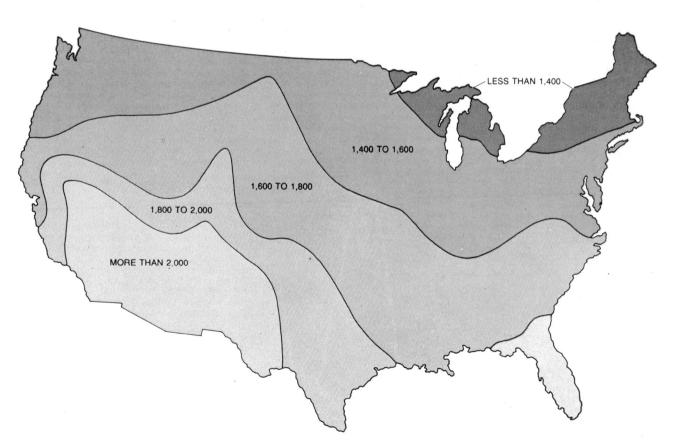

AMOUNT OF SOLAR RADIATION received each year in different areas in the U.S. is given in terms of kilowatt-hours per square meter. The amount received in the New England states is about two-thirds the amount received in the southwestern states. In every area, of course, the amount received varies with the weather, the time of day and the season. For that reason any large-scale solar electric-power system would have to be integrated with conventional power systems or would have to have a substantial storage facility of its own.

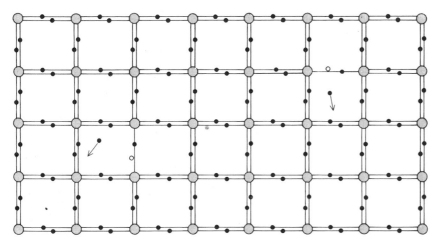

CRYSTAL OF PURE SILICON has a cubic structure, shown here in two dimensions for simplicity. The silicon atom (*gray*) has four valence electrons. Each atom is firmly held in the crystal lattice by sharing two electrons (*black*) with each of four neighbors at equal distances from it. Occasionally thermal vibrations or a photon of light will spontaneously provide enough energy to promote one of the electrons into the energy level known as the conduction band, where the electron is free to travel through the crystal and conduct electricity. When the electron moves from its bonding site, it leaves a "hole" (*white*), a local region of net positive charge.

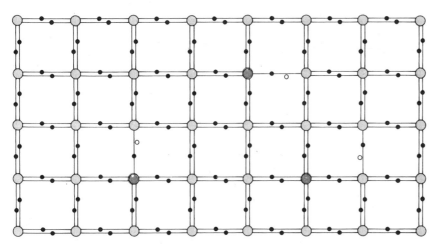

CRYSTAL OF P-TYPE SILICON can be created by doping the silicon with trace amounts of boron. Each boron atom (*dark color*) has only three valence electrons, so that it shares two electrons with three of its silicon neighbors and one electron with the fourth. Hence the *p*-type crystal has same structure as pure silicon, but it contains more holes than conduction electrons.

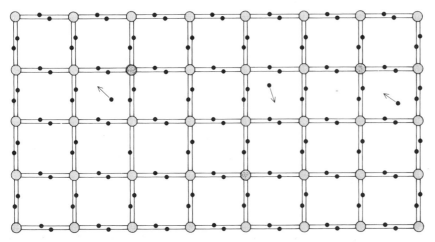

CRYSTAL OF N-TYPE SILICON can be created by doping the silicon with trace amounts of phosphorus. Each phosphorus atom (*light color*) has five valence electrons, so that not all of them are taken up in the crystal lattice. Hence *n*-type crystal has an excess of free electrons.

crystal is left to itself and no external voltage is applied to it, both the electrons and the holes wander through it randomly. When in its travels a conduction electron encounters a hole, it "falls" into it. The electron and the hole annihilate each other as the electron drops out of the conduction band and is again bound into the structure of the crystal. The energy that had been absorbed in originally creating the electron-hole pair is released as heat that slightly raises the temperature of the crystal.

In order to employ the electrons and holes as a source of electricity, it is necessary to arrange matters so that an electron cannot recombine with a hole until it has traveled through an external circuit, doing some useful work along the way. That can be achieved by taking advantage of the contrasting properties of *n*-type and *p*-type crystals. Consider a crystal that is composed of a layer of *n*-type silicon and a layer of *p*-type silicon. If pairs of electrons and holes are created near the junction of the two types of crystal, the affinity of the *n*-type crystal for electrons and that of the *p*-type crystal for holes will reduce the randomness of their movements through the crystal so that there is a net flow of electrons from the *p*-type crystal to the *n*-type one and a net flow of holes in the opposite direction.

In a silicon solar cell the layer of *n*-type silicon is of a thickness such that light falling on the surface penetrates far enough into the crystal to create electron-hole pairs in the vicinity of the junction with the *p*-type silicon. That thickness is typically half a micron. Therefore when light falls on the cell, electrons will collect in the *n*-type layer and holes will collect in the *p*-type layer until there is a voltage built up within the crystal sufficient to push any further electrons back into the *p*-type layer. In a silicon solar cell that voltage is about .65 volt.

A current can be drawn from the cell through a circuit that makes electrical contact with both the front surface of the cell and the back one. It is through the external circuit that the electrons trapped in the *n*-type layer find their way back to the *p*-type layer and can then recombine with holes. If the external circuit has a very low resistance, the current that flows through it is a measure of the rate at which the electrons are separating from the holes. That rate depends on the intensity of the light falling on the surface of the cell and on the rate at which electrons and holes are lost through their recombination.

What is the efficiency of such a silicon photovoltaic device? Under optimum conditions solar energy arrives at the earth's surface at the rate of about 1.15 kilowatts per square meter. How much of this energy can be converted into electrical energy? In more precise terms, what efficiency can we expect in photovoltaic devices? As in determining the efficiency of a fuel-burning power station there are three component questions to be considered: (1) What would the efficiency of the device be if it had the theoretically ideal characteristics? (2) To what extent does a photovoltaic device that can actually be built fall short of the performance of the ideal? (3) How closely can a device manufactured in large numbers at an acceptable cost approach the best device that can be made when cost is no object?

In considering the theoretical efficiency of a photovoltaic device let us look more closely at the process by which a photon is absorbed by an electron in a semiconductor. That process is the transfer of the quantum of energy to a single electron. If the photon is sufficiently energetic, the energy received by the electron will release it from its normal function as a bond between two neighboring atoms in the crystal; it becomes free to move in the crystal as a conduction electron, and a hole is simultaneously created.

The energy required to promote an electron from its bound state in a crystal into the conduction band is known as the band gap. The band gap varies considerably from one semiconductor to another; it can be expressed as electron volts or as the wavelength of the light that has the required quantum energy. Silicon has a band gap of 1.12 electron volts; thus in a silicon photovoltaic device an electron needs an energy of 1.12 electron volts in order to be moved into the conduction band. That energy corresponds to a wavelength of 1.1 microns, which is in the infrared region of the spectrum.

Radiation with a wavelength longer than 1.1 microns does not have enough energy to move electrons into the conduction band of a silicon photovoltaic device. Nearly half of the energy the earth receives from the sun arrives in the form of radiation with wavelengths longer than 1.1 microns. Hence that energy can play no part in generating electricity in a silicon photovoltaic cell. If the energy of a photon is greater than the band gap, that is, if in the case of silicon the wavelength of the photon is

shorter than 1.1 microns, the energy is entirely absorbed by an electron. The electron cannot retain more than 1.12 electron volts of the photon's energy, however, and so any excess energy is immediately converted into heat. In principle if the photon had enough energy, it could promote two electrons into the conduction band. In actuality all such energetic photons from the sun are absorbed by the earth's atmosphere before they reach the ground.

The result of these physical limitations is that only about half of the photons reaching the ground can create pairs of electrons and holes in a silicon photovoltaic cell, and a substantial fraction of the energy of many of these photons is converted into heat instead of electrical energy. To be somewhat more precise, the fundamental limitations set by quantum physics restrict the maximum possible efficiency of a silicon solar cell on the ground to about 21 percent. Other semiconductors have different band gaps, and their maximum theoretical efficiencies are therefore different.

A curve can be drawn relating theoretical efficiency and band gap. The de-

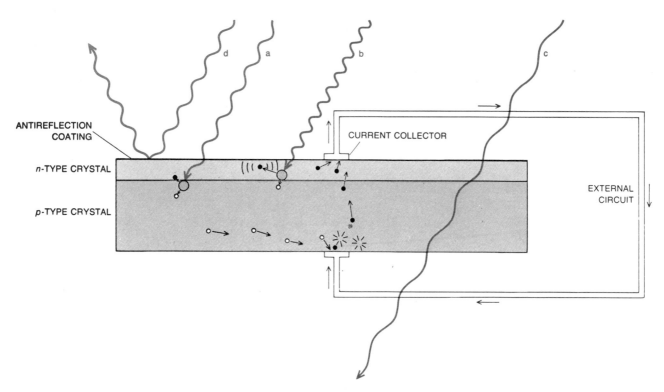

SILICON SOLAR CELL is a wafer of *p*-type silicon with a thin layer of *n*-type silicon on one side. When a photon of light with the appropriate amount of energy penetrates the cell near the junction of the two types of crystal and encounters a silicon atom (*a*), it dislodges one of the electrons, which leaves behind a hole. The energy required to promote the electron into the conduction band is known as the band gap. The electron thus promoted tends to migrate into the layer of *n*-type silicon, and the hole tends to migrate into the layer of *p*-type silicon. The electron then travels to a current collector on the front surface of the cell, generates an electric current in the external circuit and then reappears in the layer of *p*-type silicon, where it can recombine with waiting holes. If a photon with an amount of energy greater than the band gap strikes a silicon atom (*b*), it again gives rise to an electron-hole pair, and the excess energy is converted into heat. A photon with an amount of energy smaller than the band gap will pass right through the cell (*c*), so that it gives up virtually no energy along the way. Moreover, some photons are reflected from the front surface of the cell even when it has an antireflection coating (*d*). Still other photons are lost because they are blocked from reaching the crystal by the current collectors that cover part of the front surface. All these losses mean that a real silicon cell cannot convert more than about 18 percent of the solar energy it receives into electrical energy.

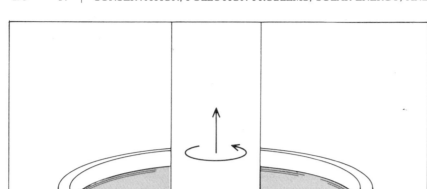

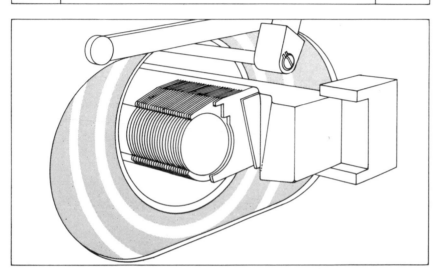

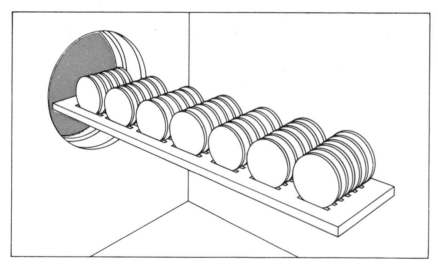

CZOCHRALSKI METHOD OF GROWING SILICON CRYSTALS is based on dipping a rotating seed crystal into a crucible of molten silicon and then slowly withdrawing it (*top*). The result is a massive cylindrical crystal three or four inches in diameter and several feet long. If an appropriate amount of boron is first added to the melt, the crystal is uniformly doped with boron. In making a silicon solar cell from such a crystal the crystal is cut into thin wafers (*middle*), a process in which a substantial portion of the crystal is lost as sawdust. One surface of each wafer, which is of *p*-type silicon, is converted into a layer of *n*-type silicon by being exposed to phosphorus (*bottom*) at a temperature high enough for the phosphorus atoms to diffuse a short distance into it. The electrical contacts are attached, the antireflection coating is applied and each cell is encapsulated in a protective skin. The finished product is a photovoltaic cell that can operate with an efficiency of between 15 and 18 percent and costs about $13.

tailed shape of such a curve depends on the distribution of energy in the spectrum of the radiation that is received; the curve on page 180 is for sunlight outside the earth's atmosphere, which is the relevant condition for space vehicles. Where devices on the ground are concerned, however, we must remember that the sunlight is filtered by the atmosphere before it reaches them. By the time it has passed through "air mass 1," which would be its vertical path through the atmosphere, it has lost a substantial part of its shortest-wavelength (ultraviolet) radiation. The curve is therefore somewhat different in detail.

Although maximum theoretical efficiency is an important characteristic of a semiconductor, there are many reasons, less fundamental than the quantum restriction I have mentioned but nonetheless real, that rule out any hope of reaching more than about four-fifths of these efficiencies. Lest it be thought that an efficiency of less than 20 percent is unworthy of our technical sophistication, it should be remembered that the epitome of advanced industrial technology, the automobile, performs its energy conversion at an efficiency of less than 20 percent, and it performs its people-moving function at a far lower figure.

The consideration that leads to a maximum theoretical efficiency of 21 percent for silicon solar cells ignores two problems that can be minimized but never eliminated completely and a third problem that can be reduced only insofar as it is economically worthwhile. The first problem lies in the fact that the figure of 21 percent refers to the utilization of the energy of radiation that has actually entered the crystal, whereas some of the sunlight is reflected from the crystal's surface. There is no way to eliminate this reflection, but it can be minimized with an antireflection coating: a layer of transparent material of a thickness such that rays of light reflected from the surface of the crystal interfere destructively with the rays reflected at the front surface of the coating. The rays that would have been reflected thus reinforce the rays that penetrate into the crystal. Even the best antireflection coating, however, can be designed to match only a single wavelength in the solar spectrum. It seems unlikely that the loss of solar energy by reflection can ever be reduced below 5 percent.

The second problem is the resistance of the semiconductor, through which the current must pass in traveling from the vicinity of the junction between the *n*-type and *p*-type layers, where the voltage is developed, to the points where the current enters the conductors of the external circuit. The ideal collector of current would be layers of metal covering the entire front and back of the solar cell and making good electrical contact with it. Such a collector presents no serious problem for the back of the cell. The conductor covering the front of the cell, however, would have to be transparent to sunlight. For fundamental reasons good conductors are not transparent, and so there is no way to cover the front surface with a layer that is sufficiently conductive and at the same time allows the light to reach the crystal. At present one must compromise by making the conductor cover as little of the cell's front surface as possible while minimizing the distance between any point where current is generated and the nearest point where it is collected.

That compromise presents its own set of problems. If one attempted to make a conductor out of a network of extremely thin metal wires in contact with the crystal's surface, only a small percentage of

the cell's area would be obscured. The resistance of the collector would be very high, however, and energy would be lost because the wires would be heated by the current. In order to reduce that resistance in an actual solar cell about 10 percent of the front surface is covered by the collector, which of course prevents 10 percent of the available light from reaching the cell itself. Hence the performance of the cell is further reduced from its ideal performance to that extent. There are still other losses attributable to the resistance the current encounters within the crystal as it travels to the collector, even though the maximum distance it must travel is typically no more than three millimeters.

If a solar cell consisted of a perfect crystal of silicon, with the ideal quantities and distribution of the n-type and p-type dopants and with no other impurities, it should be about 18 percent effi-

cient, and cells with that efficiency have been reported. Any departure from perfection, in terms of chemical purity or of the regularity of the structure of the crystal, leads to decreased efficiency. Impurities can cause leakage of current through the junction in the reverse direction, and crystallographic imperfections can act as traps where electrons and holes can recombine (and so be lost) and as highly conductive paths through which short-circuiting can occur.

Solar cells of the type I have been describing were developed for the space program, and they provide the electric power for all our space vehicles. They are a completely satisfactory solution to the problem of supplying energy without fuel of any kind, but they are expensive. In comparing the cost of electric power generated by solar cells with the cost of the power generated by the methods employed by electric utilities, the first criterion is the capital cost per kilowatt of output. It is reported that in 1959 the capital cost of solar cells was $200,000 per kilowatt. They can now be bought for nonspace purposes for about $20,000 per kilowatt, but production is very limited (amounting to the equivalent of 100 kilowatts per year). The cost of building a fuel-burning power station is about $500 per kilowatt of electric-power output. Obviously the cost of solar cells is far too high to outweigh the fact that no fuel is needed.

Why is the cost of photovoltaic devices so high? The reason is the chemical purity and crystallographic perfection that are needed to enable the solar cell to perform at its highest efficiency. The processes currently employed for attaining chemical purity and crystallographic perfection are inherently expensive. They are the same processes by which crystals are made for many purposes, such as microelectronic devices, where the highest possible quality is an absolute requirement. If photovoltaic cells are ever to be harnessed on a large scale for converting solar energy into electricity, good solar cells will have to be made cheaply. In the light of that prerequisite it is reasonable to ask: Is the manufacture of a solar cell intrinsically expensive, or is there any possibility of reducing the cost by a factor of 40, which would bring it into a competitive price range?

The process by which solar cells are currently made requires as a starting material silicon of "semiconductor grade," a very expensive material, costing about $70 per kilogram. This is not because of any scarcity of the element (silicon is one of the most abundant elements in the earth's crust) but because of the complexity of the purification process. Perfect crystals of silicon can be grown by the Czochralski method, a

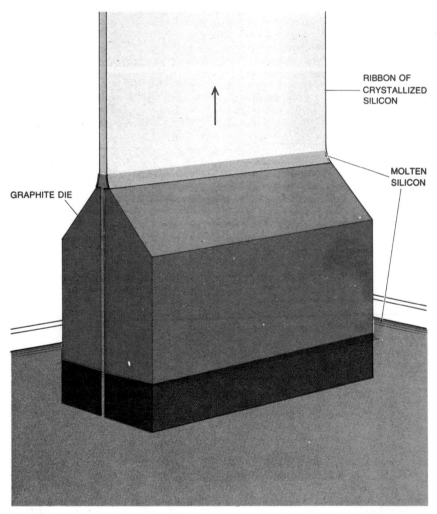

RIBBON OF
CRYSTALLIZED
SILICON

MOLTEN
SILICON

GRAPHITE DIE

EFG PROCESS FOR GROWING SILICON CRYSTALS is shown in greater detail. The graphite die (*dark gray*) rests in a pool of molten silicon (*color*). The silicon rises by capillary action through a narrow slit in the die and forms a layer of liquid silicon on the top of the die. A seed crystal (*not shown*) is dipped into the liquid at the top of the die and is drawn upward at the same rate as that at which the crystal (*light gray*) is growing downward, so that interface between crystal and liquid silicon remains a fraction of an inch above top of die. Silicon crystal grown in this way crystallizes into ribbon that can be cut into wafers with minimum of waste.

process in which a seed crystal is rotated while being slowly withdrawn from a crucible of molten silicon. If the temperature and the rates of rotation and pulling are controlled with sufficient precision, the silicon in the crucible is converted into a single perfect crystal. If a suitable quantity of boron has been added to the melt, the crystal is uniformly doped to the required extent. Such a silicon crystal is quite massive: it is typically a cylinder three or four inches in diameter and several feet long.

The next step in manufacturing a solar-cell is to slice the silicon crystal into thin wafers, each of which will eventually be an individual cell. To avoid damaging the crystal the wafers are sawed from it quite slowly. The typical wafer is a quarter of a millimeter thick, and at this point it is a uniform crystal of *p*-type silicon. A thin layer of it is converted into *n*-type silicon by exposing one surface of the wafer to phosphorus at a temperature high enough for the phosphorus atoms to diffuse into the wafer to a depth of about half a micron. Finally the electrical contacts are attached to the front and back of the wafer, the antireflection coating is applied and the entire cell is encapsulated in a protective skin.

The product of this series of operations is a cell that will generate between 10 and 15 milliwatts per square centimeter. Since the area of a typical cell is 45 square centimeters, if the cell operated at an efficiency of, say, 15 percent, it would generate up to two-thirds of a watt of electricity when it was exposed to full sunlight. At current prices such a silicon solar cell would cost about $13. The high cost is partly due to the fact that a large proportion of the initial crystal of silicon is wasted as sawdust when the wafers are cut. The principal reason the cell is so expensive, however, is that such a precise manufacturing process calls for much time and labor on the part of the skilled workers who grow the crystals, slice the wafers and fabricate the finished cell.

The output of a solar cell of a given size is proportional to the intensity of the radiation falling on it. It might therefore be economically sound to concentrate sunlight onto a relatively small area of solar cells. That could be done by focusing the light with a parabolic mirror. Such a mirror, however, has two disadvantages. First, it would focus the radiation in such a way that it would strike the cells at various angles, whereas the cells perform best if the radiation strikes them at right angles to their front surface. Second, it would be necessary to rotate the reflector to keep it pointing at the sun throughout the day, which would require a fairly precise automatic tracking device.

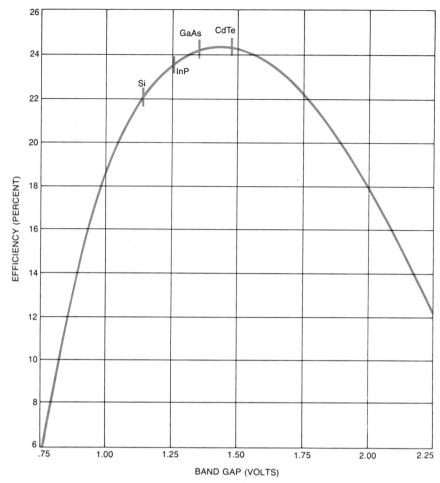

THEORETICAL MAXIMUM EFFICIENCY of a photovoltaic cell depends on the band gap of the semiconductor of which it is made. The band gap is given in terms of electron volts. In this diagram the efficiency of four different semiconductors is shown: silicon (*Si*), indium phosphorus (*InP*), gallium arsenide (*GaAs*) and cadmium tellurium (*CdTe*). Fundamental restraints of physics limit efficiency of even an ideal photovoltaic cell to less than 25 percent.

A promising alternative to a parabolic mirror is the trough-shaped reflector known as a Winston collector. The collector is designed to reflect a high proportion of incident sunlight onto a small area even when the sun is not directly in front of it, and so it is not necessary to have the collector track the sun. It has been demonstrated that a concentration ratio of eight to one is feasible, and that the performance of the cells does not suffer as long as they are not allowed to get too hot. It is therefore possible to increase the output of a given area of solar cells by a factor of at least eight, reducing the capital cost of the cells per kilowatt by the same factor. Whether or not this is economically desirable will depend on the cost of Winston collectors in relation to the cost of solar cells.

Is it necessary to maintain such extremely high standards of quality in the silicon crystals for solar cells? It is true that if a crystal is intended for a microelectronic device, even the slightest imperfection can render it useless. The requirements for a solar cell, however, are much less stringent, and the desire for high efficiency must be weighed against the high cost of perfection. As we have seen, a perfect crystal would have an efficiency of about 18 percent. Crystals of somewhat lower quality could have an efficiency of between 10 and 12 percent for a fraction of the cost. The capital cost per kilowatt of output could thus be much lower. A substantial effort, financed partly by the Energy Research and Development Administration (ERDA) and partly by private companies, is being concentrated on the development of less costly methods of making sufficiently pure silicon and of converting it into wafers from which solar cells can be made.

One of the most promising methods calls for continuously growing a silicon ribbon of the width and thickness required for solar cells. The process, known as edge-defined film-fed growth (EFG), was invented at Tyco Laboratories and is being developed by the Mobil Tyco Solar Energy Corporation. In this

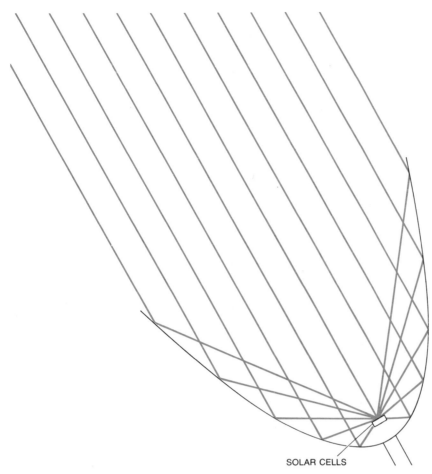

PARABOLIC REFLECTOR is one device for concentrating sunlight onto an array of solar cells in order to cut down the number of cells. Such a reflector has the disadvantage that in order for it to work well it must have an automatic driving mechanism to keep it pointed at sun.

SOLAR CELLS

national energy program of the U.S. The more important questions are whether photovoltaic devices can provide enough energy to reduce the amount of oil the U.S. imports or the number of nuclear power stations it builds, and if they can, when it might come about.

There are three quite distinct possibilities. The first is that very large solar power stations could be built, presumably in the deserts of the U.S. Southwest, where sunshine is plentiful. At an efficiency of 10 percent a collecting area of 10 square kilometers (a square 3.16 kilometers on a side) could generate a peak power of 1,000 megawatts, the typical output of a large fuel-burning power station. Although the output from solar cells is in the form of direct current, it could readily be converted into alternating current and fed into the distribution network of an electric utility.

Electricity supplied in this way could not, however, replace other means of generating electricity because it could not be relied on to provide power at all times when it is needed. Solar power stations could nonetheless reduce the demand on fuel-burning stations, and hence conserve fuel, by taking on part of the demand when the sun is shining. Such a function is sometimes called negative storage.

Alternatively, solar power stations could generate electricity when the sun is shining and store it for use when sunlight is unavailable. The energy could be stored by pumping water into elevated basins; the energy would be retrieved when the water was allowed to return to its original level through turbogenerators. It may eventually be possible to store solar energy by manufacturing hydrogen through the electrolysis of water. The hydrogen could be stored and subsequently burned instead of natural gas, or it could be fed into fuel cells to generate electricity.

Another way in which photovoltaic devices could supplement more conventional sources of power would be to deploy large panels of solar cells on space stations in stationary orbit around the earth. Energy gathered from the sun could be transmitted to terrestrial receiving stations by microwave beam. This scheme would have the advantage of providing energy continuously (if more than one space station were used), so that there would be no need for storage. Moreover, the solar cells would collect solar energy in space, where it is twice as intense as it is at the earth's surface. There is no doubt that such a power-generating space station could be built, but it would require an enormous investment. There might also be objections on the basis that the microwaves could be hazardous to life on

process the molten silicon rises by capillary action through a slot in a die made of graphite. A film of liquid silicon at the top of the die feeds the growing crystal, whose cross section is defined by the shape of the top of the die. The ribbon of silicon, .15 centimeter thick, can be grown at rates of more than 2.5 centimeters per minute; it is then simply cut with a minimum of wasted material into rectangular wafers 2.5 centimeters wide and 10 centimeters long. The crystals grown by the EFG process approach perfection far less closely than those grown by the Czochralski process, but they can nonetheless be made into solar cells with an efficiency of between 10 and 12 percent.

The EFG process is not the only way to produce cheap silicon crystals that are good enough for solar cells, and as I have mentioned silicon is not the only possible material for such cells. For example, solar cells with an efficiency of 18 percent have been made out of gallium arsenide. Moreover, thin films of other semiconductors that can be deposited from a vapor onto a substrate may eventually compete with silicon wafers in terms of dollars per kilowatt. The

EFG process has not yet been automated to the point where it can produce cheap cells. Close analysis indicates, however, that it has the potential to provide crystals for solar cells that will cost less than the magic figure of $500 per kilowatt.

Exactly how would such cells be employed to convert into electric power the intermittent and somewhat unpredictable energy of sunlight? In the first place it should be kept in mind that the cost per kilowatt of the cells will not immediately drop to an economic level, because low cost cannot be achieved until there is production on a very large scale. There are, however, smaller-scale demands that might assist the transition. Pumping water for irrigation or powering television sets (with battery storage) in remote and sunny parts of the world are applications for which even quite expensive solar cells might be the cheapest power supply. The market could be large enough to support the early years of growth needed for a new solar-cell industry.

Such an intermediate stage, however, is relatively insignificant in terms of the

the ground, and that the system would be vulnerable to an enemy.

The electric-utility industry has evolved in an era where the economics of power generation have demanded highly centralized power production. This requirement has resulted from the fact that it is highly inefficient to generate electricity in one-kilowatt amounts by burning fuel. The cheapest way to generate electricity by present methods is with power stations that operate at the level of 1,000 megawatts. Such considerations do not, however, hold for solar cells. For them the economy of scale applies to the production of the cells but not to their utilization, since they have the same efficiency whether the amount of electricity is large or small.

It can be argued that it might be cheaper for the electricity needed in the home to be generated by solar cells on the roof and stored in batteries. The solar cells could be integrated into a system that would provide heating, cooling and hot water by utilizing part of the 88 to 90 percent of the solar energy the cells are unable to convert into electricity. Furthermore, if the electricity were generated at the site where it is needed, the cost to the consumer for distribution would be eliminated. It is not likely that such a photovoltaic system could supply all the power needed in the home. Nevertheless, if the system supplied two-thirds of the power at a reasonable cost, the need for new power stations would be reduced, and the community would be less dependent on supplies of oil, natural gas, coal and uranium.

ing capacity would be possible, but it would require much more capital and much more engineering effort than are likely to be deployed. Demonstrating the feasibility of generating electricity on a large scale with solar cells will not in itself, however, guarantee their widespread adoption; that will depend on who can sell the product of his electric-generating system at the lowest price.

The capital cost of $500 per peak kilowatt for photovoltaic systems is considered a realistic aim, but how valid a criterion is it for whether such systems will be economically competitive? At first $500 per peak kilowatt looks attractive, but it must be remembered that a conventional power station (also costing $500 per peak kilowatt) can generate electricity and therefore also generate a return on its capital investment, day and night, rain or shine. Official statistics show that on the average the power stations of the U.S. are "on line" for about 50 percent of the time. That means they generate about 4,400 kilowatt-hours per installed kilowatt per year. For a solar generator, on the other hand, the amount of electricity generated per kilowatt of peak output would be about 2,100 kilowatt-hours per year in Texas and about 1,400 kilowatt-hours per year in the northeastern states. Therefore if a solar installation is to provide as much energy as a 1,000-megawatt fossil-fuel or nuclear power station, it would need to have a peak output of between 2,000 and 3,000 megawatts. The future price of fuel will determine whether a factor

of two or three in capital cost will be outweighed by the zero fuel cost of a solar installation.

The economic factors are considerably more favorable to solar generators in those places where the peak demand for electricity comes because air conditioners are going full blast. At such times, of course, solar energy is at its most abundant. Having solar generators as standby equipment for service under those conditions, however, could never make more than a small contribution to the overall demand for electric power. Still another set of economic considerations must be applied to the solar generation of electric power at the point of use. Since a substantial part of the cost of electricity to the consumer is represented by the distribution, as distinct from the generation, of electricity, at $500 per kilowatt of peak output the cost per kilowatt-hour to the homeowner would be well below what the electric utilities charge.

At the present time a photovoltaic power station costing $500 per kilowatt of peak output (if it could be built) would not be competitive with a station running on fossil fuel or uranium. As the price of such fuels increases, however, and as environmental restrictions further increase the cost of generating electricity with them, solar energy will become more attractive. The photovoltaic generation of electricity might find its place in the sun sooner than some now expect.

How soon might we hope to reap the benefits of this new technology? It depends partly on how quickly devices that work in the laboratory can be manufactured on an industrial scale, and that is a question of how fast the necessary engineering development can be achieved. No new scientific breakthrough is needed, only a great deal of meticulous engineering to put proved ideas into practice. Even if we had today a process complete in all details for making solar cells at a cost of $500 per kilowatt of peak output, however, it would still take five years to build up an industry that could produce the solar cells and the associated equipment to generate electricity on the megawatt scale. And it might take another five years beyond that before a significant fraction of new electric-generating capacity could be photovoltaic.

ERDA's goal is several 10-megawatt photovoltaic demonstration systems by the early 1980's and a significant contribution, perhaps exceeding 1 percent, to our total energy consumption (equivalent to nearly a million barrels of oil a day) by the turn of the century. A much faster buildup of photovoltaic generat-

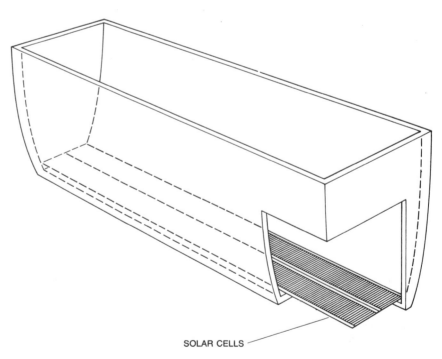

SOLAR CELLS

WINSTON COLLECTOR, invented by Roland Winston of the University of Chicago, is in the shape of a long trough. It concentrates sunlight on a strip of solar cells and has the advantage that it does not need to be rotated in order to follow sun across sky in course of the day.

Energy in the Universe

by Freeman J. Dyson
September 1971

The energy flows on the earth are embedded in the energy flows in the universe. A delicate balance among gravitation, nuclear reactions and radiation keeps the energy from flowing too fast

*Man has no Body distinct from his Soul;
for that called Body is a portion
of Soul
discern'd by the five Senses,
the chief
inlets of Soul in this age.*

*Energy is the only life and is
from the Body;
and Reason is the bound or outward
circumference of Energy.*

Energy is Eternal Delight.

—WILLIAM BLAKE,
The Marriage of Heaven and Hell, 1793

One need not be a poet or a mystic to find Blake's definition of energy more satisfying than the definitions given in textbooks on physics. Even within the framework of physical science energy has a transcendent quality. On many occasions when revolutions in thought have demolished old sciences and created new ones, the concept of energy has proved to be more valid and durable than the definitions in which it was embodied. In Newtonian mechanics energy was defined as a property of moving masses. In the 19th century energy became a unifying principle in the construction of three new sciences: thermodynamics, quantitative chemistry and electromagnetism. In the 20th century energy again appeared in fresh disguise, playing basic and unexpected roles in the twin intellectual revolutions that led to relativity theory and quantum theory. In the special theory of relativity Einstein's equation $E = mc^2$, identifying energy with mass, threw a new light on our view of the astronomical universe, a light whose brilliance no amount of journalistic exaggeration has been able to obscure. And in quantum mechanics Planck's equation $E = h\nu$, restricting the

energy carried by any oscillation to a constant multiple of its frequencies, transformed in an even more fundamental way our view of the subatomic universe. It is unlikely that the metamorphoses of the concept of energy, and its fertility in giving birth to new sciences, are yet at an end. We do not know how the scientists of the next century will define energy or in what strange jargon they will discuss it. But no matter what language the physicists use they will not come into contradiction with Blake. Energy will remain in some sense the lord and giver of life, a reality transcending our mathematical descriptions. Its nature lies at the heart of the mystery of our existence as animate beings in an inanimate universe.

The purpose of this article is to give an account of the movement of energy in the astronomical world, insofar as we understand it. I shall discuss the genesis of the various kinds of energy that are observed on the earth and in the sky, and the processes by which energy is channeled in the evolution of stars and galaxies. This overall view of the sources and flow of energy in the cosmos is intended to put in perspective the preceding articles, which deal with the problems of the use of energy by mankind on the earth. In looking to our local energy resources it is well to consider how we fit into the larger scheme of things. Ultimately what we can do here on the

earth will be limited by the same laws that govern the economy of astronomical energy sources. The converse of this statement may also be true. It would not be surprising if it should turn out that the origin and destiny of the energy in the universe cannot be completely understood in isolation from the phenomena of life and consciousness. As we look out into space we see no sign that life has intervened to control events anywhere except precariously on our own planet. Everywhere else the universe appears to be mindlessly burning up its reserves of energy, inexorably drifting toward the state of final quiescence described imaginatively by Olaf Stapledon: "Presently nothing was left in the whole cosmos but darkness and the dark whiffs of dust that once were galaxies." It is conceivable, however, that life may have a larger role to play than we have yet imagined. Life may succeed against all the odds in molding the universe to its own purposes. And the design of the inanimate universe may not be as detached from the potentialities of life and intelligence as scientists of the 20th century have tended to suppose.

The cosmos contains energy in various forms, for example gravitation, heat, light and nuclear energy. Chemical energy, the form that plays the major role in present-day human activities, counts for very little in the universe as a whole.

CELESTIAL ENERGY SOURCE is represented by the computer-generated display on the following page, which is based on data gathered by means of a rocket-borne X-ray detection device. The display, known as a correlation map, was used to locate with high precision a strong X-ray source (designated GX 5-1) in the constellation Sagittarius near the direction of the galactic center. The experiment was carried out by a team of investigators from the Massachusetts Institute of Technology; the details of the experimental procedure are described in the September 1970 issue of *The Astrophysical Journal.* The mechanism responsible for the large energy fluxes emanating from such sources is unknown, but it is believed to play an important role in the overall energy flow of the universe.

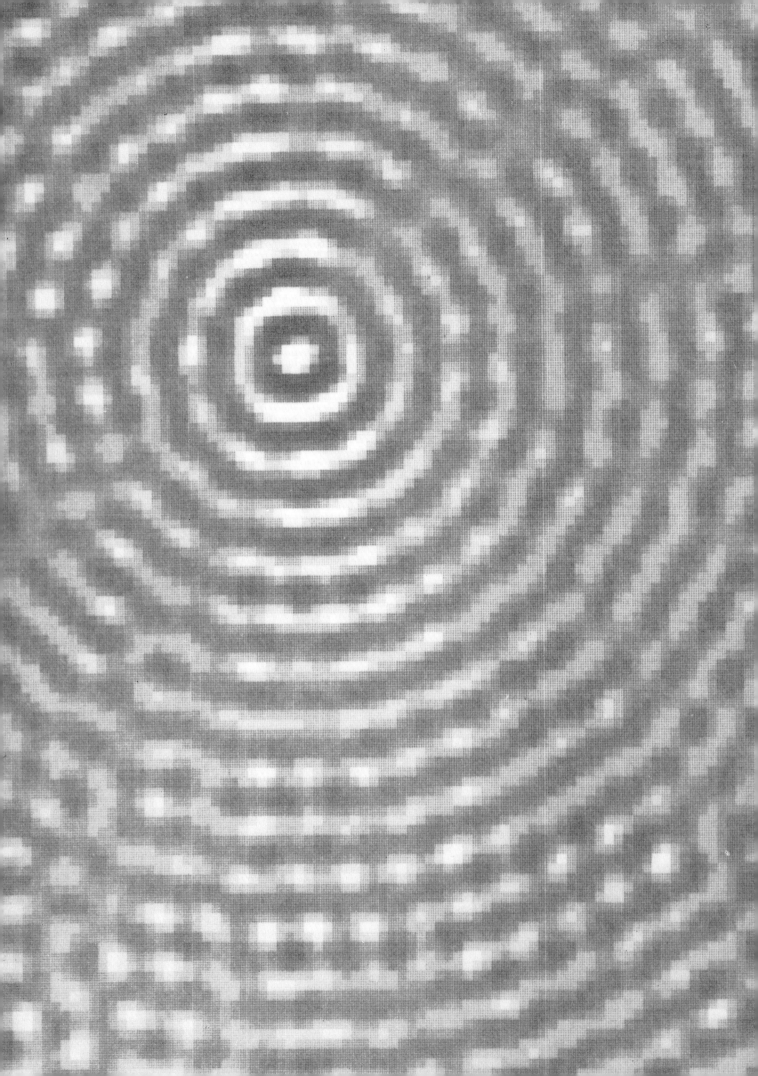

In the universe the predominant form of energy is gravitational. Every mass spread out in space possesses gravitational energy, which can be released or converted into light and heat by letting the mass fall together. For any sufficiently large mass this form of energy outweighs all others.

The laws of thermodynamics decree that each quantity of energy has a characteristic quality called entropy associated with it. The entropy measures the degree of disorder associated with the energy. Energy must always flow in such a direction that the entropy increases. Thus we can arrange the different forms of energy in an "order of merit," the highest form being the one with the least disorder or entropy [see illustration below]. Energy of a higher form can be degraded into a lower form, but a lower form can never be wholly converted back into a higher form. The basic fact determining the direction of energy flow in the universe is that gravitational energy is not only predominant in quantity but also highest in quality. Gravitation carries no entropy and stands first in the order of merit. It is for this reason that a hydroelectric power station converting the gravitational energy of water to electricity can have an efficiency close to 100 percent, which no chemical or nuclear power station can approach. In the universe as a whole the main theme of energy flow is the gravitational contraction of massive objects, the gravitational energy released in contraction being converted into energy of motion, light and heat. The flow of water from a reservoir to a turbine situated a little closer to the center of the earth is in essence a controlled gravitational contraction of the earth, only on a more modest scale than astronomers are accustomed to consider. The universe evolves by the gravitational contraction of objects of all sizes, from clusters of galaxies to planets.

When one views the universe in broad outline in this way, a set of paradoxical questions at once arises. Since thermodynamics favors the degradation of gravitational energy to other forms, how does it happen that the gravitational energy of the universe is still predominant after 10 billion years of cosmic evolution? Since large masses are unstable against gravitational collapse, why did they not all collapse long ago and convert their gravitational energy into heat and light in a quick display of cosmic fireworks? Since the universe is on a one-way slide toward a state of final death in which energy is maximally degraded, how does it manage, like King Charles, to take such an unconscionably long time a-dying? These questions are not easy to answer. The further one goes in answering them, the more remarkable and paradoxical becomes the apparent stability of the cosmos. It turns out that the universe as we know it survives not by any inherent stability but by a succession of seemingly accidental "hangups." By a hangup I mean an obstacle, usually arising from some quantitative feature of the design of the universe, that arrests the normal processes of degradation of energy. Psychological hangups are generally supposed to be bad for us, but cosmological hangups are absolutely necessary for our existence.

The first and most basic hangup built into the architecture of the universe is the size hangup. A naïve person looking at the cosmos has the impression that the whole thing is extravagantly, even irrelevantly, large. This extravagant size is our primary protection against a variety of catastrophes. If a volume of space is filled with matter with an average density d, the matter cannot collapse gravitationally in a time shorter than the "free-fall time" t, which is the time it would take to fall together in the absence of any other hangups. The formula relating d with t is $Gdt^2 = 1$, where G is the constant in Newton's law of gravitation. The effect of this formula is that when we have an extravagantly small density d, and therefore an extravagantly big volume of space, the free-fall time t can become so long that gravitational collapse is postponed to a remote future.

If we take for d the average density of mass in the visible universe, which works out to about one atom per cubic meter, the free-fall time is about 100 billion years. This is longer than the probable age of the universe (10 billion years), but only by a factor of 10. If the matter in the universe were not spread out with such an exceedingly low density, the free-fall time would already have ended and our remote ancestors would long ago have been engulfed and incinerated in a universal cosmic collapse.

The matter inside our own galaxy has an average density about a million times higher than that of the universe as a whole. The free-fall time for the galaxy is therefore about 100 million years. Within the time span of life on the earth the galaxy is not preserved from gravitational collapse by size alone. Our survival requires other hangups besides the hangup of size.

Another form of degradation of gravitational energy, one less drastic than gravitational collapse, would be the disruption of the solar system by close encounters or collisions with other stars. Such a degradation of the orbital motions of the earth and planets would be just as fatal to our existence as a complete collapse. We have escaped this catastrophe only because the distances between stars in our galaxy are also extravagantly large. Again a calculation shows that our galaxy is barely large enough to make the damaging encounters unlikely. So even within our galaxy

FORM OF ENERGY	ENTROPY PER UNIT ENERGY
GRAVITATION	0
ENERGY OF ROTATION	0
ENERGY OF ORBITAL MOTION	0
NUCLEAR REACTIONS	10^{-6}
INTERNAL HEAT OF STARS	10^{-3}
SUNLIGHT	1
CHEMICAL REACTIONS	1–10
TERRESTRIAL WASTE HEAT	10–100
COSMIC MICROWAVE RADIATION	10^4

"ORDER OF MERIT" of the major forms of energy in the universe ranks the various energy forms roughly according to their associated entropy per unit energy, expressed in units of inverse electron volts. The entropy, which measures the degree of disorder associated with a particular form of energy, varies approximately inversely with the temperature associated with that energy form. In the cases of gravitation, rotation and orbital motion there is no associated temperature and hence the entropy is zero. Energy generally flows from higher levels to lower levels in the table, that is, in such a direction that the entropy increases. The cosmic microwave background radiation appears to be an ultimate heat sink; no way is known in which this energy could be further degraded or converted into any other form. The universe survives not by any inherent stability but by a succession of seemingly accidental "hangups," or obstacles, usually arising from some quantitative feature of the design of the universe, that act to arrest the normal processes of the degradation of energy.

SIZE HANGUP, the first and most basic hangup that is built into the architecture of the universe, is symbolized by this photograph of a large cluster of galaxies in the constellation Hercules. A cluster may contain anywhere from two galaxies to several thousand. It typically occupies approximately 10^{20} cubic light-years of space and maintains an average distance between galaxies of about a million light-years. It is the extravagantly large volume of space that is the primary protection against a variety of cosmic catastrophes. By making the "free-fall time" of the universe so long, for example, the size hangup postpones the ultimate gravitational collapse of the universe to a remote future. The photograph was made with the 200-inch Hale reflecting telescope on Palomar Mountain.

the size hangup is necessary to our preservation, although it is not by itself sufficient.

The second on the list of hangups is the spin hangup. An extended object cannot collapse gravitationally if it is spinning rapidly. Instead of collapsing, the outer parts of the object settle into stationary orbits revolving around the inner parts. Our galaxy as a whole is preserved by this hangup, and the earth is preserved by it from collapsing into the sun. Without the spin hangup no planetary system could have been formed at the time the sun condensed out of the interstellar gas.

The spin hangup has produced ordered structures with an impressive appearance of permanence, not only galaxies and planetary systems but also double stars and the rings of Saturn. None of these structures is truly permanent. Given sufficient time all will be degraded by slow processes of internal energy dissipation or by random encounters with other objects in the universe. The solar system seems at first to be a perfect perpetual motion machine, but in reality its longevity is dependent on the combined action of the spin hangup and the size hangup.

The third hangup is the thermonuclear hangup. This hangup arises from the fact that hydrogen "burns" to form helium when it is heated and compressed. The thermonuclear burning (actually fusion reactions between hydrogen nuclei) releases energy, which opposes any further compression. As a result any object such as a star that contains a large proportion of hydrogen is unable to collapse gravitationally beyond a certain point until the hydrogen is all burned up. For example, the sun has been stuck on the thermonuclear hangup for 4.5 billion years and will take about another five billion years to burn up its hydrogen before its gravitational contraction can be resumed [*see top illustration on page 189*]. Ultimately the supply of nuclear energy in the universe is only a small fraction of the supply of gravitational energy. But the nuclear energy acts as a delicately adjusted regulator, postponing the violent phases of gravitational collapse and allowing stars to shine peacefully for billions of years.

There is good evidence that the universe began its existence with all the matter in the form of hydrogen, with perhaps some admixture of helium but few traces of heavier elements. The evidence comes from the spectra of stars moving in our galaxy with very high velocities with respect to the sun. The high velocities mean that these stars do not take part in the general rotation of the galaxy. They are moving in orbits that are oblique to the plane of the galaxy, and therefore their velocity and the sun's combine to give a relative velocity of the order of hundreds of kilometers per second. Such a velocity is in contrast to that of common stars, which orbit with the sun in the central plane of the galaxy and show relative velocities of the order of

SPIN HANGUP is exemplified by this photograph of a typical galaxy of the "open spiral" type. Galaxies, planetary systems, double stars and the rings of Saturn are among the celestial objects that are spared temporarily from the inevitable gravitational collapse by the spin hangup. This particular galaxy, designated M 101, illustrates the mechanism of star formation that is probably still at work in our own galaxy. Each spiral arm consists of clumps of bright, newly formed stars left behind by the passage of a rotating hydrodynamic-gravitational wave in the galactic disk. Photograph was also made by the 200-inch telescope on Palomar Mountain.

TWO STELLAR POPULATIONS are characteristically present in spiral galaxies, the older stars forming a roughly spherical halo cloud and the younger stars forming a comparatively thin central disk. This photograph of the spiral galaxy M 104, seen almost edge on, gives a particularly clear view of the two types of stellar population. In our own galaxy the arrangement is the same but the proportions are different. If our galaxy were seen this way, the disk would look much brighter than that of M 104, the halo much fainter.

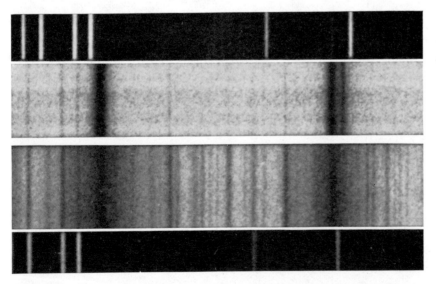

SPECTROGRAMS of two stars in our galaxy, a high-velocity halo star (*second from top*) and a normal disk star (*third from top*) of approximately the same spectral type, were made with the 120-inch telescope at the Lick Observatory. The spectrum of the high-velocity star, designated HD 140283, shows comparatively weak absorption lines for all the elements except hydrogen. The spectrum of the normal star, our own sun, contains numerous lines associated with the heavier elements, particularly carbon and iron. Since hydrogen burns to make carbon and iron but carbon and iron cannot burn to make hydrogen, the objects in our galaxy with the least contamination of hydrogen by heavier elements must be the oldest. Spectra at top and bottom provide bright lines for reference purposes.

tens of kilometers per second. The high-velocity stars form a "halo," or spherical cloud, that is bisected by the rotating galactic disk, which contains the bulk of the ordinary stars [*see top illustration at left*].

The obvious explanation of this state of affairs is that the high-velocity stars are the oldest. They condensed out of the primeval galaxy while it was still in a state of free fall, before it encountered the spin hangup. After the spin hangup the galaxy settled down into a disk, and the ordinary stars were formed in orbits within the disk, where they have remained ever since. This picture of the history of the galaxy is dramatically confirmed by the spectroscopic evidence [*see bottom illustration at left*]. The spectra of the extreme high-velocity stars show extremely weak absorption lines for all the elements except hydrogen. These stars are evidently composed of less than a tenth—sometimes less than a hundredth—as much of the common elements carbon, oxygen and iron as we find in the sun. Such major deficiencies of the common elements are almost never found in low-velocity stars. Since hydrogen burns to make carbon and iron, but carbon and iron cannot burn to make hydrogen, the objects with the least contamination of hydrogen by heavier elements must be the oldest. We can still see a few stars in our neighborhood dating back to a time so early that the contamination by heavier elements was close to zero.

The discovery that the universe was originally composed of rather pure hydrogen implies that the thermonuclear hangup is a universal phenomenon. Every mass large enough to be capable of gravitational collapse must inescapably pass through a prolonged hydrogen-burning phase. The only objects exempt from this rule are masses of planetary size or smaller, in which gravitational contraction is halted by the mechanical incompressibility of the material before the ignition point of thermonuclear reactions is reached. The preponderance of hydrogen in the universe ensures that our night sky is filled with well-behaved stars like our own sun, placidly pouring out their energy for the benefit of any attendant life-forms and giving to the celestial sphere its historic attribute of serene immobility. It is only by virtue of the thermonuclear hangup that the heavens have appeared to be immobile. We now know that in corners of the universe other than our own violent events are the rule rather than the exception. The prevalence of catastrophic outbursts of energy was revealed to us through the rapid

hydrogen gently for billions of years instead of blowing up like a bomb? To answer this question it is necessary to invoke yet another hangup.

The crucial difference between the sun and a bomb is that the sun contains ordinary hydrogen with only a trace of the heavy hydrogen isotopes deuterium and tritium, whereas the bomb is made mainly of heavy hydrogen. Heavy hydrogen can burn explosively by strong nuclear interactions, but ordinary hydrogen can react with itself only by the weak-interaction process. In this process two hydrogen nuclei (protons) fuse to form a deuteron (a proton and a neutron) plus a positron and a neutrino. The proton-proton reaction proceeds about 10^{18} times more slowly than a strong nuclear reaction at the same density and temperature. It is this weak-interaction hangup that makes ordinary hydrogen useless to us as a terrestrial source of energy. The hangup is essential to our existence, however, in at least three ways. First, without this hangup we would not have a sufficiently long-lived and stable sun. Second, without it the ocean would be an excellent thermonuclear high explosive and would constitute a perennial temptation to builders of "doomsday machines." Third and most important, without the weak-interaction hangup it is unlikely that any appreciable quantity of hydrogen would have survived the initial hot, dense phase of the evolution of the universe. Essentially all the matter in the universe would have been burned to helium before the first galaxies started to condense, and no normally long-lived stars would have had a chance to be born.

If one looks in greater detail at the theoretical reasons for the existence of the weak-interaction hangup, our salvation seems even more providential. The hangup depends decisively on the nonexistence of an isotope of helium with a mass number of 2, the nucleus of which would consist of two protons and no neutrons. If helium 2 existed, the proton-proton reaction would yield a helium-2 nucleus plus a photon, and the helium-2 nucleus would in turn spontaneously decay into a deuteron, a positron and a neutrino. The first reaction being strong, the hydrogen would burn fast to produce helium 2. The subsequent weak decay of the helium 2 would not limit the rate of burning. It happens that there does exist a well-observed state of the helium-2 nucleus, but the state is unbound by about half a million volts. The nuclear force between two protons is attractive and of the order of 20 million volts, but it just barely fails to produce a bound state. If

STAGE	DURATION	RANGE OF OUTPUT OF ENERGY
1. GRAVITATIONAL CONTRACTION	143 MILLION YEARS	INITIALLY 600, DECREASING RAPIDLY TO .7
2. HYDROGEN BURNING	10.3 BILLION YEARS	INITIALLY .7, INCREASING SLOWLY TO 3
3. RESUMED GRAVITATIONAL CONTRACTION OF CORE	500 MILLION YEARS	3−10
4. HELIUM AND CARBON BURNING	500 MILLION YEARS	10−1,000, FLUCTUATING IN COMPLICATED FASHION
5. FINAL GRAVITATIONAL CONTRACTION	13 MILLION YEARS	1,000−.01
6. WHITE-DWARF PHASE	INFINITE	.01, COOLING SLOWLY TO ZERO

EFFECT OF THERMONUCLEAR HANGUP on the life of the sun is evident in this table, which gives the sun's energy output at various stages in its evolution in units of the present solar luminosity (2.10^{33} ergs per second). Only the first three stages are known well enough to be accurately computed. The details of stages *4* and *5* are uncertain because the mechanisms of convective instability in the sun's interior and of mass loss at the surface are not completely understood. During stage *4* the sun will probably pass through a "red giant" phase, and during stage *5* through a "planetary nebula" phase. What is certain is that in stages *4* and *5* the energy output will be high and the duration short compared with the energy output and duration of stage *2*. If it were not for the thermonuclear hangup of stage *2*, the sun would have squandered all its energy and reached stage *6* long ago, probably in less than a billion years. As matters stand, the sun is only halfway through stage *2*.

progress of radio astronomy over the past 30 years. These outbursts are still poorly understood, but it seems likely that they occur in regions of the universe where the thermonuclear hangup has been brought to an end by the exhaustion of hydrogen.

It may seem paradoxical that the thermonuclear hangup has such benign and pacifying effects on extraterrestrial affairs in view of the fact that, so far at least, our terrestrial thermonuclear devices are neither peaceful nor particularly benign. Why does the sun burn its

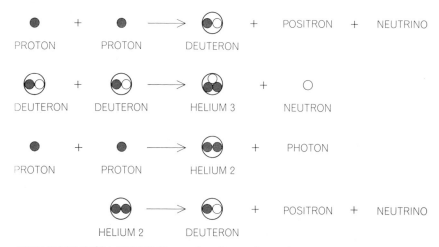

SOME FUSION REACTIONS discussed in this article are depicted schematically here. In the sun (*top*) ordinary hydrogen nuclei (protons) fuse to form a deuteron (a proton and a neutron) plus a positron and a neutrino. In a thermonuclear bomb (*second from top*) two heavy hydrogen isotopes, in this case both deuterons, fuse by the strong interaction process to form a helium-3 nucleus plus a neutron. The proton-proton reaction proceeds about 10^{18} times more slowly than the corresponding deuteron-deuteron reaction. If a helium-2 nucleus could exist, the proton-proton reaction would yield a helium-2 nucleus plus a photon (*third from top*), and the helium-2 nucleus would in turn spontaneously decay into a deuteron, a positron and a neutrino (*fourth from top*). As a consequence there would be no weak-interaction hangup, and essentially all of the hydrogen existing in the universe would have been burned to helium even before the first galaxies had started to condense.

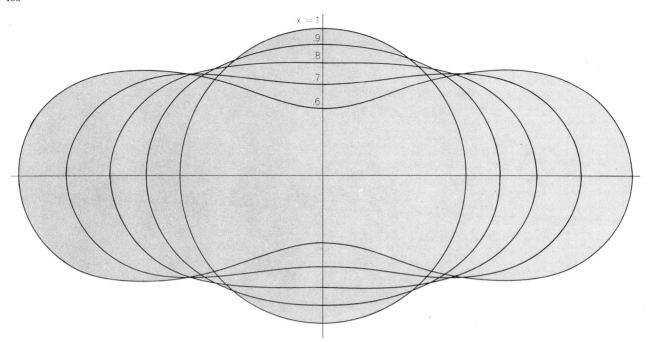

SURFACE-TENSION HANGUP has enabled fissionable nuclei such as uranium to survive in the earth's crust for aeons. Before these nuclei can split spontaneously their surface must be stretched into a nonspherical shape, and this stretching is opposed by an extremely powerful force of surface tension. This diagram shows the shapes of various nuclei when they go "over the hump" during fission; the shapes were computed according to the liquid-drop model of the nucleus. The nuclei are labeled by a parameter x, which is the ratio of electrostatic energy to surface tension. Nuclei from thorium to plutonium all have values of x between .7 and .8. The larger x is, the more unstable the nucleus is and the smaller the deformation required before fission occurs.

the force were a few percent stronger, there would be no weak-interaction hangup.

I have discussed four hangups: size, spin, thermonuclear and weak-interaction. The catalogue is by no means complete. There is an important class of transport or opacity hangups, which arise because the transport of energy by conduction or radiation from the hot interior of the earth or the sun to the cooler surface takes billions of years to complete. It is the transport hangup that keeps the earth fluid and geologically active, giving us such phenomena as continental drift, earthquakes, volcanoes and mountain uplift. All these processes derive their energy from the original gravitational condensation of the earth four billion years ago, supplemented by a modest energy input from subsequent radioactivity.

Last on my list is a special surface-tension hangup that has enabled the fissionable nuclei of uranium and thorium to survive in the earth's crust until we are ready to use them. These nuclei are unstable against spontaneous fission. They contain so much positive charge and so much electrostatic energy that they are ready to fly apart at the slightest provocation. Before they can fly apart, however, their surface must be stretched into a nonspherical shape, and this stretching is opposed by an extremely powerful force of surface tension. A nucleus is kept spherical in exactly the same way a droplet of rain is kept spherical by the surface tension of water, except that the nucleus has a tension about 10^{18} times as strong as that of the raindrop. In spite of this surface tension a nucleus of uranium 238 does occasionally fission spontaneously, and the rate of the fissioning can be measured. Nonetheless, the hangup is so effective that less than one in a million of the earth's uranium nuclei has disappeared in this way during the whole of geological history.

No hangup can last forever. There are times and places in the universe at which the flow of energy breaks through all hangups. Then rapid and violent transformations occur, of whose nature we are still ignorant. Historically it was physicists and not astronomers who recorded the first evidence that the universe is not everywhere as quiescent as traditional astronomy had pictured it. The physicist Victor Hess discovered 60 years ago that even our quiet corner of the galaxy is filled with a uniform cloud of the extremely energetic particles now called cosmic rays. We still do not know in detail where these particles come from, but we do know that they represent an important channel in the overall energy flow of the universe. They carry on the average about as much energy as starlight.

The cosmic rays must certainly originate in catastrophic processes. Various attempts to explain them as by-products of familiar astronomical objects have proved quantitatively inadequate. In the past 30 years half a dozen strange new types of object have been discovered, each of which is violent and enigmatic enough to be a plausible parent of cosmic rays. These include the supernovas (exploding stars), the radio galaxies (giant clouds of enormously energetic electrons emerging from galaxies), the Seyfert galaxies (galaxies with intensely bright and turbulent nuclei), the X-ray sources, the quasars and the pulsars. All these objects are inconspicuous only because they are extremely distant from us. And once again only the size hangup —the vastness of the interstellar spaces— has diluted the cosmic rays enough to save us from being fried or at least sterilized by them. If sheer distance had not effectively isolated the quiet regions of the universe from the noisy ones, no type of biological evolution would have been possible.

The longest-observed and least mysterious of the violent objects are the supernovas. These appear to be ordinary stars, rather more massive than the sun, that have burned up their hydrogen and passed into a phase of gravitational collapse. In various ways the rapid release of gravitational energy can cause the star to explode. There may in some cases be a true thermonuclear detonation, with the core of the star, composed mainly of carbon and oxygen, burning instantaneously to iron. In other cases the collapse may cause the star to spin so rapidly that hydrodynamic instability disrupts it. A third possibility is that a spinning magnetic field becomes so intensified by gravitational collapse that it can drive off the surface of the star at high velocity. Probably several different kinds of supernova exist, each with a different mechanism of energy transfer. In all cases the basic process must be a gravitational collapse of the core of the star. By one means or another some fraction of the gravitational energy released by the collapse is transferred outward and causes the outer layers of the star to explode. The outward-moving energy appears partly as visible light, partly as the energy of motion of the debris and partly as the energy of cosmic rays. In addition a small fraction of the energy may be converted into the nuclear energy of the unstable nuclear species thorium and uranium, and small amounts of these elements may be injected by the explosion into the interstellar gas. As far as we know no other mechanism can create the special conditions required for the production of fissionable nuclei.

We have firm evidence that a locally violent environment existed in our galaxy immediately before the birth of the solar system. It is likely that the violence and the origin of the sun and the earth were part of the same sequence of events. The evidence for violence is the existence in certain very ancient meteorites of xenon gas with an isotopic composition characteristic of the products of spontaneous fission of the nucleus plutonium 244. Supporting evidence is provided by radiation damage in the form of fission-fragment tracks that can be made visible by etching in pieces of other meteorites [*see illustration below*]. The meteorites do not contain enough uranium or thorium to account for either the xenon or the fission tracks. Plutonium 244, although it is the longest-lived isotope of plutonium, has a half-life of only 80 million years which is very short compared with the age of the earth. Therefore the meteorites must be coeval with the solar system, and the plutonium must have been made close to, in both time and space, the event that gave birth to the sun.

We are only beginning to understand the way stars and planets are born. It seems that stars are born in clusters of a few hundred or a few thousand at a time rather than singly. There is perhaps a cyclical rhythm in the life of a galaxy. For 100 million years the stars and the interstellar gas in any particular sector of a galaxy lie quiet. Then some kind of shock or gravitational wave passes by, compressing the gas and triggering gravitational condensation. Various hangups are overcome, and a large mass of gas condenses into new stars in a limited region of space. The most massive stars shine brilliantly for a few million years and die spectacularly as supernovas. The brief blaze of the clusters of short-lived massive stars makes the shock wave visible, from a distance of millions of light-years, as a bright spiral arm sweeping around the galaxy. After the massive new stars are burned out the less massive stars continue to condense, partially contaminated with plutonium. These more modest stars continue their quiet and frugal existence for billions of years after the spiral arm that gave them birth has passed by. In some such rhythm as this, 4.5 billion years ago, our solar system came into being.

Whether some similar rhythms, on an even more gigantic scale, are involved in the birth of the radio galaxies, the quasars and the nuclei of Seyfert galaxies we simply do not know. Each of these objects pours out quantities of energy millions of times greater than the output of the brightest supernova. We know nothing of their origins, and we know nothing of their effects on their surroundings. It would be strange if their effects did not ultimately turn out to be of major importance, both for science and for the history of life in the universe.

The main sources of energy available to us on the earth are chemical fuels, uranium and sunlight. In addition we hope one day to learn how to burn in a controlled fashion the deuterium in the oceans. All these energy stores exist here by virtue of hangups that have temporarily halted the universal processes of energy degradation. Sunlight is sustained by the thermonuclear, the weak-interaction and the opacity hangups. Urani-

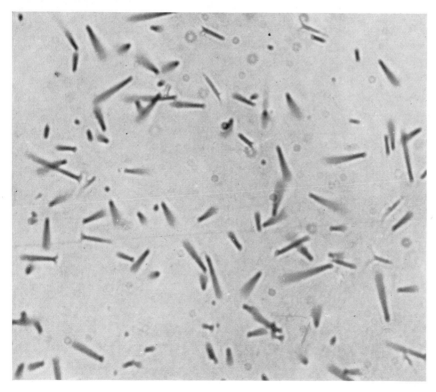

ANCIENT EVIDENCE that a locally violent environment existed in our galaxy immediately before the birth of the solar system is provided by photographs such as this one, which was made by P. B. Price of the University of California at Berkeley. The photograph shows radiation damage in the form of fission-fragment tracks made visible by etching in a crystal from a very ancient meteorite. Meteorites of this type do not contain enough uranium or thorium to account for the fission tracks. Instead the tracks appear to be the products of the spontaneous fission of the nucleus plutonium 244, which has a half-life of only 80 million years, a period that is very short compared with the age of the earth. Therefore the meteorites must be coeval with the solar system, and the plutonium must have been made close to, in both time and space, the event that gave birth to the solar system.

um is preserved by the surface-tension hangup. Coal and oil have been buried in the ground and saved from oxidation by various biological and chemical hang-ups, the details of which are still under debate. Deuterium has been preserved in low abundance, after almost all of it was burned to form helium in the earliest stages of the history of the universe, because no thermonuclear reaction ever runs quite to completion.

Humanity is fortunate in having such a variety of energy resources at its disposal. In the very long run we shall need energy that is absolutely pollution-free; we shall have sunlight. In the fairly long run we shall need energy that is inexhaustible and moderately clean; we shall have deuterium. In the short run we shall need energy that is readily usable and abundant; we shall have uranium. Right now we need energy that is cheap and convenient; we have coal and oil. Nature has been kinder to us than we had any right to expect. As we look out into the universe and identify the many accidents of physics and astronomy that have worked together to our benefit, it almost seems as if the universe must in some sense have known that we were coming.

Since the Apollo voyages gave us a closeup view of the desolate landscape of the moon, many people have formed an impression of the earth as a uniquely beautiful and fragile oasis in a harsh and hostile universe. The distant pictures of the blue planet conveyed this impression most movingly. I wish to assert the contrary view. I believe the universe is friendly. I see no reason to suppose that the cosmic accidents that provided so abundantly for our welfare here on the earth will not do the same for us wherever else in the universe we choose to go.

Ko Fung was one of the great natural philosophers of ancient China. In the fourth century he wrote: "As for belief, there are things that are as clear as the sky, yet men prefer to sit under an upturned barrel." Some of the current discussions of the resources of mankind on the earth have a claustrophobic quality that Ko Fung's words describe very accurately. I hope that with this article I may have persuaded a few people to come out from under the barrel, and to look to the sky with hopeful eyes. I began with a quotation from Blake. Let me end with another from him, this time echoing the thought of Ko Fung: "If the doors of perception were cleansed every thing would appear to man as it is, infinite. For man has closed himself up, till he sees all things thro' narrow chinks of his cavern."

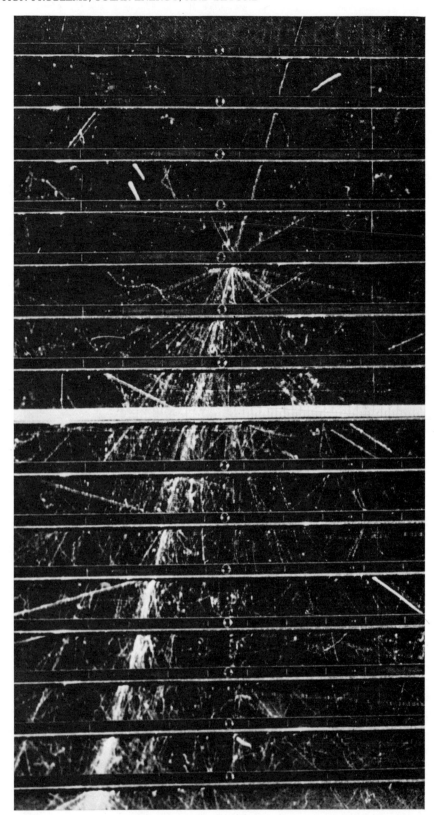

RECENT EVIDENCE of violent events in parts of the universe other than our own is contained in this cloud-chamber photograph of a primary cosmic ray track, obtained at an altitude of 17,200 feet on Mount Chacaltaya in Bolivia by Alfred Z. Hendel and his colleagues at the University of Michigan. The cloud chamber contained 17 iron plates, each half an inch thick. The cosmic ray, in this case a high-energy proton, entered the chamber from the top and passed through five plates before colliding with an iron nucleus in the sixth plate, producing a shower of secondary reaction products, mainly pi mesons. The energy of the incoming proton, approximately 1,100 billion electron volts, was measured by means of a detector mounted below the cloud chamber. Although it is not understood in detail where cosmic rays come from, they are known to carry about as much energy as starlight.

APPENDIX

APPENDIX

INTRODUCTION

The following figures and tables are taken from the textbook *Energy: An Introduction to Physics* by Robert H. Romer (W. H. Freeman and Company, San Francisco, 1976). This self-consistent set of data will facilitate discussions of energy technologies and energy policies.

Note (in Figure 1.1) the close correspondence between per capita power consumption and per capita GNP. The decrease in the ratio around 1930 is due to increases in technical efficiencies. The most recent decrease in the ratio (not shown), since the price rise of 1974, is due to increased efforts in conservation.

Figure 1.2 shows the leveling off in the consumption of firewood (before 1850) and in coal (around 1910). Oil and gas take over around World War II as the most important fuels in the United States. But the reliance on oil imports grows steadily (Figure 1.3). Growth in electrical generating capacity and production appear to increase dramatically around 1950 (Figures 1.4 and 1.5), but a logarithmic plot (Figure 1.6) shows reasonably constant growth rates until the mid-seventies, but a much higher rate than for direct uses of fuels (Figure 1.7).

Figure 2.1 is well worth studying in detail. It shows energy flows in 1973, just before the large price increase for oil and other fuels. (It will be important to contrast this figure with a similar one for, say, 1983). Table 2.1 recapitulates the sources of energy in 1973, and Figure 2.2 shows the detailed consumption uses of fuels and electricity for the same year. Tables 2.2, 2.3, and 2.5 present more detailed data and also give average annual percentage changes during the period 1970-73, just before the price increase. Note especially the large increase in oil for electric generation; this increase was caused by peak demands that were supplied by oil-fueled turbines, by increased electricity demands for space heating and for water pollution treatment plants, by the rather sudden requirements of the Clean Air Act which discouraged the expansion of coal-fired plants, by delays in nuclear plants, and by shortages of natural gas. As a consequence, oil imports jumped dramatically during 1970–73.

Tables 2.6, 2.7, 2.8, and 2.9 present detailed breakdowns of energy consumption for residential, commercial, industrial, and transportation uses, respectively. Table 2.10 recaps the data according to the type of fossil fuel used.

Table 3.1 lists the energy content of fuels in various sets of units. For comparison, bread and butter are listed also (they compare to the energy content of the worst and of the best coal, respectively. Interestingly, the energy content of TNT is about half that of bread). Nuclear "fuels" have energy contents that are typically a million times greater than those of fossil fuels; this ratio reflects of course the ratio of nuclear binding energies to molecular binding energies.

A subject of perpetual controversy is the estimate of recoverable resources of fuels, especially of oil and gas. A large part of the problem is the fact that "recoverable" implies some judgment about future prices and future technologies. Table 4.1 gives estimates of *eventual* total production *including* what has already been produced; these exceed the so-called proven reserves, especially for U.S. crude oil. On the other hand, oil shale and tar sands may provide a greater resource than shown, considering the tar deposits in Venezuela and elsewhere. Other hydrocarbon resources are peat and geopressured methane (found, e.g., in deep deposits near the Gulf Coast).

Table 4.2 shows cumulative production in relation to the total resource; the remaining recoverable resource (not shown directly) would be the difference. Production over past decades is shown in Figure 4.1 through 4.6 for coal, oil, and gas. Projected production for oil and coal is indicated in Figures 4.7 and 4.8.

General data on solar energy are shown in Table 5.1, with Figure 5.1 showing the annual average insolation over the United States. Figure 5.2 allows an estimate of the required collection area in order to meet U.S. energy needs, for various assumed conversion efficiencies. One disadvantage of solar energy is its diluteness: The energy delivered to 1 m^2 over 1 year equals that contained in 1 barrel of crude oil. If converted with an efficiency of 10%, then it has the energy content of about 4 gallons of gasoline.

Table 6.1 gives some ideas about energy use and conservation potentials in the household. The big electricity users are air conditioning, refrigerator-freezer, clothes dryer, and of course electric lights. The big users of energy, not necessarily electricity, are space heating and hot water.

Data shown for energy requirements for passenger transportation (Table 7.1) range all the way from bicycles (one-third the energy consumption of walking) to supersonic transport aircraft and ocean liners. Table 7.2 shows energy consumption in freight transportation, ranging from pipelines to airplanes.

History of Energy Production and Consumption in the United States

FIGURE 1.1
Per capita average power consumption and per capita gross national product in the United States. (Gross national product figures are given in 1973 dollars; dollar figures for previous years have been increased to take account of inflation.)

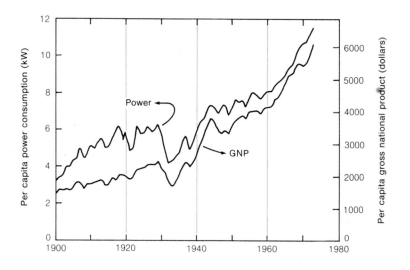

FIGURE 1.2
Annual United States consumption of energy from various sources since 1850. Energy from fuel wood is included here, both as a separate item and as part of the total; wood is *not* included in other graphs and tables.

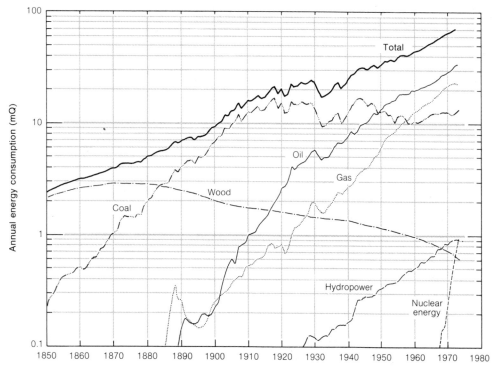

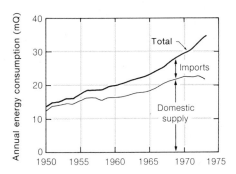

FIGURE 1.3
Foreign and domestic supplies of oil. Annual
United States consumption of energy from
oil (natural gas liquids included).

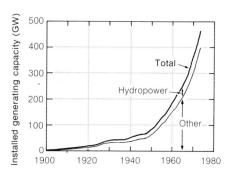

FIGURE 1.4
Installed electrical generating capacity
of the United States.

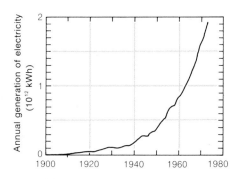

FIGURE 1.5
Generation of electrical energy in the United
States.

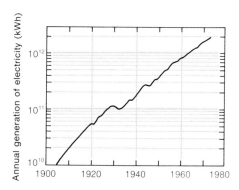

FIGURE 1.6
Generation of electrical energy in the United
States (semilogarithmic graph).

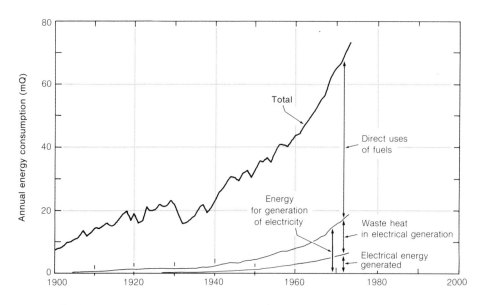

FIGURE 1.7
Electrical energy in the United States in relation to overall energy consumption.

Sources and Uses of Energy in the United States, 1973

FIGURE 2.1
Energy flow in the United States (1973).

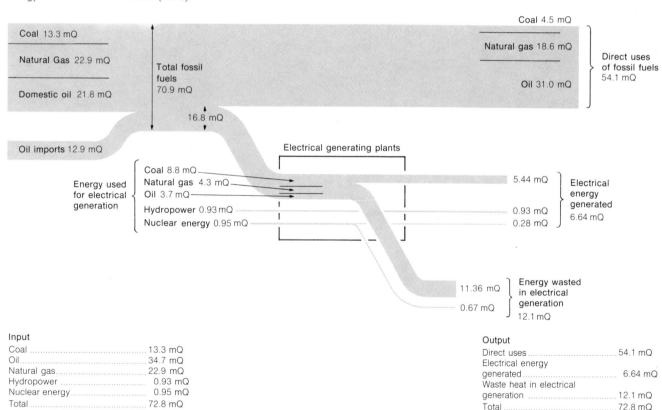

Input

Coal	13.3 mQ
Oil	34.7 mQ
Natural gas	22.9 mQ
Hydropower	0.93 mQ
Nuclear energy	0.95 mQ
Total	72.8 mQ

Output

Direct uses	54.1 mQ
Electrical energy generated	6.64 mQ
Waste heat in electrical generation	12.1 mQ
Total	72.8 mQ

FIGURE 2.2
Energy consumption in the United States (1973).

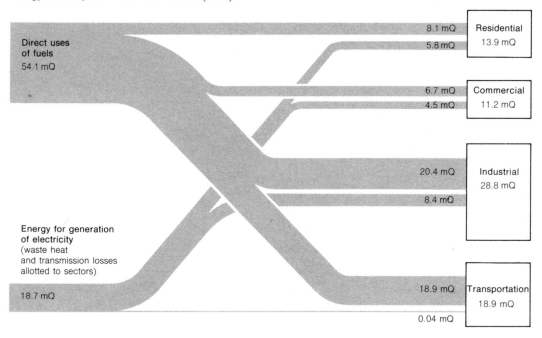

TABLE 2.1   Sources of energy, 1973.

Source	Energy* (mQ)	Percentage of national total
Coal	13.3 (+1.7%)	18.3%
Oil†	34.7 (+5.5%)	47.7%
Natural gas	22.9 (+1.2%)	31.4%
Total fossil fuels	70.9 (+3.3%)	97.4%
Hydroelectric power	0.93 (+3.1%)	1.3%
Nuclear power	0.95 (+56%)	1.3%
Total	72.8 (+3.7%)	100%

*Figures in parentheses are average annual percentage rates of change during the period 1970–1973.
†Analysis of the oil supply.

	Energy (mQ)	Percentage of total oil supply
Domestic crude oil	19.3 (−1.4%)	55.6%
Domestic natural gas liquids	2.5 (+0.9%)	7.2%
Total domestic supply	21.8 (−1.1%)	62.8%
Imported oil (crude oil and refined products) (For trends in oil imports, see Figure 1.3)	12.9 (+22.8%)	37.2%
Total	34.7 (+5.5%)	100%

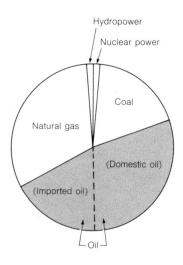

Chart fot TABLE 2.1

TABLE 2.2   Electrical Generation, 1973.

Source of energy	Average efficiency	Electrical energy generated*,† (mQ)	Percentage of electrical energy generated	Energy used for generation* (mQ)	Percentage of energy used for generation
Coal	33%	2.95 (+5.9%)	44.4%	8.82 (+5.6%)	47.2%
Oil	31%	1.16 (+18%)	17.5%	3.71 (+16%)	19.8%
Natural gas	31%	1.33 (−3.7%)	20.0%	4.28 (−3.3%)	22.9%
Hydroelectric power**	100%	0.93 (+3.1%)	14.0%	0.93 (+3.1%)	5.0%
Nuclear power	30%	0.28 (+56%)	4.2%	0.95 (+56%)	5.1%
Total	35.6%††	6.64 (+5.9%)	100%	18.7 (+6.0%)	100%

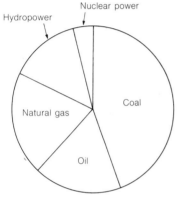

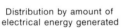

Distribution by amount of electrical energy generated

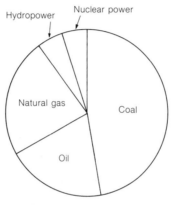

Distribution by amount of energy consumed in generation

Charts for TABLE 2.2

*Figures in parentheses are average annual percentage rates of change during the period 1970–1973.

†Total amount of electrical energy generated: 6.64 mQ = 1.95 × 10¹² kWh. Approximately 95% of this energy is generated by public and private utilities; the remainder is generated by industries for internal use.

**Generation of electrical energy with hydropower is taken to be 100% efficient; the actual efficiency is somewhat smaller, but allowance for this would make only slight changes in the totals.

††Average efficiency *not* including hydropower: 32.2%.

TABLE 2.3   Sources of energy (1973), divided according to direct uses and use for generation of electrical energy.

Source of energy	Direct uses* (mQ)	Use for generation of electrical energy* (mQ)	Total* (mQ)
Coal	4.54 (−4.5%)	8.82 (+5.6%)	13.3 (+1.7%)
Oil	31.0 (+4.5%)	3.71 (+16%)	34.7 (+5.5%)
Natural gas	18.6 (+2.4%)	4.28 (−3.3%)	22.9 (+1.2%)
Hydroelectric power	0	0.93 (+3.1%)	0.93 (+3.1%)
Nuclear power	0	0.95 (+56%)	0.95 (+56%)
Total	54.1 (+2.9%)	18.7 (+6.0%)	72.8 (+3.7%)

*Figures in parentheses are average annual percentage rates of change during the period 1970–1973.

TABLE 2.4   Direct uses of fuels, and use of energy for generation of electricity, 1973.

Uses	Energy* (mQ)	Percentage of national total
Direct uses of fossil fuels		
Coal	4.54 (−4.5%)	6.2%
Oil	31.0 (+4.5%)	42.6%
Natural gas	18.6 (+2.4%)	25.5%
Total direct use	54.1 (+2.9%)	74.3%
Electrical generation		
Electrical energy generated	6.64 (+5.9%)	9.1%
Energy wasted in electrical generation	12.06 (+6.1%)	16.5%
Total for electrical generation	18.7 (+6.0%)	25.7%
Total energy consumption	72.8 (+3.7%)	100%

*Figures in parentheses are average annual percentage rates of change during the period 1970–1973.

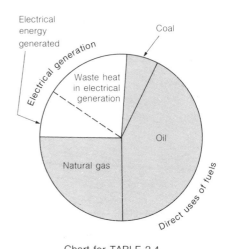

Chart for TABLE 2.4

TABLE 2.5   Energy consumption by sector, 1973.

Sector	Direct uses of fuels* (mQ)	Electricity (including waste heat from generation)* (mQ)	Total energy consumption* (mQ)	Percentage of national total
Residential	8.1 (+0.8%)	5.8 (+7.5%)	13.9 (+3.3%)	19.1%
Commercial	6.7 (+3.0%)	4.5 (+7.0%)	11.2 (+4.6%)	15.4%
Industrial	20.4 (+2.3%)	8.4 (+4.6%)	28.8 (+3.0%)	39.6%
Transportation	18.9 (+4.4%)	0.04 (0%)	18.9 (+4.4%)	26.0%
Total	54.1 (+2.9%)	18.7 (+6.0%)	72.8 (+3.7%)	100%

*Figures in parentheses are average annual percentage rates of change during the period 1970–1973.

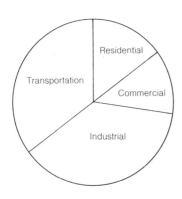

Distribution of energy from direct uses of fossil fuels

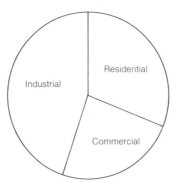

Distribution of energy used for generation of electricity (transportation sector omitted)

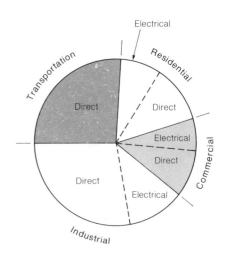

Distribution of total energy

Charts for TABLE 2.5

TABLE 2.6   Energy consumption in the residential sector, 1973.

Use	Direct uses of fuels* (mQ)	Electricity (including waste heat from generation)* (mQ)	Total energy consumption* (mQ)	Percentage of residential sector	Percentage of national total
Space heating	6.4 (+0.7%)	1.26 (+24.7%)	7.7 (+3.5%)	55.4%	10.6%
Hot water	1.2 (+1.3%)	0.62 (+2.5%)	1.82 (+1.6%)	13.1%	2.5%
Lights	0	1.09 (+0.3%)	1.09 (+0.3%)	7.8%	1.5%
Refrigeration	0	0.76 (+4.8%)	0.76 (+4.8%)	5.5%	1.0%
Cooking	0.36 (0%)	0.28 (+3.6%)	0.64 (+1.6%)	4.6%	0.9%
Air conditioning	0	0.56 (+8.3%)	0.56 (+8.3%)	4.0%	0.8%
Television	0	0.42 (+5.9%)	0.42 (+5.9%)	3.0%	0.6%
Clothes drying	0.08 (+2.6%)	0.19 (+8.8%)	0.27 (+6.8%)	1.9%	0.4%
Freezers	0	0.29 (+8.3%)	0.29 (+8.3%)	2.1%	0.4%
Misc. appliances	0	0.22 (+2.5%)	0.22 (+2.5%)	1.6%	0.3%
Dish washers	0	0.06 (+10.6%)	0.06 (+10.6%)	0.4%	0.08%
Washing machines	0	0.04 (0%)	0.04 (0%)	0.3%	0.05%
Total	8.1 (+0.8%)	5.8 (+7.5%)	13.9 (+3.3%)	100%	19.1%

*Figures in parentheses are average annual percentage rates of change during the period 1970–1973.

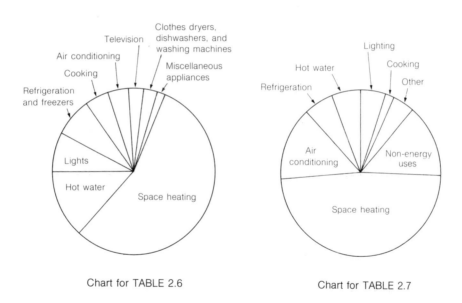

Chart for TABLE 2.6   Chart for TABLE 2.7

TABLE 2.7   Energy consumption in the commercial sector, 1973.

Use	Direct uses of fuels* (mQ)	Electricity (including waste heat from generation)* (mQ)	Total energy consumption* (mQ)	Percentage of commercial sector	Percentage of national total
Space heating	4.4 (+1.2%)	1.0 (+30%)	5.4 (+4.7%)	48.2%	7.4%
Non-energy uses†	1.6 (+9.2%)	0	1.6 (+9.2%)	14.3%	2.2%
Air conditioning	0.15 (+8.1%)	1.47 (+7.3%)	1.6 (+7.3%)	14.3%	2.2%
Refrigeration	0	0.68 (0%)	0.68 (0%)	6.1%	0.9%
Hot water	0.43 (+1.0%)	0.22 (+1.0%)	0.65 (+1.0%)	5.8%	0.9%
Lights (except street and highway)	0	0.47 (+0.4%)	0.47 (+0.4%)	4.2%	0.6%
Cooking	0.13 (+1.4%)	0.08 (+4.0%)	0.21 (+2.5%)	1.9%	0.3%
Street and highway lights	0	0.10 (+0.3%)	0.10 (+0.3%)	0.9%	0.14%
Other	0	0.48 (+0.4%)	0.48 (+0.4%)	4.3%	0.7%
Total	6.7 (+3.0%)	4.5 (+7.0%)	11.2 (+4.6%)	100%	15.4%

*Figures in parentheses are average annual percentage rates of change during the period 1970–1973.
†Largely use of petroleum products for highway paving, etc.

TABLE 2.8   Energy consumption in the industrial sector, 1973.

Use	Direct uses of fuels* (mQ)	Electricity (including waste heat from generation)* (mQ)	Total energy consumption* (mQ)	Percentage of industrial sector	Percentage of national total
Chemicals (production of basic chemicals, synthetic fibers, drugs, fertilizers, etc.)	3.6 (+2.7%)	2.4 (+5.0%)	6.0 (+3.6%)	20.8%	8.2%
Metals (smelting and refining, manufacture of basic metal products)	3.9 (+0.5%)	2.0 (+4.7%)	5.9 (+1.9%)	20.5%	8.1%
Petroleum refining, manufacture of paving and roofing materials, lubricants, etc.	3.0 (+4.0%)	0.3 (+5.1%)	3.3 (+4.1%)	11.5%	4.5%
Food manufacturing and processing	1.0 (+2.4%)	0.5 (+4.0%)	1.5 (+2.9%)	5.2%	2.1%
Paper products	1.0 (+1.4%)	0.5 (+4.8%)	1.5 (+2.5%)	5.2%	2.1%
Glass, concrete, asbestos, etc.	1.0 (+1.7%)	0.4 (+4.1%)	1.4 (+2.4%)	4.9%	1.9%
Other industries	6.9 (+2.6%)	2.3 (+4.2%)	9.2 (+3.0%)	31.9%	12.6%
Total	20.4 (+2.3%)	8.4 (+4.6%)	28.8 (+3.0%)	100%	39.6%

*Figures in parentheses are average annual percentage rates of change during the period 1970–1973.

TABLE 2.9   Energy consumption in the transportation sector, 1973.

	Energy consumption*,† (mQ)	Percentage of transportation sector	Percentage of national total
Automobiles (intercity travel)	3.61 (+2.3%)	19.1%	5.0%
Automobiles (local travel)	6.80 (+5.9%)	36.0%	9.3%
Buses (intercity travel)	0.04 (0%)	0.21%	0.05%
Buses (local travel)	0.06 (+4.2%)	0.32%	0.08%
Buses (school buses)	0.03 (−5.0%)	0.16%	0.04%
Trucks (intercity)	1.18 (+0.4%)	6.2%	1.6%
Trucks (local)	3.20 (+5.0%)	16.9%	4.4%
Subways	0.03 (−1.0%)	0.16%	0.04%
Railroads (passenger)	0.02 (−9%)	0.11%	0.03%
Railroads (freight)	0.52 (+1.0%)	2.8%	0.7%
Planes (commercial passenger)	0.77 (0%)	4.1%	1.1%
Planes (private)	0.10 (−1%)	0.5%	0.14%
Planes (air freight)	0.18 (+6%)	1.0%	0.25%
Planes (military)	0.52 (−6%)	2.8%	0.7%
Waterways	0.47 (+3.5%)	2.5%	0.6%
Pipelines	0.22 (+4.2%)	1.2%	0.3%
Other transportation	1.04 (+4%)	5.5%	1.4%
Non-energy uses of fossil fuels (motor oils, greases, etc.)	0.16 (+1.8%)	0.8%	0.2%
Total	18.9 (+4.4%)	100%	26.0%

*Figures in parentheses are average annual percentage rates of change during the period 1970–1973.

†Electrical energy uses (approximately 0.2% of energy used for transportation) are not shown separately.

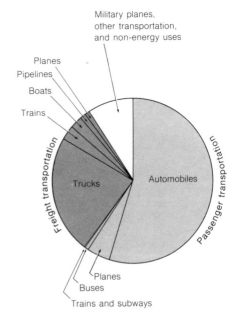

Chart for TABLE 2.8

Chart for TABLE 2.9

TABLE 2.10 Use of fossil fuels, 1973.

Use	Coal* (mQ)	Oil* (mQ)	Natural gas* (mQ)	Total* (mQ)
Direct uses				
Residential sector	0	3.26 (+1.4%)	4.81 (+0.4%)	8.1 (+0.8%)
Commercial sector	0.30 (−11%)	3.86 (+4.6%)	2.56 (+2.8%)	6.7 (+3.0%)
Industrial sector	4.24 (−4%)	5.74 (+5.9%)	10.47 (+3.4%)	20.4 (+2.3%)
Transportation sector	0.003 (−20%)	18.16 (+4.6%)	0.75 (+0.2%)	18.9 (+4.4%)
Total direct uses	4.54 (−4.5%)	31.0 (+4.5%)	18.6 (+2.4%)	54.1 (+2.9%)
Use for generating electrical energy	8.82 (+5.6%)	3.71 (+16%)	4.28 (−3.3%)	16.8 (+4.9%)
Total	13.3 (+1.7%)	34.7 (+5.5%)	22.9 (+1.2%)	70.9 (+3.3%)

*Figures in parentheses are average annual percentage rates of change during the period 1970–1973.

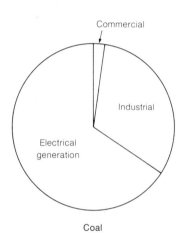

Coal

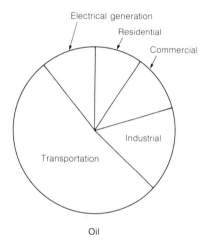

Oil

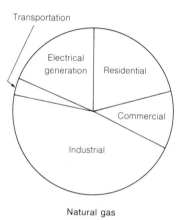

Natural gas

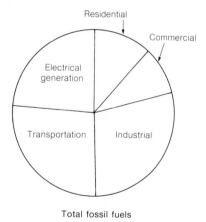

Total fossil fuels

Charts for TABLE 2.10

3   Energy Content of Fuels

TABLE 3.1   Energy content.[a]

Fuel	(Commonly used units)	Values (BTU/ton)	(J/kg)
Coal (bituminous and anthracite)		25×10^6	29×10^6
Lignite		10×10^6	12×10^6
Peat		3.5×10^6	4×10^6
Crude oil	5.6×10^6 BTU/barrel	37×10^6	43×10^6
Gasoline	5.2×10^6 BTU/barrel	38×10^6	44×10^6
NGL's (Natural gas liquids)	4.2×10^6 BTU/barrel	37×10^6	43×10^6
Natural gas[b]	1030 BTU/ft^3	47×10^6	55×10^6
Hydrogen gas[b]	333 BTU/ft^3	107×10^6	124×10^6
Methanol (methyl alcohol)	6×10^4 BTU/gal	17×10^6	20×10^6
Charcoal		24×10^6	28×10^6
Wood	20×10^6 BTU/cord	12×10^6	14×10^6
Miscellaneous farm wastes		12×10^6	14×10^6
Dung		15×10^6	17×10^6
Assorted garbage and trash		10×10^6	12×10^6
Bread	1100 kcal/lb	9×10^6	10×10^6
Butter	3600 kcal/lb	29×10^6	33×10^6
Fission	200 MeV/fission	7×10^{13} [c]	8×10^{13} [c]
		5×10^{11} [d]	5.8×10^{11} [d]
D-D Fusion (deuterium)	7 MeV/deuteron	2.9×10^{14} [e]	3.3×10^{14} [e]
		8.6×10^{10} [f]	10^{11} [f]
D-T Fusion (lithium)[g]	7 MeV/Li nucleus	8.4×10^{13}	9.7×10^{13}
Complete "mass-energy conversion"[h]	931 MeV/amu	7.7×10^{16}	9×10^{16}

[a]These data are only intended for use in making estimates of available energy. Various types of wood, for example, have energy values covering a rather wide range; different samples of coal, oil, and other fuels also have varying energy values. Various fuels obtained from the processing of oil (for example residual oil, kerosene, various types of gasoline) have energy values per unit mass within about 20% of those listed for crude oil and gasoline.

[b]Quantities of natural gas are usually reported in cubic feet, the volume of gas at a pressure of 1 atmosphere and a temperature of 60°F, or in thousands of cubic feet (often abbreviated Mcf).

[c]per ton or kilogram of nuclei undergoing fission.

[d]per ton or kilogram of uranium metal, when only the U^{235} (abundance 0.72%) is used.

[e]per ton or kilogram of pure deuterium.

[f]per ton or kilogram of hydrogen, containing 0.015% deuterium.

[g]The data for D-T fusion are based on the assumption that deuterium is available in unlimited quantities, that tritium is produced and that energy production is limited by the availability of lithium.

[h]The data given for complete mass-energy conversion are given for purposes of comparison; no practical "fuel" is known that would yield this much energy.

Fossil Fuels—Resources and Production

4

TABLE 4.1 Fossil fuel resources: estimates of eventual total production.

NOTE: Such estimates are subject to considerable uncertainty and should only be considered as giving the order of magnitude of the eventual total production. Estimates of production of energy from sources already in use are much more reliable than they are for tar sands and oil shales. The amount of oil that *exists* in the world's oil shales is estimated to be about 2×10^{15} barrels, 10,000 times larger than the figure given here; most of this oil is probably unobtainable, but the situation could be changed either by new technological developments or by changing economic conditions.

Fuel	United States		World (United States included)	
	Physical units	Approximate energy content (Q)	Physical units	Approximate energy content (Q)
Coal and lignite	1.6×10^{12} tons	37	8.4×10^{12} tons	170
Crude oil	200×10^9 barrels	1.1	2100×10^9 barrels	12
Natural gas liquids (NGL's)	40×10^9 barrels	0.17	400×10^9 barrels	1.7
Natural gas	1.1×10^{15} ft³	1.1	12×10^{15} ft³	12
Canadian tar sands	–	–	300×10^9 barrels	1.7
Oil shales	80×10^9 barrels	0.45	190×10^9 barrels	1.1
Total	–	40	–	200

TABLE 4.2 Major fossil fuels: world and United States consumption rates in relation to resources.

	World (United States included)		
	Coal and lignite	Petroleum liquids (crude oil and NGL's)	Natural gas
Resources*	8.4×10^{12} tons	2500×10^9 barrels	12×10^{15} ft^3
Cumulative production (through 1973)	0.16×10^{12} tons (1.9% of resources)	312×10^9 barrels (12% of resources)	0.6×10^{15} ft^3 (5% of resources)
Production rate (1973)	3270×10^6 tons/yr	21.2×10^9 barrels/yr	44×10^{12} ft^3/yr
Time remaining until resources would be exhausted, if production were to continue at the 1973 rate.†	2500 yr	100 yr	260 yr

	United States		
	Coal and lignite	Petroleum liquids (crude oil and NGL's)	Natural gas
Resources*	1.6×10^{12} tons	240×10^9 barrels	1.1×10^{15} ft^3
Cumulative production through 1973)	0.04×10^{12} tons (2.5% of resources)	117×10^9 barrels (49% of resources)	0.41×10^{15} ft^3 (37% of resources)
Production rate (1973)	600×10^6 tons/yr	4.0×10^9 barrels/yr	21.7×10^{12} ft^3/yr
Time remaining until resources would be exhausted, if production were to continue at the 1973 rate.	2600 yr	30 yr	32 yr

*See Table 4.1
†See also Figures 4.7 and 4.8.

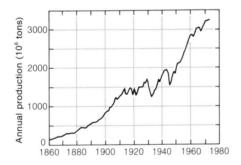

FIGURE 4.1
World production of coal and lignite.
Cumulative production through 1973:
16×10^{10} tons.

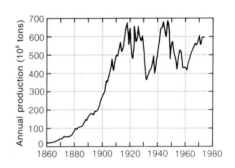

FIGURE 4.2
United States production of coal and lignite.
Cumulative production through 1973:
4×10^{10} tons.

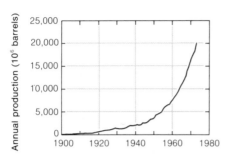

FIGURE 4.3
World production of crude oil. Cumulative
production through 1973: 300 × 10⁹ barrels.

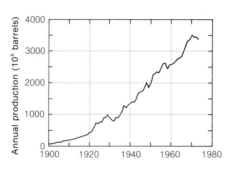

FIGURE 4.4
United States production of crude oil.
Cumulative production through 1973:
106 × 10⁹ barrels.

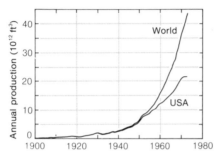

FIGURE 4.5
World and United States production of natural
gas. Cumulative production through 1973:
World, 600 × 10¹² ft³; United States,
410 × 10¹² ft³.

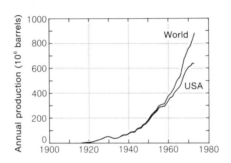

FIGURE 4.6
World and United States production of NGL's
(natural gas liquids). Cumulative production
through 1973: World, 13 × 10⁹ barrels;
United States, 11 × 10⁹ barrels.

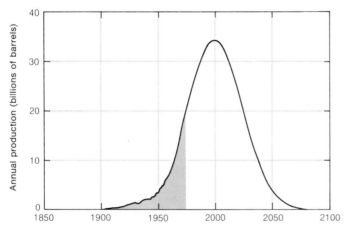

FIGURE 4.7
Projected cycle of world oil production (total
production = 2100 × 10⁹ barrels). For the
United States, peak production probably
occurred during the early 1970s.

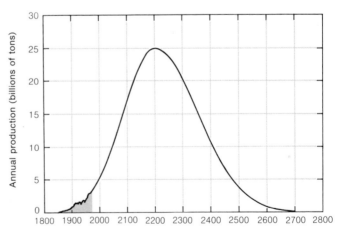

FIGURE 4.8
Projected cycle of world coal production (total
production = 8.4 × 10¹² tons). The projected
cycle for the United States would have a
similar appearance; the size would be reduced
because of the lower total production.

5   Solar Energy

TABLE 5.1   Solar energy data.

	Values
Total power radiated from the sun in all directions	3.8×10^{26} W
Power incident on the earth at the top of the atmosphere*	1.73×10^{17} W $= 5.45 \times 10^{24}$ J/yr $= 5174$ Q/yr
Solar constant (S_o)—power per unit area at the top of the earth's atmosphere, for a surface directly facing the sun*	$S_o = 1.353$ kW/m² (approximate rounded value: 1.4 kW/m²)
Amount incident at ground level per unit area, for a surface directly facing the sun (this amount varies with weather conditions and with the amount of atmosphere in the path; the value given here is typical for a time near noon on a clear and cloudless day)	1 kW/m²
Energy delivered to a horizontal surface (approximate average rate for the 48 contiguous states, averaged over all hours of the day and night and averaged over a full year)†	200 W/m²

NOTE: The langley. Solar energy data are frequently reported in *langleys* per day, where 1 langley is defined as 1 cal/cm². The langley is a unit of energy per unit area, and therefore the number of langleys per unit time is the average power per unit area: 1 langley/day $= 0.484$ W/m². El Paso, for instance, receives during June an average of 730 langleys/day, an average power per unit area of $730 \times 0.484 = 353$ W/m².

*These are average values for a whole year. During the year, the rate at which energy is received varies by 3.4% above and below these values, the highest value occurring near January 1 when the earth is closest to the sun, and the lowest value on about July 1 when the earth is farthest from the sun.

†See Figure 5.1 for more detailed data.

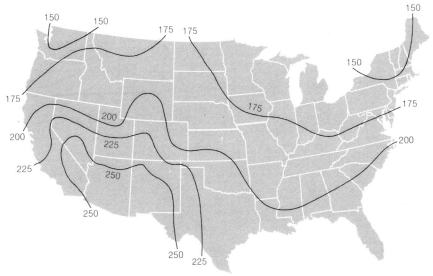

FIGURE 5.1
Solar energy zones in the United States. The numbers give the annual average radiation on a horizontal surface, in watts per square meter (averaged over a full year and over all hours of the day and night). *Note*: solar radiation may vary significantly between nearby locations because of local variations in cloudiness. Some of these variations are not shown here, and caution should be used if this map is used to estimate available solar energy for a particular location.

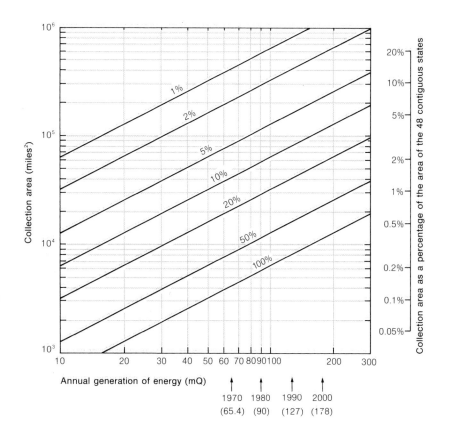

FIGURE 5.2
Potentially available solar energy in the United States. The graph shows the required collection area versus the annual amount of energy derived from solar energy for various possible conversion efficiencies. (For example, at an efficiency of 2%, an area of 10^5 square miles could supply 30 mQ/yr.) This figure is based on an average value of 200 W/m² on a horizontal surface. The arrows show *total* annual consumption of energy in the United States for 1970 and projected values for later years.

6   Energy Requirements for Electrical Appliances

TABLE 6.1   Average wattage and estimated energy consumption of various appliances.

Appliance	Average power required (W)	Estimated electrical energy used per year (kWh)	Appliance	Average power required (W)	Estimated electrical energy used per year (kWh)
Air conditioner (window, 5000 BTU/hr)	1565	1390	Humidifier	175	163
Blanket	177	147	Iron	1000	144
Broiler	1436	100	Microwave oven	1450	190
Carving knife	92	8	Radio	71	86
Clock	2	17	Radio-phonograph	110	110
Clothes dryer	4855	990	Razor	14	2
Coffee maker	895	105	Refrigerator (12 ft^3)	240	730
Deep-fat fryer	1450	83	Refrigerator (12 ft^3, frostless)	320	1215
Dishwasher	1200	363	Refrigerator-freezer (14 ft^3)	325	1135
Egg cooker	516	14	Refrigerator-freezer (14 ft^3, frostless)	615	1830
Fan (attic)	370	290	Roaster	1333	205
Fan (circulating)	88	43	Sewing machine	75	11
Fan (window)	200	170	Stove	12,200	1175
Floor polisher	305	15	Sun lamp	280	16
Food blender	385	15	Television		
Food freezer (15 ft^3)	340	1200	Black & white, vacuum tubes	160	350
Food freezer (15 ft^3, frostless)	440	1760	Black & white, solid state	55	120
Food mixer	127	13	Color, vacuum tubes	300	660
Food waste disposer	445	30	Color, solid state	200	440
Frying pan	1200	185	Toaster	1145	40
Grill (sandwich)	1160	33	Tooth brush	7	0.5
Hair dryer	380	14	Trash compactor	400	50
Heat lamp	250	13	Vacuum cleaner	630	46
Heater (portable)	1320	175	Vibrator	40	2
Heating pad	65	10	Waffle iron	1115	22
Hot plate	1260	90	Washing machine (automatic)	512	103

Source: Electrical Energy Association, New York, N.Y.

Energy Requirements for Transportation

TABLE 7.1   Energy requirements for passenger transportation.

Mode of transport	Maximum capacity (no. of passengers)*	Vehicle mileage (miles/gal)†	Passenger mileage (passenger-miles/gal)†	Energy consumption (BTU/passenger-mile)
Bicycle	1	1560	1560	80
Walking	1	470	470	260
Intercity bus	45	5	225	550
Commuter train (10 cars)	800	0.2	160	775
Subway train (10 cars)	1000	0.15	150	825
Volkswagen sedan**	4	30	120	1030
Local bus	35	3	105	1180
Intercity train (4 coaches)	200	0.4	80	1550
Motorcycle	1	60	60	2060
Automobile††	4	12	48	2580
747 jet plane	360	0.1	36	3440
727 jet plane	90	0.4	36	3440
707 jet plane	125	0.25	31	3960
United States SST (proposed)	250	0.1	25	4950
Light plane (2 seat)	2	12	24	5160
Executive jet plane	8	2	16	7740
Concorde SST	110	0.12	13	9400
Snowmobile	1	12	12	10,300
Ocean liner	2000	0.005	10	12,400

*The relative effectiveness of various modes of transportation can be drastically altered if a smaller number of passengers is carried.
†Miles per gallon of gasoline or the equivalent in food or in other fuel; all values must be regarded as approximate.
**Long-distance intercity travel.
††Typical American automobile, used partly for local travel and partly for long-distance driving.

TABLE 7.2   Energy requirements for freight transportation.

Mode of transport	Mileage (ton-miles/gal)	Energy consumption (BTU/ton-mile)
Oil pipelines	275	450
Railroads	185	670
Waterways	182	680
Truck	44	2800
Airplane	3	42000

BIBLIOGRAPHIES

I ENERGY USE, CONVERSION, TRANSPORTATION, AND STORAGE

1. Energy and Power

ENERGY IN THE FUTURE. Palmer Cosslett. Putnam D. Van Nostrand Company, Inc., 1953.

ENERGY IN THE AMERICAN ECONOMY, 1850–1975: AN ECONOMIC STUDY OF ITS HISTORY AND PROSPECTS. Sam H. Schurr and Bruce C. Netschert. The Johns Hopkins Press, 1960.

RESOURCES IN AMERICA'S FUTURE: PATTERNS OF REQUIREMENTS AND AVAILABILITIES 1960–2000. Hans H. Landsberg, Leonard L. Fischman and Joseph L. Fisher. The Johns Hopkins Press, 1963.

DIRECT USE OF THE SUN'S ENERGY. Farrington Daniels. Yale University Press, 1964.

THE WORLD ELECTRIC POWER INDUSTRY. N. B. Guyol. University of California Press, 1969.

CONTROLLED NUCLEAR FUSION: STATUS AND OUTLOOK. David J. Rose in *Science*, Vol. 172, No. 3985, pages 797–808; May 21, 1971.

2. The Conversion of Energy

EFFICIENCY OF THERMOELECTRIC DEVICES. Eric T. B. Gross in *American Journal of Physics*, Vol. 29, No. 1, pages 729–731; November, 1961.

ELECTRICAL ENERGY BY DIRECT CONVERSION. Claude M. Summers. Publication No. 147, The Office of Engineering Research, Oklahoma State University, March, 1966.

APPROACHES TO NONCONVENTIONAL ENERGY CONVERSION EDUCATION. Eric T. B. Gross in *IEEE Transactions on Education*, Vol. E-10, No. 2, pages 98–99; June, 1967.

3. The Economic Geography of Energy

A HISTORY OF TECHNOLOGY. Edited by Charles Singer, E. J. Holmyard and A. R. Hall. Oxford University Press, 1954–1958.

NATIONAL POWER SURVEY: GUIDELINES FOR GROWTH OF ELECTRIC POWER INDUSTRY. Federal Power Commission, U.S. Government Printing Office, 1964.

INTERNATIONAL PETROLEUM ENCYCLOPEDIA. The Petroleum Publishing Co., 1968.

II COAL

4. Oil and Gas from Coal

CLEAN FUELS FROM COAL: SYMPOSIUM PAPERS, PRESENTED SEPTEMBER 10–14, 1973, CHICAGO, ILLINOIS. Institute of Gas Technology, December, 1973.

PROCEEDINGS OF SIXTH SYNTHETIC PIPELINE GAS SYMPOSIUM: CHICAGO, ILLINOIS, OCTOBER 28–30, 1974. American Gas Association, 1974.

FEDERAL ENERGY ADMINISTRATION PROJECT INDEPENDENCE BLUEPRINT FINAL TASK FORCE REPORT. Prepared by the Interagency Task Force on Synthetic Fuels from Coal, under the direction of the U.S. Department of the Interior. U.S. Government Printing office, November, 1974.

FISCHER-TROPSCH PLANT DESIGN CRITERIA: PRESENTED AT 68TH ANNUAL MEETING OF AMERICAN INSTITUTE OF CHEMICAL ENGINEERS, LOS ANGELES, NOVEMBER 19, 1975. J. B. O'Hara, A. Bela, N. E. Jentz and S. K. Khaderi. The Ralph M. Parsons Company, 1975.

OIL/GAS PLANT DESIGN CRITERIA: PRESENTED AT 68TH ANNUAL MEETING OF AMERICAN INSTITUTE OF CHEMICAL ENGINEERS, LOS ANGELES, NOVEMBER 19, 1975. J. B. O'Hara, G. H. Hervey, S. M. Fass and E. A. Mills. The Ralph M. Parsons Company, 1975.

5. The Strip-Mining of Western Coal

LEASED AND LOST: A STUDY OF PUBLIC AND INDIAN COAL LEASING IN THE WEST. Council on Economic Priorities in *Economic Priorities Report*, Vol. 5, No. 2; 1974.

PROJECT INDEPENDENCE. Federal Energy Administration. U.S. Government Printing Office, 1974.

REHABILITATION POTENTIAL OF WESTERN COAL LANDS. National Academy of Sciences and National Academy of Engineering. Ballinger Publishing Co., 1974.

EFFECTS OF COAL DEVELOPMENT IN THE NORTHERN GREAT PLAINS. Northern Great Plains Resource Program. Denver Federal Center, 1975.

RESULTS OF A STUDY OF OVERBURDEN HANDLING TECHNIQUES AND RECLAMATION PRACTICES AT WESTERN U.S. SURFACE COAL MINES. U.S. Bureau of Mines, 1975.

SURFACE MINING CONTROL AND RECLAMATION ACT OF 1975. U.S. House of Representatives, Committee on Interior and Insular Affairs. U.S. Government Printing Office, 1975.

III NUCLEAR ENERGY

6. The Necessity of Fission Power

THE NUCLEAR CONTROVERSY. Ralph E. Lapp. Fact Systems/Reddy Kilowatt, Inc., 1974.

PROJECT INDEPENDENCE REPORT. Federal Energy Administration, U. S. Government Printing Office, 1974.

WASH-1400, REACTOR SAFETY STUDY: AN ASSESSMENT OF ACCIDENT RISKS IN U.S. COMMERCIAL NUCLEAR POWER PLANTS. U. S. Atomic Energy Commission. U.S. Government Printing Office, 1975.

ERDA-48. Energy Research and Development Agency. U. S. Government Printing Office, 1975.

AMERICA'S ENERGY FUTURE. Ralph E. Lapp. Fact Systems/Reddy Kilowatt, Inc.

HIGH-LEVEL RADIOACTIVE WASTE FROM LIGHT-WATER REACTORS. Bernard L. Cohen in *Reviews of Modern Physics*, Vol. 49, No. 1, pages 1–20; January, 1977.

7. The Disposal of Radioactive Wastes from Fission Reactors

HIGH-LEVEL RADIOACTIVE WASTE MANAGEMENT ALTERNATIVES. Edited by K. J. Schneider and A. M. Platt. Battelle Memorial Institute, Pacific Northwest Laboratories, 1974.

ALTERNATIVE PROCESSES FOR MANAGING EXISTING COMMERCIAL HIGH LEVEL RADIOACTIVE WASTES. Nuclear Regulatory Commission Report NUREG-0043, 1976.

ALTERNATIVES FOR MANAGING WASTES FROM REACTORS AND POST-FISSION OPERATIONS IN THE LWR FUEL CYCLE. Division of Nuclear Fuel Cycle and Production. Energy Research and Development Administration Report ERDA-76-43, 1976.

ENVIRONMENTAL SURVEY OF THE REPROCESSING AND WASTE MANAGEMENT PORTIONS OF THE LWR FUEL CYCLE. Nuclear Regulatory Commission Report NUREG-0116, 1976.

HIGH-LEVEL RADIOACTIVE WASTE FROM LIGHT-WATER REACTORS. Bernard L. Cohen in *Reviews of Modern Physics*, Vol. 49, No. 1, pages 1n20; January, 1977.

8. The Reprocessing of Nuclear Fuels

NUCLEAR CHEMICAL ENGINEERING. Manson Benedict and Thomas H. Pigford. McGraw-Hill Book Company, 1957.

ENGINEERING FOR NUCLEAR FUEL REPROCESSING. Justin T. Long. Gordon and Breach Science Publishers, Inc., 1968.

ENVIRONMENT EFFECT OF A COMPLEX NUCLEAR FACILITY. W. P. Bebbington in *Chemical Engineering Progress*, Vol. 70, No. 3, pages 85–86; March, 1974.

REPROCESSING—WHAT WENT WRONG? Simon Rippon in *Nuclear Engineering International*, Vol. 21, No. 239, pages 21–31; February, 1976.

9. Nuclear Power, Nuclear Weapons and International Stability

MOVING TOWARD LIFE IN A NUCLEAR ARMED CROWD? Albert Wohlstetter, Thomas A. Brown, Gregory Jones, David McGarvey, Henry Rowen, Vincent Taylor and Roberta Wohlstetter. Pan Heuristics, April 22, 1976.

NUCLEAR POWER ISSUES AND CHOICES. Nuclear Energy Policy Study Group. Ballinger Publishing Company, 1977.

THE NATIONAL ENERGY PLAN. Executive Office of the President: Energy Policy and Planning. Government Printing Office, April 29, 1977.

NUCLEAR PROLIFERATION AND SAFEGUARDS. Office of Technology Assessment, Congress of the United States, June, 1977.

10. Superphénix: A Full-Scale Breeder Reactor

FAST BREEDER REACTORS. Glenn T. Seaborg and Justin L. Bloom in *Scientific American*, Vol. 223, No. 5, pages 13–21; November, 1970.

THE NECESSITY OF FISSION POWER. H. A. Bethe in *Scientific American*, Vol. 234, No. 1, pages 21–31; January, 1976.

THE BREEDER: WHEN AND WHY. *EPRI Journal*, Vol. 1, No. 2; March, 1976.

11. The Prospects of Fusion Power

PROGRESS IN CONTROLLED THERMONUCLEAR RESEARCH. R. W. Gould, H. P. Furth, R. F. Post and F. L. Ribe in Presidentation Made before the President's Science Advisory Committee, December 15, 1970, and AEC's General Advisory Committee, December 16, 1970.

WORLD SURVEY OF MAJOR FACILITIES IN CONTROLLED FUSION. *Nuclear Fusion*, special Supplement 1970, STI/Pub/23. International Atomic Energy Agency, 1970.

WHY FUSION? William C. Gough in *Proceedings of the Fusion Reactor Design Symposium*, Held at Texas Tech University, Lubbock, Texas, on June 2–5, 1970.

IV   CONSERVATION, POLLUTION PROBLEMS, SOLAR ENERGY, AND BEYOND

12. The Fuel Consumption of Automobiles

CALTECH SEMINAR SERIES ON ENERGY CONSUMPTION IN PRIVATE TRANSPORTATION: TECHNICAL REPORT. John R. Pierce. California Institute of Technology, 1974.

FUEL ECONOMY OF THE 1975 MODELS: SAE PAPER NO. 740970. Thomas C. Austin and Karl H. Hellman. Society of Automotive Engineers, 1974.

POTENTIAL FOR MOTOR VEHICLE FUEL ECONOMY IMPROVEMENT: REPORT TO THE CONGRESS. U.S. Department of Transportation and the U.S. Environmental Protection Agency, October 24, 1974.

13. Human Energy Production as a Process in the Biosphere

NATIONAL RESOURCES FOR U.S. GROWTH: A LOOK AHEAD TO THE YEAR 2000. Hans H. Landsberg. Resources for the Future, Inc., 1964.

CLEANING OUR ENVIRONMENT: THE CHEMICAL BASIS FOR ACTION. Subcommittee on Environmental Improvement, Committee on Chemistry and Public Affairs. American Chemical Society, 1969.

RESOURCES AND MAN: A STUDY AND RECOMMENDATIONS. The Committee on Resources and Man. W. H. Freeman and Company, 1969.

THE CHANGING GLOBAL ENVIRONMENT. Edited by S. F. Singer. D. Reidel Publishing Company, 1975.

POWER GENERATION AND ENVIRONMENTAL CHANGE: SYMPOSIUM OF THE COMMITTEE ON ENVIRONMENTAL AFFAIRS, AMERICAN ASSOCIATION FOR THE ADVANCEMENT OF SCIENCE, BOSTON, DECEMBER 28, 1969. Edited by David A. Berkowitz and Arthur M. Squires. The M.I.T. Press, 1971.

14. The Photovoltaic Generation of Electricity

DIRECT USE OF THE SUN'S ENERGY. Farrington Daniels. Yale University Press, 1964.

SATELLITE SOLAR POWER STATION, P. E. Glaser in *Solar Energy*, Vol. 12, No. 3, pages 353–361; 1969.

GROWTH OF CONTROLLED PROFILE CRYSTALS FROM THE MELT, PART III: THEORY. Bruce Chalmers, H. E. LaBelle, Jr., and A. I. Mlavsky in *Materials Research Bulletin*, Vol. 6, No. 8, pages 681–690; August, 1971.

ENERGY: THE SOLAR-HYDROGEN ALTERNATIVE. J. O' M. Bockris. John Wiley & Sons, 1975.

SOLAR CELLS. Harold J. Hovel in *Semiconductors and Semimetals: Vol. 11*. Academic Press, 1975.

15. Energy in the Universe

GRAVITATION THEORY AND GRAVITATIONAL COLLAPSE. B. Kent Harrison, Kip S. Thorne, Masami Wakano and John Archibald Wheeler. The University of Chicago Press, 1965.

THE DYNAMICS OF DISK-SHAPED GALAXIES. C. C. Lin in *Annual Review of Astronomy and Astrophysics,* Vol. 5, pages 453–464; 1967.

GRAVITATION AND THE UNIVERSE: JAYNE MEMORIAL LECTURE FOR 1969. Robert H. Dicke. American Philosophical Society, 1970.

INDEX

The author gratefully acknowledges the assistance of Charles A. Cravotta III in compiling the index.